APPLIED FINITE
ELEMENT ANALYSIS

APPLIED FINITE
ELEMENT ANALYSIS
Second Edition

LARRY J. SEGERLIND
Agricultural Engineering Department
Michigan State University

JOHN WILEY AND SONS
New York · Chichester · Brisbane · Toronto · Singapore

Library of Congress Cataloging in Publication Data :

Segerlind, Larry J., 1937–
  Applied finite element analysis.

  Bibliography: p. 411.
  Includes index.
  1. Finite element method.   I. Title.
TA347.F5S43   1984   620'.001'515353   84-7455
ISBN 0-471-80662-5

Printed in the United States of America

10 9 8 7 6 5 4 3 2 1

To my parents T. J. and Bessie and my grade school teacher Alice Brunger

Preface

The finite element method is a widely accepted numerical procedure for solving the differential equations of engineering and physics and is the computational basis of many computer-aided design systems. The teaching of the fundamentals of the finite element method is rapidly becoming a necessity in those curricula which solve problems in the general areas of structural analysis, continuum mechanics, heat transfer, and fluid flow.

This is an introductory textbook covering the basic concepts of the finite element method and their application to the analysis of plane structures and two-dimensional continuum problems in heat transfer, irrotational fluid flow, and elasticity. The topics can be handled by advanced senior and beginning graduate students and have been taught in a course at this level for 10 years. No prior knowledge of structural analysis or the finite element method is assumed.

The major differences between this book and the first edition are the organization, the inclusion of five new chapters that introduce the analysis of plane structures, an increase in the number of homework problems, and the use of Galerkin's method in conjunction with the solution of field problems.

The organization of the material is unique and makes the book more useful as a teaching tool. The book consists of four parts: basic concepts, field problems, structural and solid mechanics problems, and linear and quadratic elements. The basic concepts cover six chapters and contain information needed to study field problem and/or structural and solid mechanics applications. Once the basic concepts have been completed, either Part Two, the field problems, or Part Three, structural and solid mechanics, can be studied. These two sections of the book are *independent*; the students can study either part or chapters from both parts depending on their interests. Part Four, linear and quadratic elements, covers general procedures for developing the element shape functions and numerically integrating the element matrices. This part can be covered after some of the continuum applications have been completed.

The organization of the book offers the instructor at least three different teaching options: (1) a general finite element course with topics selected from both field and solid mechanics applications, (2) a course emphasizing the solution of field problems, or (3) a course emphasizing the solution of solid mechanics problems. The author teaches the first type of course and covers Chapters 1 through 11, 17, 18, 23, and 24 in approximately 40 lectures.

The chapters that introduce the analysis of plane structures are included for agricultural and mechanical engineering students who need an introduction to the matrix analysis of structures but who do not have time to take several civil engineering courses to get the material. The displacement method of structural analysis based on the principle of minimum potential energy is presented. This formulation is consistent with the finite element analysis of elasticity problems.

Galerkin's method has been used to solve the field problems because it is an approach that is more readily accepted by seniors and beginning graduate students. These students have not had any variational calculus; hence, the concept of a functional is often quite mysterious to them. Galerkin's method offers two primary advantages. The inclusion of derivative boundary conditions is a straightforward procedure, and the student will find that he or she has studied a method that can be used to formulate differential equations with a first-derivative term. The disadvantage of Galerkin's method is the surfacing of the interelement requirements, which are never used and are somewhat difficult to explain away. It should be noted, however, that the interelement requirements also occur in the variational formulation when it is correctly applied. The variational method is not presented correctly in most finite element books.

The number of homework problems has been increased to over 300 with most chapters having at least 10 and many having 15 or more. The problems are a mixture of numerical calculation, analytical derivations, or the evaluation of important integrals and problems requiring a computer solution. A complete solution manual is available from the publisher.

The finite element method must be implemented on a digital computer; therefore, four computer programs have been included. These programs are written specifically for the beginning user and contain diagnostic checks to detect errors made by first-time users. These checks are very efficient and eliminate most of the student-instructor contact relative to finding data input errors.

I would like to thank the many students in MMM809 who used the original handwritten pages of this book as well as several typed versions. Their questions and suggestions were an invaluable contribution and have influenced the complete organization of the book as well as specific paragraphs within the text. I would also like to thank my wife Donna for her patience during the writing of this manuscript.

East Lansing, Michigan   **Larry J. Segerlind**

Contents

APPLIED FINITE
ELEMENT ANALYSIS

PART ONE
BASIC CONCEPTS

The first six chapters cover concepts basic to all applications of the finite element method. These chapters should be covered before going to the application chapters.

Chapter 1

INTRODUCTION

The finite element method is a numerical procedure for obtaining solutions to many of the problems encountered in engineering analysis. It has two primary subdivisions. The first utilizes discrete elements to obtain the joint displacements and member forces of a structural framework. The second uses the continuum elements to obtain approximate solutions to heat transfer, fluid mechanics, and solid mechanics problems. The formulation using the discrete elements is referred to as the "matrix analysis of structures" and yields results identical with the classical analysis of structural frameworks. The second approach is the true finite element method. It yields approximate values of the desired parameters at specific points called nodes. A general finite element computer program, however, is capable of solving both types of problems and the name "finite element method" is often used to denote both the discrete element and the continuum element formulations.

The finite element method combines several mathematical concepts to produce a system of linear or nonlinear equations. The number of equations is usually very large, anywhere from 20 to 20,000 or more and requires the computational power of the digital computer. The method has little practical value if a computer is not available.

It is impossible to document the exact origin of the finite element method because the basic concepts have evolved over a period of 150 or more years. The method as we know it today is an outgrowth of several papers published in the 1950s that extended the matrix analysis of structures to continuum bodies. The space exploration of the 1960s provided money for basic research, which placed the method on a firm mathematical foundation and stimulated the development of multiple-purpose computer programs that implemented the method. The design of airplanes, missiles, space capsules, and the like, provided application areas.

Although the origin of the method is vague, its advantages are clear. The method is easily applied to irregular-shaped objects composed of several different materials and having mixed boundary conditions. It is applicable to steady-state and time-dependent problems as well as problems involving nonlinear material properties. General computer programs that are user-independent can be, and have been, developed. User-assisting programs that generate a grid from a limited number of shape-defining points are available as well as programs that analyze the results and display them in graphic form for further study.

The finite element method is the computational basis of many computer-assisted design programs. The increased use of computer-assisted design makes it imperative that the practicing engineer have a knowledge of how the finite element method works.

3

1.1 SOLUTION OF BOUNDARY VALUE PROBLEMS

The best way to solve any physical problem governed by a differential equation is to obtain the analytical solution. There are many situations, however, where the analytical solution is difficult to obtain. The region under consideration may be so irregular that it is mathematically impossible to describe the boundary. The configuration may be composed of several different materials whose regions are mathematically difficult to describe. Problems involving anisotropic materials are usually difficult to solve analytically, as are equations having nonlinear terms.

A numerical method can be used to obtain an approximate solution when an analytical solution cannot be developed. All numerical solutions produce values at discrete points for one set of the independent parameters. The complete solution procedure is repeated each time these parameters change. Numerical solutions, however, are more desirable than no solution at all. The calculated values provide important information about the physical process even though they are at discrete points.

There are several procedures for obtaining a numerical solution to a differential equation. The methods can be separated into three basic groupings: (1) the finite difference method, (2) the variational method, and (3) the methods that weight a residual. These methods are briefly discussed in the following paragraphs.

Finite Difference Method

The finite difference method approximates the derivatives in the governing differential equation using difference equations. This method is useful for solving heat transfer and fluid mechanics problems and works well for two-dimensional regions with boundaries parallel to the coordinate axes. The method, however, is rather cumbersome when regions have curved or irregular boundaries, and it is difficult to write general computer programs for the method.

Variational Method

The variational approach involves the integral of a function that produces a number. Each new function produces a new number. The function that produces the lowest number has the additional property of satisfying a specific differential equation. To help clarify this concept, consider the integral

$$\Pi = \int_0^H \left[\frac{D}{2} \left(\frac{dy}{dx} \right)^2 - Qy \right] dx \tag{1.1}$$

The numerical value of Π can be calculated given a specific equation $y = f(x)$. The calculus of variations shows, however, that the particular equation $y = g(x)$, which yields the lowest numerical value for Π, is the solution to the differential equation

$$D \frac{d^2 y}{dx^2} + Q = 0 \tag{1.2}$$

with the boundary conditions $y(0) = y_0$ and $y(H) = y_H$.

The process can be reversed. Given a differential equation, an approximate solution can be obtained by substituting different trial functions into the appropriate functional. The trial function that gives the minimum value of Π is the approximate solution.

The variational method is the basis for many finite element formulations, but it has a major disadvantage: It is not applicable to any differential equation containing a first derivative term.

Weighted Residual Methods

The weighted residual methods also involve an integral. In these methods, an approximate solution is substituted into the differential equation. Since the approximate solution does not satisfy the equation, a residual or error term results. Suppose that $y=h(x)$ is an approximate solution to (1.2). Substitution gives

$$D \frac{d^2 h(x)}{dx^2} + Q = R(x) \neq 0 \tag{1.3}$$

since $y=h(x)$ does not satisfy the equation. The weighted residual methods require that

$$\int_0^H W_i(x) R(x)\, dx = 0 \tag{1.4}$$

The residual $R(x)$ is multiplied by a weighting function $W_i(x)$, and the integral of the product is required to be zero. The number of weighting functions equals the number of unknown coefficients in the approximate solution. There are several choices for the weighting functions, and some of the more popular choices have been assigned names.

Collocation Method. Impulse functions $W_i(x) = \delta(x - X_i)$ are selected as the weighting functions. This selection is equivalent to requiring the residual to vanish at specific points. The number of points selected equals the number of undetermined coefficients in the approximate solution.

Subdomain Method. Each weighting function is selected as unity, $W_i(x) = 1$, over a specific region. This is equivalent to requiring the integral of the residual to vanish over an interval of the region. The number of integration intervals equals the number of undetermined coefficients in the approximate solution.

Galerkin's Method. Galerkin's method uses the same functions for $W_i(x)$ that were used in the approximating equation. This approach is the basis of the finite element method for problems involving first-derivative terms. This method yields the same result as the variational method when applied to differential equations that are self-adjoint. Galerkin's method is used to develop the finite element equations for the field problems discussed in this book.

Least Squares Method.   The least squares method utilizes the residual as the weighting function and obtains a new error term defined by

$$Er = \int_0^H [R(x)]^2 \, dx \tag{1.5}$$

This error is minimized with respect to the unknown coefficients in the approximate solution. The least squares method has been utilized to formulate finite element solutions, but it is not as popular as Galerkin's method and the variational approach.

The variational method and the weighted residual methods each involve an integral. These methods can be grouped under the heading of integral formulations. A numerical solution based on an integral formulation is a new concept for many people; therefore, some of the more common methods are illustrated by solving a simple problem.

1.2   INTEGRAL FORMULATIONS FOR NUMERICAL SOLUTIONS

The immediate objective is to illustrate how each of the integral methods discussed in the previous section can be used to obtain an approximate solution to a physical problem. The example is a simply supported beam subjected to concentrated moments at each end. The beam and its bending moment diagram are shown in Figure 1.1.

The governing differential equation is

$$EI \frac{d^2y}{dx^2} - M(x) = 0 \tag{1.6}$$

with the boundary conditions

$$y(0) = 0 \qquad \text{and} \qquad y(H) = 0 \tag{1.7}$$

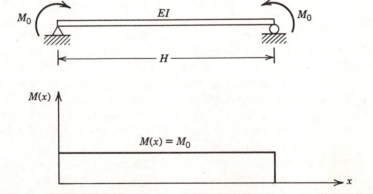

Figure 1.1.   A simply supported beam with concentrated end moments.

The coefficient EI represents the resistance of the beam to deflection and $M(x)$ is the bending moment equation. In this example, $M(x)$ has the constant value M_0.

An approximate equation for the beam deflection is

$$y(x) = A \sin \frac{\pi x}{H} \tag{1.8}$$

where A is an undetermined coefficient. This solution is an acceptable candidate because it satisfies the boundary conditions $y(0) = y(H) = 0$ and has a shape similar to the expected deflection curve. The exact solution of the differential equation is

$$y(x) = \frac{M_0 x}{2EI}(x - H) \tag{1.9}$$

1.2.1 Variational Method

The integral for the differential equation (1.6) is

$$\Pi = \int_0^H \left[\frac{EI}{2}\left(\frac{dy}{dx}\right)^2 + M_0 y \right] dx \tag{1.10}$$

The value of A that makes (1.8) the best approximation to the deflection curve is the value that makes Π a minimum. To evaluate A, Π must be written as a function of A and then minimized with respect to A. Noting that

$$\frac{dy}{dx} = \frac{A\pi}{H}\cos\frac{\pi x}{H}$$

we find that Π becomes

$$\Pi = \int_0^H \left[\frac{EI}{2}\left(\frac{A\pi}{H}\cos\frac{\pi x}{H}\right)^2 + M_0 A \sin\frac{\pi x}{H} \right] dx$$

or

$$\Pi = \left(\frac{EI\pi^2}{4H}\right)A^2 + \left(\frac{2M_0 H}{\pi}\right)A \tag{1.11}$$

Minimizing Π yields

$$\frac{\partial \Pi}{\partial A} = 2\left(\frac{EI\pi^2}{4H}\right)A + \frac{2M_0 H}{\pi} = 0$$

and

$$A = -\frac{4M_0 H^2}{\pi^3 EI} \tag{1.12}$$

The approximate solution is

$$y(x) = -\frac{4M_0 H^2}{\pi^3 EI}\sin\frac{\pi x}{H} \tag{1.13}$$

1.2.2 Collocation Method

The collocation method requires that the residual equation for the approximate solution be zero at as many points as there are undetermined coefficients. The residual is obtained by substituting (1.8) into (1.6). The result is

$$R(x) = -EI \frac{A\pi^2}{H^2} \sin \frac{\pi x}{H} - M_0 \tag{1.14}$$

since

$$\frac{d^2 y}{dx^2} = -\frac{A\pi^2}{H^2} \sin \frac{\pi x}{H}$$

There is only one undetermined coefficient; therefore, $R(x)$ is equated to zero at one point between 0 and H. Selecting $x = H/2$ for convenience, we obtain

$$R\left(\frac{H}{2}\right) = -EI \frac{A\pi^2}{H^2} \sin \frac{\pi}{2} - M_0 = 0$$

and

$$A = -\frac{M_0 H^2}{EI\pi^2} \tag{1.15}$$

The approximate solution is

$$y(x) = -\frac{M_0 H^2}{EI\pi^2} \sin \frac{\pi x}{H} \tag{1.16}$$

Had a point other than $x = H/2$ been selected, a different approximate solution would have been obtained.

1.2.3 Subdomain Method

The subdomain method requires that $\int R(x) \, dx = 0$ over as many subintervals as there are undetermined coefficients. The user can choose how long to make each subinterval. In the present example there is only one unknown coefficient; thus the interval must be $[0, H]$. The residual equation is (1.14); thus

$$\int_0^H R(x) \, dx = \int_0^H \left[-EI \frac{A\pi^2}{H^2} \sin \frac{\pi x}{H} - M_0 \right] dx = 0$$

Integration yields

$$-\left(\frac{2EI\pi}{H}\right) A - M_0 H = 0$$

and

$$A = -\frac{M_0 H^2}{2\pi EI} \tag{1.17}$$

The approximate solution is

$$y(x) = -\frac{M_0 H^2}{2\pi EI} \sin \frac{\pi x}{H} \tag{1.18}$$

1.2.4 Galerkin's Method

When using Galerkin's method, $\int W_i(x)R(x)\,dx$ is evaluated using the same functions for $W_i(x)$ that were used in the approximate solution. In this example there is only one weighting function, $W_i(x) = \sin \pi x/H$. The residual equation is (1.14) and the integral is

$$\int_0^H \sin \frac{\pi x}{H} \left[-EI \frac{A\pi^2}{H^2} \sin \frac{\pi x}{H} - M_0 \right] dx = 0$$

Integrating yields

$$-\frac{EI\pi^2 A}{2H} + \frac{2M_0 H}{\pi} = 0$$

Solving gives

$$A = -\frac{4M_0 H^2}{\pi^3 EI} \tag{1.19}$$

and the approximate solution is

$$y(x) = -\frac{4M_0 H^2}{\pi^3 EI} \sin \frac{\pi x}{H} \tag{1.20}$$

This solution is identical to the solution obtained using the variational method.

1.2.5 Least Squares Method

A new error term, $Er = \int [R(x)]^2\,dx$, is formed when using the least squares method. Substitution of the residual equation gives

$$Er = \int_0^H \left[-\frac{EI\pi^2}{H^2} A \sin \frac{\pi x}{H} - M_0 \right]^2 dx$$

Integration gives

$$Er = \frac{A^2 H}{2} \left(\frac{EI\pi^2}{H^2} \right)^2 + \frac{4M_0 EI\pi}{H} A + M_0^2 H$$

The error is minimized with respect to A producing

$$\frac{\partial Er}{\partial A} = AH \left(\frac{EI\pi^2}{H^2} \right)^2 + \frac{4M_0 EI\pi}{H} = 0 \tag{1.21}$$

After solving for A, the approximate solution is

$$y(x) = -\frac{4M_0 H^2}{\pi^3 EI} \sin \frac{\pi x}{H} \tag{1.22}$$

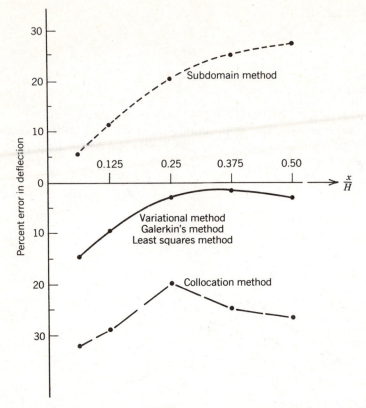

Figure 1.2.   Percentage error for the five approximate solutions for the simply supported beam.

and is identical to the solutions obtained using the variational method and Galerkin's method, (1.13) and (1.20).

It is not possible to state which method is the most accurate. The error depends on the approximating function and the equation being solved. Percentage error curves for the different methods are given in Figure 1.2. It appears that equation (1.22) is more accurate than either equation (1.16) or (1.18). However, it is possible to find a collocation point which produces a maximum deflection that agrees with the exact value (see Problem 1.16). The collocation points or subregions selected affect the accuracy of the approximate solutions.

The important point to be gained from these examples is that the numerical solution of a differential equation can be formulated in terms of an integral. The integral formulation is a basic characteristic of the finite element method.

1.3 POTENTIAL ENERGY FORMULATIONS

The solution of solid mechanics problems, which includes the solution of two- and three-dimensional elasticity problems as well as plate and shell structures, can be approached in several ways. The classical approach is to formulate the governing differential equation and obtain the analytical solution. This does not work for many problems because of difficulties in mathematically describing the structural geometry and/or the boundary conditions. A popular alternative to the classical approach is a numerical procedure based on a principle which states that the displacements at the equilibrium position occur such that the potential energy of a stable system is a minimum value.

A contributing term to the potential energy is the strain energy. This is the energy stored during the deformation process. The strain energy is a volume integral involving products of the stress and strain components. For example, the strain energy in an axial force member is

$$\Lambda = \int_V \frac{\sigma_{xx}\varepsilon_{xx}}{2}\,dV \tag{1.23}$$

More will be said about the principle of minimum potential energy and strain energy in later chapters. The important concept to realize now is that the displacement analysis of structural and solid mechanics problems combines the strain energy integral with a minimization process. Computationally, the analysis of a truss or plate structure looks very similar to the variational and Galerkin approaches of the previous section. The similarity should be apparent by the time the reader has completed this book.

1.4   THE FINITE ELEMENT METHOD

The finite element method is a numerical procedure for solving physical problems governed by a differential equation or an energy theorem. It has two characteristics that distinguish it from other numerical procedures:

1. The method utilizes an integral formulation to generate a system of algebraic equations.
2. The method uses continuous piecewise smooth functions for approximating the unknown quantity or quantities.

The second characteristic distinguishes the finite element method from other numerical procedures that utilize an integral formulation. Recall the approximate solution used in the previous section, $y = A \sin \pi x/H$. This function is continuous and has an infinite number of continuous derivatives. The finite element method uses a continuous function but a function with only enough continuity in the derivatives to allow the integrals to be evaluated. For an integral formulation such as the variational method (1.10), no continuity is required in the first derivative. The integral can be evaluated when the first derivative is piecewise continuous. An equation composed of several linear segments can be used as the approximating equation.

A finite element model for the beam deflection problem considered in the previous section might appear as shown in Figure 1.3. It could consist of several linear segments defined in terms of the nodal values, as shown in Figure 1.3*a*.

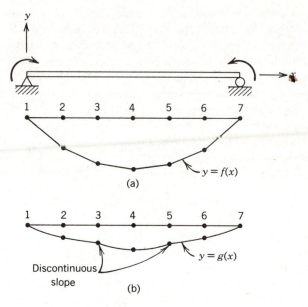

Figure 1.3.   (a) A linear finite element model. (b) A quadratic finite element model.

The interval between each node would be considered an element, and the deflection is approximated by straight-line segments. An alternate grid could consist of three elements each defined by three node points as shown in Figure 1.3*b*. In this case a quadratic equation is defined over each set of three points. In either case, the equations $y = f(x)$ or $y = g(x)$ would not have a continuous first derivative between any pair of adjacent elements.

Functions without continuous first-derivative terms can also be used with Galerkin's method. The second-derivative term, d^2y/dx^2, is modified using integration by parts.

The finite element method can be subdivided into five basic steps. These steps are listed here and illustrated in the next two chapters.

1. Discretize the region. This includes locating and numbering the node points, as well as specifying their coordinates values.
2. Specify the approximation equation. The order of the approximation, linear or quadratic, must be specified and the equations must be written in terms of the unknown nodal values. An equation is written for each element.
3. Develop the system of equations. When using Galerkin's method, the weighting function for each unknown nodal value is defined and the weighted residual integral is evaluated. This generates one equation for each unknown nodal value. In the potential energy formulation, the potential energy of the system is written in terms of the nodal displacements and then is minimized. This gives one equation for each of the unknown displacements.

4. Solve the system of equations.

5. Calculate quantities of interest. These quantities are usually related to the derivative of the parameter and include the stress components, and heat flow and fluid velocities.

1.5 OBJECTIVE AND ORGANIZATION

The objective of this book is to provide a basic introduction to the finite element method as it is used to obtain solutions to heat transfer, irrotational flow and elasticity problems, and the analysis of two-dimensional structural frameworks. This book is not a comprehensive textbook on the finite element method. You should, however, be able to read and understand more advanced books and the technical literature once the material in this book has been completed.

This textbook contains many example problems and problem assignments. The organization of the material has evolved over several years of teaching the material to seniors and graduate students. The book is divided into four parts: basic concepts, field problems, structural and solid mechanics problems, and numerically integrated elements. Once the basic concepts have been covered, the reader may turn to either the group of chapters covering the field problems or those chapters discussing the structural and solid mechanics problems. These two parts of the book are independent. The numerically integrated elements can be covered after either the field problems or the mechanics problems have been studied. Initial emphasis is placed on the linear elements because the matrices for these elements can be evaluated using a hand calculator. The numerically integrated elements should be covered only after the linear elements are thoroughly understood.

A good knowledge of undergraduate mathematics, including some linear algebra, is all that is necessary to handle the material in the first three parts. It is assumed that the reader has a background knowledge in some of the application areas. A limited knowledge of advanced calculus is needed to understand the numerically integrated elements.

A comment on the matrix notation used in this book appears in Appendix I.

PROBLEMS

1.1–1.5 Obtain an approximate displacement equation for the simply supported beam of length H and section property EI shown in Figure P1.1. Assume that the trial displacement equation is $y(x) = A \sin \pi x/H$. Compare the deflection at the center with the theoretical value $y = -5WH^4/384EI$. The governing differential equation is

$$EI \frac{d^2y}{dx^2} - \frac{Wx(H-x)}{2} = 0$$

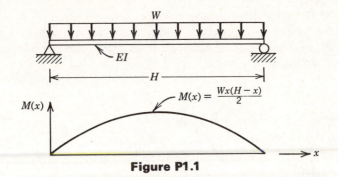

Figure P1.1

1.1   Evaluate A by minimizing the integral

$$\Pi = \int_0^H \left[\frac{EI}{2} \left(\frac{dy}{dx} \right)^2 + \left(\frac{Wx(H-x)}{2} \right) y \right] dx$$

1.2   Evaluate A by requiring that the residual vanish at (a) $x=H/3$, and (b) $x=H/2$.

1.3   Evaluate A using the subdomain method.

1.4   Evaluate A using Galerkin's method.

1.5   Evaluate A using the least squares method.

1.6–1.9   Obtain an approximate displacement equation for the simply supported beam shown in Figure P1.6 using the trial solution $y(x)=A \sin \pi x/H$. Compare the deflection at the center with the theoretical value $y = -0.06415 M_0 H^2/EI$. The governing differential equation is

$$EI \frac{d^2 y}{dx^2} - \frac{M_0 x}{H} = 0$$

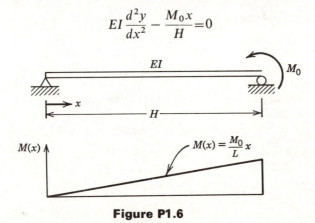

Figure P1.6

1.6   Evaluate A by minimizing the integral·

$$\Pi = \int_0^H \left[\frac{EI}{2} \left(\frac{dy}{dx} \right)^2 + \frac{M_0 x}{H} y \right] dx$$

1.7   Evaluate A by requiring the residual to vanish at (a) $x=H/2$, and (b) $x=0.577H$.

1.8 Evaluate A using the subdomain method.

1.9 Evaluate A using Galerkin's method.

1.10–1.13 Obtain an approximate displacement equation for the simply supported beam shown in Figure P1.10 using the trial solution $y(x) = A \sin \pi x/H$. Compare the deflection at the center with the theoretical value of $y = PH^3/48EI$. The governing differential equations are

$$EI\frac{d^2y}{dx^2} - \frac{Px}{2} = 0 \qquad 0 \leqslant x \leqslant \frac{H}{2}$$

$$EI\frac{d^2y}{dx^2} - \frac{P}{2}(H-x) = 0 \qquad \frac{H}{2} \leqslant x \leqslant H$$

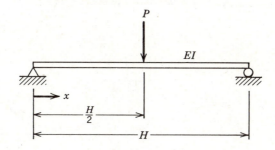

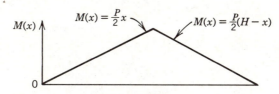

Figure P1.10

1.10 Evaluate A by minimizing the integrals

$$\Pi = \int_0^H \frac{EI}{2}\left(\frac{dy}{dx}\right)^2 dx + \int_0^{H/2} \frac{Pxy}{2} dx + \int_{H/2}^H \frac{P}{2}(H-x)y\,dx$$

1.11 Evaluate A by requiring the residual to vanish at $x = H/2$.

1.12 Evaluate A using the subdomain method.

1.13 Determine the collocation point for which A is equal to the deflection at the center of the beam, that is, $A = -PH^3/48EI$.

1.14 Obtain an approximate displacement equation for the simply supported beam in Figure 1.1 using the equation

$$y(x) = A \sin \frac{\pi x}{H} + B \sin \frac{3\pi x}{H}$$

in conjunction with the variational method. *Hint:* Π must be minimized with respect to both A and B.

1.15 Do Problem 1.14 by requiring that the residual vanish at $x = H/4$ and

$x = H/2$. Both points must be used because there are two unknown coefficients, A and B.

1.16 Determine the collocation points x/H for the beam in Figure 1.1 that yields $A = -M_0 H^2/8EI$ (the correct value of the maximum deflection). Use $y(x) = A \sin \pi x/H$ as the approximate equation.

1.17 Evaluate A using $y(x) = A(x^2 - xH)$ for an approximate solution for the beam shown in Figure 1.1. Use the variational method.

1.18 Evaluate A for the equation and beam in Problem 1.17 by applying the collocation method at $x = H/2$.

1.19 Calculate the value of Π for the beam in Figure 1.1 using (1.10) and the exact solution (1.9). Verify that the value for Π for the exact solution is less than the value for the approximate solution (1.13).

Chapter 2

ONE-DIMENSIONAL LINEAR ELEMENT

Our immediate objective is to discuss the division of a one-dimensional region into linear elements and to develop an element equation. The element equation is then generalized so that a continuous piecewise smooth equation can be written for the region. The linear element is used to obtain an approximate solution to

$$D\frac{d^2\phi}{dx^2}+Q=0 \tag{2.1}$$

in the next chapter. This element is also used to calculate the displacements in a system of axial force members, Chapter 17.

2.1 DIVISION OF THE REGION INTO ELEMENTS

The one-dimensional region is a line segment and the division into subregions or elements is quite straightforward. The line segment is divided into shorter segments by using nodes (Figure 2.1). The nodes are usually numbered consecutively from left to right as are the elements. The element numbers are enclosed in parentheses to distinguish them from the node numbers.

There are some rules to guide the placement of the nodes when obtaining an approximate solution to a differential equation.

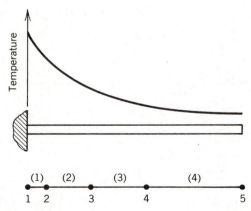

Figure 2.1   Division of a one-dimensional region into elements.

1. Place the nodes closer together in the regions where the unknown parameter changes rapidly and further apart where the unknown is relatively constant.
2. Place a node wherever there is a stepped change in the value of the coefficients D and Q of (2.1).
3. Place a node wherever the numerical value of ϕ in (2.1) is needed.

The first rule requires that the user have some knowledge of how the unknown parameter varies. This is where engineering experience enters the solution process. The second rule is necessary because integrals that include the parameters D and Q of (2.1) must be evaluated. The integrals are easier to evaluate if the coefficients do not experience a stepped change within the interval of integration.

2.2   THE LINEAR ELEMENT

The one-dimensional linear element is a line segment with a length L and two nodes, one at each end (Figure 2.2). The nodes are denoted by i and j and the nodal values by Φ_i and Φ_j. The origin of the coordinate system is to the left of node i.

The parameter ϕ* varies linearly between the nodes, and the equation for ϕ is

$$\phi = a_1 + a_2 x \tag{2.2}$$

The coefficients a_1 and a_2 can be determined by using the nodal conditions

$$\begin{aligned}
\phi &= \Phi_i \quad \text{at} \quad x = X_i \\
\phi &= \Phi_j \quad \text{at} \quad x = X_j
\end{aligned} \tag{2.3}$$

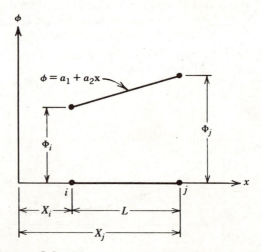

Fugure 2.2   The one-dimensional linear element.

---

*The symbol ϕ is used throughout this text to denote a general scalar quantity. Uppercase symbols, such as X, Y, Φ, and U, denote nodal values.

to develop the pair of equations

$$\Phi_i = a_1 + a_2 X_i$$
$$\Phi_j = a_1 + a_2 X_j \tag{2.4}$$

which yield a_1 and a_2 as

$$a_1 = \frac{\Phi_i X_j - \Phi_j X_i}{X_j - X_i}$$

$$a_2 = \frac{\Phi_j - \Phi_i}{X_j - X_i} \tag{2.5}$$

Substitution of (2.5) into (2.2) and rearranging give

$$\phi = \left(\frac{X_j - x}{L}\right)\Phi_i + \left(\frac{x - X_i}{L}\right)\Phi_j \tag{2.6}$$

where $X_j - X_i$ has been replaced by the element length L.

Equation (2.6) is in a standard finite element form. The nodal values are multiplied by linear functions of x, which are called shape functions or interpolation functions. These functions are denoted by N with a subscript to indicate the node with which a specific shape function is associated. The shape functions in (2.6) are denoted by N_i and N_j with

$$N_i = \frac{X_j - x}{L} \quad \text{and} \quad N_j = \frac{x - X_i}{L} \tag{2.7}$$

Equation (2.6) can be rewritten as

$$\phi = N_i\Phi_i + N_j\Phi_j \tag{2.8}$$

and also as

$$\phi = [N]\{\Phi\} \tag{2.9}$$

where $[N] = [N_i \quad N_j]$ is a row vector of shape functions and

$$\{\Phi\} = \begin{Bmatrix} \Phi_i \\ \Phi_j \end{Bmatrix}$$

is a column vector containing the element nodal values.

A few comments about the shape functions are in order. Each shape function has a value of one at its own node and zero at the other node and the two shape functions sum to one. A third characteristic is that the shape functions are always polynomials of the same type as the original interpolation equation. Equation (2.2) is a linear equation and the shape functions are linear equations. If the interpolation equation had been a quadratic model defined by three nodes, the resulting shape functions would have also been quadratic equations. Another characteristic is that the derivatives of the shape functions with respect to x sum to zero. The shape functions are shown in Figure 2.3.

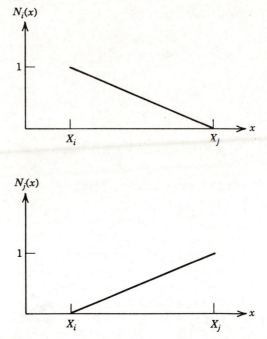

Figure 2.3   The linear shape functions N_i and N_j.

ILLUSTRATIVE EXAMPLE

A one-dimensional linear element has been used to approximate the temperature
distribution in a fin. The solution indicates that the temperatures at nodes i and
j are 120 and 90°C, respectively. Determine the temperature at a point 4 cm from
the origin and the temperature gradient within the element. Nodes i and j are
located at 1.5 and 6 cm from the origin in Figure 2.4.

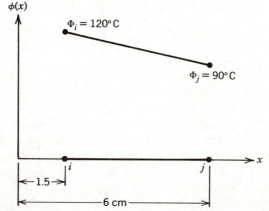

Figure 2.4.   Nodal values for the example problem.

The temperature, ϕ, within the element is given by (2.6)

$$\phi=\left(\frac{X_j-x}{L}\right)\Phi_i+\left(\frac{x-X_i}{L}\right)\Phi_j$$

The element data are

$$X_i=1.5\text{ cm} \qquad X_j=6.0\text{ cm}$$
$$\Phi_i=120°\text{C} \qquad \Phi_j=90°\text{C}$$
$$x=4.0\text{ cm} \qquad L=4.5\text{ cm}$$

Substitution yields

$$\phi=\left(\frac{6-4}{4.5}\right)120+\left(\frac{4-1.5}{4.5}\right)90$$

$$\phi=103.3°\text{C}$$

The temperature gradient is the derivative of (2.6)

$$\frac{d\phi}{dx}=\frac{\Phi_j-\Phi_i}{L} \tag{2.10}$$

Substituting the nodal values, we obtain

$$\frac{d\phi}{dx}=\left(\frac{90-120}{4.5}\right)=-6.67°\text{C/cm}$$

2.3 A CONTINUOUS PIECEWISE SMOOTH EQUATION

A continuous piecewise smooth equation for a one-dimensional region can be constructed by connecting several linear equations with the properties developed in the previous section. Each of these equations can be written as

$$\phi^{(e)}=N_i^{(e)}\Phi_i+N_j^{(e)}\Phi_j \tag{2.11}$$

where

$$N_i^{(e)}=\frac{X_j-x}{X_j-X_i} \qquad \text{and} \qquad N_j^{(e)}=\frac{x-X_i}{X_j-X_i} \tag{2.12}$$

The superscript (e) indicates an element quantity. All that is needed to complete the process is to insert the correct values of i, j, and e for each element. The values of i and j for a corresponding e are obtained from the grid. Node i is the left-hand node of an element. The element information for the grid in Figure 2.1 is

e	i	j
1	1	2
2	2	3
3	3	4
4	4	5

The equation for each element in Figure 2.1 is

$$\phi^{(1)} = N_1^{(1)}\Phi_1 + N_2^{(1)}\Phi_2$$
$$\phi^{(2)} = N_2^{(2)}\Phi_2 + N_3^{(2)}\Phi_3$$
$$\phi^{(3)} = N_3^{(3)}\Phi_3 + N_4^{(3)}\Phi_4 \tag{2.13}$$
$$\phi^{(4)} = N_4^{(4)}\Phi_4 + N_5^{(4)}\Phi_5$$

Note that $N_2^{(1)}$ and $N_2^{(2)}$ are different equations even though both involve node two. The equations for these two quantities are

$$N_2^{(1)} = \frac{x - X_1}{X_2 - X_1} \quad \text{and} \quad N_2^{(2)} = \frac{X_3 - x}{X_3 - X_2}$$

Realize that each equation of (2.13) is for a single element and is not applicable outside the element. The first equation should be written as

$$\phi^{(1)} = N_1^{(1)}\Phi_1 + N_2^{(1)}\Phi_2 \qquad X_1 \leqslant x \leqslant X_2$$

but the range of x is deleted in most of the finite element literature and is deleted in this book. Whenever an element equation is given, the implication is that it is valid for only a single element.

2.4   A COMMENT ON NOTATION

The need to denote element quantities occurs on most of the pages of this book. The following notation is used so that a superscript (e) does not have to be placed on every coefficient.

1.   When brackets or parentheses have a superscript (e), that is, $(G\phi + Q)^{(e)}$, then every term within the brackets or parentheses should be interpreted on an element basis.
2.   A quantity on the left-hand side of an equal sign with a superscript (e) implies that the quantities on the right-hand side of the sign are for a particular element. For example,

$$\phi^{(e)} = N_i\Phi_i + N_j\Phi_j$$

implies that N_i and N_j are really $N_i^{(e)}$ and $N_j^{(e)}$, and that Φ_i and Φ_j are the element nodal values.

PROBLEMS

2.1   The nodal coordinates X_i and X_j and the nodal values of Φ_i and Φ_j for several linear elements are given below. Evaluate ϕ at the given value of x. The x values are in centimeters, and Φ_i and Φ_j are in degrees Celsius.

x	X_i	X_j	Φ_i	Φ_j
(a) 0.8	0.0	1.5	60	43
(b) 3.6	3.0	4.5	27	33
(c) 7.1	6.5	7.5	63	51
(d) 1.8	0.5	3.0	0	−15
(e) 2.2	1.0	3.0	60	67

2.2 (a–e)   Evaluate $d\phi/dx$ for the corresponding element in Problem 2.1.

2.3   The shape functions for the quadratic element shown in Figure P2.3 are

$$N_i = \frac{2}{L^2}(x - X_j)(x - X_k)$$

$$N_j = -\frac{4}{L^2}(x - X_i)(x - X_k)$$

$$N_k = \frac{2}{L^2}(x - X_i)(x - X_j)$$

(a) Show that these shape functions equal one at their own node and are zero at the other two nodes. Also show that the shape functions sum to one.

(b) Show that the derivatives of N_i, N_j, and N_k with respect to x sum to zero.

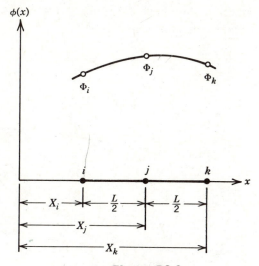

Figure P2.3

2.4   The implementation of the finite element method requires the evaluation of integrals that contain the shape functions or their derivatives. Evaluate

(a) $\displaystyle\int_{X_i}^{X_j} N_i\, dx$ (b) $\displaystyle\int_{X_i}^{X_j} \frac{dN_i}{dx}\frac{dN_j}{dx}\, dx$ (c) $\displaystyle\int_{X_i}^{X_j} N_j^2\, dx$

for the linear element.

2.5 The coordinate s shown in Figure P2.5 has its origin at node i and a value of L at node j. Develop the shape functions $N_i(s)$ and $N_j(s)$ starting with $\phi(s)=a_1+a_2 s$ and solving for a_1 and a_2.

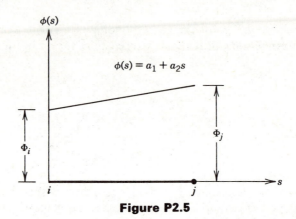

Figure P2.5

2.6 The coordinate ξ shown in Figure P2.6 is a natural coordinate whose origin is at the center of the element. The value of ξ at nodes i and j is 1 and -1, respectively. Develop the shape functions $N_i(\xi)$ and $N_j(\xi)$ starting with $\phi(\xi)=a_1+a_2\xi$ and solving for a_1 and a_2.

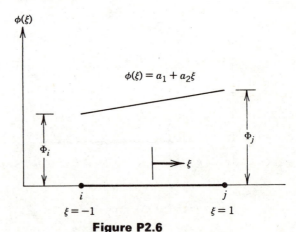

Figure P2.6

2.7 The beam element shown in Figure P2.7 has two vertical displacements, v_i and v_j, and two rotations, θ_i and θ_j, defined at the end points. The displacement equation is

$$v = a_1 + a_2 x + a_3 x^2 + a_4 x^3$$

where

$$a_1 = v_i, \quad a_3 = \frac{3}{L^2}(v_j - v_i) - \frac{1}{L}(2\theta_i + \theta_j)$$

$$a_2 = \theta_i, \quad a_4 = \frac{2}{L^3}(v_i - v_j) + \frac{1}{L^2}(\theta_i + \theta_j)$$

Develop the shape function equations for the interpolation equation

$$v(x) = N_1 v_i + N_2 \theta_i + N_3 v_j + N_4 \theta_j$$

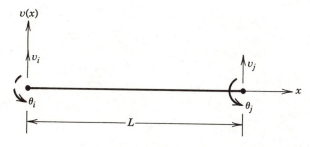

Figure P2.7

2.8 The equation for $\phi(x, y)$ in a two-dimensional rectangular element shown in Figure P2.8 is

$$\phi(x, y) = C_1 + C_2 s + C_3 t + C_4 st$$

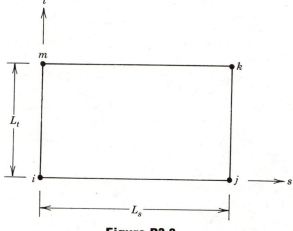

Figure P2.8

The coefficients are

$$C_1 = \Phi_1, \ C_3 = \frac{1}{L_t}(\Phi_m - \Phi_i)$$

$$C_2 = \frac{1}{L_s}(\Phi_j - \Phi_i), \ C_4 = \frac{1}{L_s L_t}(\Phi_i - \Phi_j + \Phi_k - \Phi_m)$$

where Φ_i, Φ_j, Φ_k, and Φ_m are the nodal values of ϕ. Develop the shape function equations for the interpolation equation

$$\phi(x, y) = N_i \Phi_i + N_j \Phi_j + N_k \Phi_k + N_m \Phi_m$$

Chapter 3

A FINITE ELEMENT EXAMPLE

The shape function information developed in Chapter 2 is general information that can be used in solving solid mechanics problems as well as differential equations. The objective in this chapter is to illustrate the finite element method by developing an approximate solution for the one-dimensional differential equation

$$D\frac{d^2\phi}{dx^2}+Q=0 \tag{3.1}$$

with the boundary conditions

$$\phi(0)=\phi_0 \quad \text{and} \quad \phi(H)=\phi_H \tag{3.2}$$

Two physical problems are embedded within (3.1): the deflection of simply supported beams when the bending moment diagram is known and heat flow through a composite wall when the surface temperatures are known.

The finite element equations are obtained using Galerkin's formulation. Evaluation of the residual integral yields a nodal equation that is applied in a recursive manner to generate a system of linear equations. The nodal equation is used to solve a beam deflection problem.

The reader whose immediate interest is the analysis of structural frameworks may substitute Chapters 17 and 18 for Chapters 3 and 4. Chapters 3 and 4, however, should be read before beginning Part Two of the book.

3.1 WEIGHTING FUNCTIONS

A system of linear equations is generated by evaluating the weighted residual integral*

$$-\int_0^H W(x)\left(D\frac{d^2\phi}{dx^2}+Q\right)dx=0 \tag{3.3}$$

using a new weighting function for each node where ϕ is unknown. Galerkin's formulation of the weighted residual method requires that the weighting functions be constructed using the shape functions N_i and N_j. The weighting functions in the Galerkin finite element formulation are defined as follows: *The weighting function for the sth node, W_s, consists of the shape functions associated with the sth node.*

---

*The integral has been multiplied by a negative one so that the results can be written in a more convenient form.

The weighting function for node three of a linear grid (Figure 3.1) consists of the shape functions for node three.

$$W_3(x) = \begin{cases} N_3^{(2)} & X_2 \leqslant x \leqslant X_3 \\ N_3^{(3)} & X_3 \leqslant x \leqslant X_4 \end{cases} \tag{3.4}$$

In general,

$$W_s(x) = \begin{cases} N_s^{(e)} & X_r \leqslant x \leqslant X_s \\ N_s^{(e+1)} & X_s \leqslant x \leqslant X_t \end{cases} \cdot \tag{3.5}$$

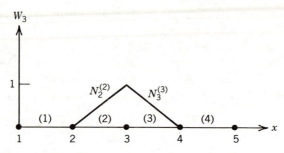

Figure 3.1. The weighting function for node three.

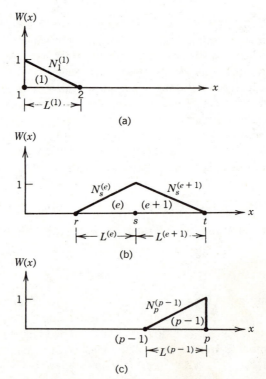

Figure 3.2. The weighting functions for (a) the first node, (b) an interior node, and (c) the last node.

This function is occasionally called a hat function for reasons apparent in Figure 3.2*b*. The weighting functions for the first and last nodes are shown in Figure 3.2*a* and 3.2*c*. The respective equations are

$$W_1(x) = N_1^{(1)} \qquad \text{and} \qquad W_p(x) = N_p^{(p-1)} \tag{3.6}$$

The weighting function for every node consists either of N_i, N_j, or a combination of the two.

3.2 THE WEIGHTED RESIDUAL INTEGRAL

After defining the weighting functions, the next step is to evaluate the residual integral (3.3). Using the sequence of nodes *r*, *s*, and *t* in Figure 3.2*b*, we find that (3.3) becomes

$$R_s = R_s^{(e)} + R_s^{(e+1)} = - \int_{X_r}^{X_s} \left[N_s \left(D \frac{d^2\phi}{dx^2} + Q \right) \right]^{(e)} dx$$

$$- \int_{X_s}^{X_t} \left[N_s \left(D \frac{d^2\phi}{dx^2} + Q \right) \right]^{(e+1)} dx = 0 \tag{3.7}$$

because $W_s = 0$ for $x < X_r$ and $x > X_t$. The integral splits into two parts because $W_s(x)$ is defined by two separate equations within the interval $X_r \leqslant x \leqslant X_t$. The terms $R_s^{(e)}$ and $R_s^{(e+1)}$ represent the contributions of elements (e) and $(e+1)$ to the residual equation for node *s*.

There is a problem associated with each integral in (3.7). The approximate solution does not have continuity in the first derivative $d\phi/dx$; therefore, the integral of $d^2\phi/dx^2$ is not defined. This difficulty can be circumvented, however, by changing $d^2\phi/dx^2$ into a new term. Consider the first integral in (3.7) and note that

$$\frac{d}{dx}\left(N_s \frac{d\phi}{dx} \right) = N_s \frac{d^2\phi}{dx^2} + \frac{dN_s}{dx} \frac{d\phi}{dx} \tag{3.8}$$

Then

$$N_s \frac{d^2\phi}{dx^2} = \frac{d}{dx}\left(N_s \frac{d\phi}{dx} \right) - \frac{dN_s}{dx} \frac{d\phi}{dx} \tag{3.9}$$

Substituting into the integral gives

$$- \int_{X_r}^{X_s} \left(N_s D \frac{d^2\phi}{dx^2} \right)^{(e)} dx = -\left(D N_s \frac{d\phi}{dx} \right)^{(e)} \Bigg|_{X_r}^{X_s}$$

$$+ \int_{X_r}^{X_s} \left(D \frac{dN_s}{dx} \frac{d\phi}{dx} \right)^{(e)} dx \tag{3.10}$$

A similar set of operations applied to the first term of the second integral in (3.7) produces

$$- \int_{X_s}^{X_t} \left(N_s D \frac{d^2 \phi}{dx^2} \right)^{(e+1)} dx = - \left(D N_s \frac{d\phi}{dx} \right)^{(e+1)} \Big|_{X_s}^{X_t}$$

$$+ \int_{X_s}^{X_t} \left(D \frac{dN_s}{dx} \frac{d\phi}{dx} \right)^{(e+1)} dx \tag{3.11}$$

The first terms in each of (3.10) and (3.11) simplify because the shape functions are either zero or one at the respective nodes. The complete residual equation is

$$R_s = R_s^{(e)} + R_s^{(e+1)} = - \int_0^H W_s \left(D \frac{d^2 \phi}{dx^2} + Q \right) dx$$

$$= - \left(D \frac{d\phi}{dx} \right)^{(e)} \Big|_{x=X_s}$$

$$+ \int_{X_r}^{X_s} \left(D \frac{dN_s}{dx} \frac{d\phi}{dx} - N_s Q \right)^{(e)} dx$$

$$+ \left(D \frac{d\phi}{dx} \right)^{(e+1)} \Big|_{x=X_s}$$

$$+ \int_{X_s}^{X_t} \left(D \frac{dN_s}{dx} \frac{d\phi}{dx} - N_s Q \right)^{(e+1)} dx = 0 \tag{3.12}$$

The pair of terms evaluated at $x = X_s$ establishes an interelement requirement. The residual cannot be zero until the difference between these two quantities is zero.

3.3   EVALUATION OF THE INTEGRAL

Evaluation of the integrals in (3.12) yields the residual equation for an interior node. The equations for the first and last nodes can be obtained from these operations and are left as exercises.

Starting with element (e)

$$\phi^{(e)} = N_r \Phi_r + N_s \Phi_s$$

$$\phi^{(e)} = \left(\frac{X_s - x}{L} \right) \Phi_r + \left(\frac{x - X_r}{L} \right) \Phi_s \tag{3.13}$$

Thus

$$N_s^{(e)} = \frac{x - X_r}{L}, \qquad \frac{dN_s^{(e)}}{dx} = \frac{1}{L} \tag{3.14}$$

and

$$\frac{d\phi^{(e)}}{dx} = \frac{1}{L} (-\Phi_r + \Phi_s) \tag{3.15}$$

Substitution of the appropriate terms and evaluating the integrals gives

$$\int_{X_r}^{X_s} D \frac{dN_s}{dx} \frac{d\phi}{dx} dx = \frac{D}{L}(-\Phi_r + \Phi_s) \tag{3.16}$$

and

$$\int_{X_r}^{X_s} Q N_s \, dx = \frac{QL}{2} \tag{3.17}$$

Equations (3.16) and (3.17) are combined with the interelement contribution for element (e) to give

$$R_s^{(e)} = -\left(D \frac{d\phi}{dx}\right)^{(e)}\Bigg|_{x=X_s} + \frac{D}{L}(-\Phi_r + \Phi_s) - \frac{QL}{2} \tag{3.18}$$

Proceeding with the second integral of (3.12)

$$\phi^{(e+1)} = N_s \Phi_s + N_t \Phi_t$$

$$\phi^{(e+1)} = \left(\frac{X_t - x}{L}\right)\Phi_s + \left(\frac{x - X_s}{L}\right)\Phi_t \tag{3.19}$$

Thus

$$N_s^{(e+1)} = \frac{X_t - x}{L}, \qquad \frac{dN_s^{(e+1)}}{dx} = -\frac{1}{L} \tag{3.20}$$

and

$$\frac{d\phi^{(e+1)}}{dx} = \frac{1}{L}(-\Phi_s + \Phi_t) \tag{3.21}$$

Evaluation of the integrals yields

$$\int_{X_s}^{X_t} D \frac{dN_s}{dx} \frac{d\phi}{dx} dx = \frac{D}{L}(\Phi_s - \Phi_t) \tag{3.22}$$

$$\int_{X_s}^{X_t} Q N_s \, dx = \frac{QL}{2} \tag{3.23}$$

and

$$R_s^{(e+1)} = D \frac{d\phi}{dx}\Bigg|_{x=X_s} + \frac{D}{L}(\Phi_s - \Phi_t) - \frac{QL}{2} \tag{3.24}$$

Combining (3.18) and (3.24) gives the residual equation for node s

$$R_s = \left(D \frac{d\phi}{dx}\right)^{(e+1)}\Bigg|_{x=X_s} - \left(D \frac{d\phi}{dx}\right)^{(e)}\Bigg|_{x=X_s}$$

$$- \left(\frac{D}{L}\right)^{(e)}\Phi_r + \left[\left(\frac{D}{L}\right)^{(e)} + \left(\frac{D}{L}\right)^{(e+1)}\right]\Phi_s - \left(\frac{D}{L}\right)^{(e+1)}\Phi_t$$

$$- \left(\frac{QL}{2}\right)^{(e)} - \left(\frac{QL}{2}\right)^{(e+1)} = 0 \tag{3.25}$$

The usual solution procedure is to generate the system of equations without the interelement terms. Once the equations have been solved,

$$\left(D\frac{d\phi}{dx}\right)^{(e+1)}\bigg|_{x=X_s} - \left(D\frac{d\phi}{dx}\right)^{(e)}\bigg|_{x=X_s} \tag{3.26}$$

can be calculated. Theoretically, the value of (3.26) can be used to evaluate the quality of the grid and to indicate where the grid should be refined. A practical way to implement this idea, however, has not been developed.

Equation (3.26) does contain some important information. If $D^{(e)} = D^{(e+1)}$, then the interelement requirement reduces to

$$\left(\frac{d\phi}{dx}\right)^{(e+1)}\bigg|_{x=X_s} - \left(\frac{d\phi}{dx}\right)^{(e)}\bigg|_{x=X_s} \tag{3.27}$$

There must be continuity in the slopes before the residual is zero. This continuity can never be attained with a linear element unless the solution is a straight line. The value of (3.27) becomes smaller as the grid is refined, but it is never zero for all of the nodes.

The interelement requirement (3.26) can be viewed as an error term similar to the one associated with finite difference approximations. The error term is not incorporated into the system of equations. It is, however, a constant reminder that the solution is only approximate.

Deleting the interelement requirement from (3.25), assuming a sequential numbering of the nodes and elements, and writing all quantities in terms of s gives the nodal residual equation

$$R_s = -\left(\frac{D}{L}\right)^{(s-1)}\Phi_{s-1} + \left[\left(\frac{D}{L}\right)^{(s-1)} + \left(\frac{D}{L}\right)^{(s)}\right]\Phi_s - \left(\frac{D}{L}\right)^{(s)}\Phi_{s+1}$$

$$-\left(\frac{QL}{2}\right)^{(s-1)} - \left(\frac{QL}{2}\right)^{(s)} = 0 \tag{3.28}$$

3.4 ANALYSIS OF A SIMPLY SUPPORTED BEAM

The general residual equation, (3.28), obtained by evaluating the weighted residual integral is now used to obtain approximate deflection values for a simply supported beam subjected to concentrated end moments.

The beam in Figure 3.3 has been reinforced over the center one-half of its span through the use of steel plates that are welded to the basic section. The beam data, its bending moment diagram, and the finite element model are given in the figure.

The governing differential equation for the deflection curve is

$$EI\frac{d^2\phi}{dx^2} - M(x) = 0 \tag{3.29}$$

where EI is a bending stiffness term composed of the elastic modulus of the material, E, N/cm^2, and the area moment of the cross section, I, cm^4. The internal bending

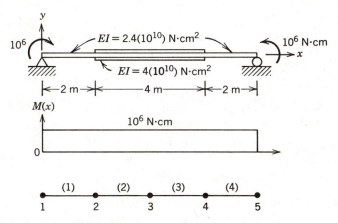

Figure 3.3 A simply supported beam subjected to concentrated end moments.

moment, $M(x)$, is in N · cm and y is the deflection, cm. The boundary conditions are $y(0)=y(800 \text{ cm})=0$.

Comparing (3.29) with (3.1) yields the relationships $D=EI$ and $Q=-M(x)=-10^6$. The element data for the beam are summarized as follows (the element length must be in centimeters).

e	D	Q	L
1	$2.4(10)^{10}$	-10^6	200
2	$4.0(10)^{10}$	-10^6	200
3	$4.0(10)^{10}$	-10^6	200
4	$2.4(10)^{10}$	-10^6	200

Since Q and L have constant values, (3.28) simplifies to

$$R_s = \frac{-D^{(s-1)}Y_{s-1}+(D^{(s-1)}+D^{(s)})Y_s-D^{(s)}Y_{s+1}}{L}-QL=0 \qquad (3.30)$$

where Y has been used to denote the nodal deflection values.

Writing the residual equation for nodes two, three, and four gives

$$R_2 = -1.2Y_1+3.2Y_2-2.0Y_3+2=0$$
$$R_3 = -2.0Y_2+4.0Y_3-2.0Y_4+2=0$$
$$R_4 = -2.0Y_3+3.2Y_4-1.2Y_5+2=0 \qquad (3.31)$$

after the 10^8 multiplier has been canceled. Incorporation of the boundary conditions $Y_1=Y_5=0$ produces

$$R_2 = \quad 3.2Y_2-2.0Y_3 \qquad\qquad =-2$$
$$R_3 = -2.0Y_2+4.0Y_3-2.0Y_4=-2$$
$$R_4 = \qquad\qquad -2.0Y_3+3.2Y_4=-2 \qquad (3.32)$$

which yields

$$Y_2 = -2.50 \text{ cm}, \qquad Y_3 = -3.0 \text{ cm}, \qquad \text{and} \qquad Y_4 = -2.50 \text{ cm}$$

when solved.

The solution of (3.32) for Y_2, Y_3, and Y_4 would appear to end the example. There are many situations, however, where there is a need to calculate quantities pertaining to a particular element once the nodal values are known. A couple of items are calculated here to illustrate the procedures.

The first calculation involves determining the deflection at $x = 300$ cm. This point occurs in element two (see Figure 3.3); thus

$$y^{(2)} = N_2^{(2)} Y_2 + N_3^{(2)} Y_3$$

$$= \left(\frac{X_3 - x}{X_3 - X_2}\right) Y_2 + \left(\frac{x - X_2}{X_3 - X_2}\right) Y_3$$

Noting that $X_2 = 200$ cm and $X_3 = 400$ cm, and using the calculated nodal values, we find that

$$y = \left(\frac{400 - 300}{400 - 200}\right)(-2.50) + \left(\frac{300 - 200}{400 - 200}\right)(-3.0)$$

$$= -\frac{1}{2}(2.50 + 3.0) = -2.75 \text{ cm}$$

For the second calculation, the slope at node one is evaluated. This node is located in element one; therefore, using (2.10), we obtain

$$\frac{dy^{(1)}}{dx} = \frac{1}{L}(-Y_1 + Y_2) = \frac{-2.50 - 0}{200} = -0.0125 \text{ cm/cm}$$

The slope is constant within the element. The constant slope property is a major disadvantage of the linear elements.

The beam deflection problem solved in this section was selected because of the governing differential equation. The reader should not leave this chapter with the idea that all beam deflection problems should be solved this way. There is a more efficient method that involves a specific beam element. This element is covered in the structural mechanics part of this book (Part Three).

3.5   MATRIX NOTATION

Some comments relative to the system of equations (3.32) are in order. It is both convenient and orderly to use matrix notation when working with a system of equations. One way of writing (3.32) in matrix notation is

$$\begin{Bmatrix} R_2 \\ R_3 \\ R_4 \end{Bmatrix} = \begin{bmatrix} 3.2 & -2 & 0 \\ -2 & 4 & -2 \\ 0 & -2 & 3.2 \end{bmatrix} \begin{Bmatrix} Y_2 \\ Y_3 \\ Y_4 \end{Bmatrix} - \begin{Bmatrix} -2 \\ -2 \\ -2 \end{Bmatrix} = \begin{Bmatrix} 0 \\ 0 \\ 0 \end{Bmatrix} \qquad (3.33)$$

This set of equations is written symbolically as

$$\{R\} = [K]\{Y\} - \{F\} = \{0\} \tag{3.34}$$

where $\{R\}$ is the vector of residual equations,

$$[K] = \begin{bmatrix} 3.2 & -2 & 0 \\ -2 & 4 & -2 \\ 0 & -2 & 3.2 \end{bmatrix} \tag{3.35}$$

is called the global stiffness matrix, and

$$\{F\} = \begin{Bmatrix} -2 \\ -2 \\ -2 \end{Bmatrix} \tag{3.36}$$

is called the global force vector. The terms *stiffness* and *force* come from the matrix analysis of structures. The stiffness matrix, $[K]$, is symmetric when the system of equations has a potential energy formulation or a Galerkin formulation of a differential equation that is self-adjoint.

PROBLEMS

3.1 (a) Obtain the final system of finite element equations for the nodal deflections of the stepped beam shown in Figure P3.1.

(b) Solve the equations in (a) and calculate the deflection at $x = 3H/16$.

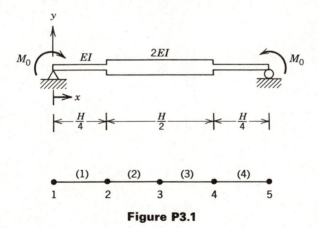

Figure P3.1

3.2 The differential equation $D^{(e)} d^2\phi/dx^2 = 0$ is applicable to each section of the composite wall shown in Figure P3.2, where $D^{(e)}$ is the thermal conductivity. Calculate the nodal temperature values within the wall and evaluate the heat flow through each material. The heat flow is given by $q = -D^{(e)} d\phi/dx$. A unit of surface area is assumed.

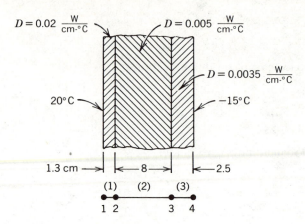

Figure P3.2

3.3 Perform the calculations in Problem 3.2 for the configuration shown in Figure P3.3.

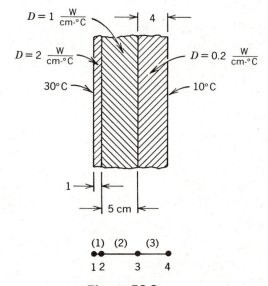

Figure P3.3

3.4 Start with (3.28) and develop and solve the system of finite element equations for an approximate solution to the differential equation $d^2\phi/dx^2 + Q = 0$ using the value for Q and the boundary conditions given in the following table. Divide the interval $[0, 2]$ into four elements, each with a length of 0.5 cm. The nodes and elements are numbered as shown in Figure P3.4.

	Q	$\phi(0)$	$\phi(2)$
(a)	4	0	3.0
(b)	6	1	2.0
(c)	-3	2	0.5
(d)	2	-1	1.5
(e)	-5	3	-2.0

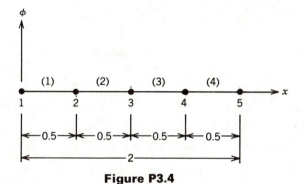

Figure P3.4

3.5 Evaluate the residual equation for node one using the weighting function shown in Figure 3.2a. Note that the answer is the same as (3.24) with $(e+1)=$ (1), $s=1$, and $t=2$.

3.6 Evaluate the residual equation for node p using the weighting function shown in Figure 3.2c. Note that the answer is the same as (3.18) with $(e)=(p-1)$, $r=p-1$, and $s=p$.

3.7 Evaluate the contribution of

$$\int_0^H WQ\,dx$$

to the Galerkin residual equation R_s when Q varies linearly over an element

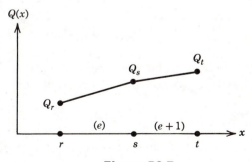

Figure P3.7

(Figure P3.7). The equation for $Q(x)$ in each element is

$$Q^{(e)} = N_r^{(e)}Q_r + N_s^{(e)}Q_s$$
$$Q^{(e+1)} = N_s^{(e+1)}Q_s + N_t^{(e+1)}Q_t$$

where Q_r, Q_s, and Q_t are the nodal values of Q. Note that $(X_a - X_b)^3 = X_a^3 - 3X_a^2 X_b + 3X_a X_b^2 - X_b^3$.

3.8 The residual equation for a uniform grid and a linear variation of $Q(x)$ between nodes is given by

$$R_s = \frac{-D^{(s-1)}Y_{s-1} + [D^{(s-1)} + D^{(s)}]Y_s - D^{(s)}Y_{s+1}}{L}$$

$$- L\frac{(Q_{s-1} + 4Q_s + Q_{s+1})}{6} = 0$$

Use this equation to obtain the nodal displacements for the beam shown in Figure P3.8. The governing differential equation is

$$EI\frac{d^2\phi}{dx^2} - M(x) = 0$$

and $M(x)$ is given in the figure. Each element is 200 cm long; $EI = 2(10^{10})\text{N}\cdot\text{cm}^2$.

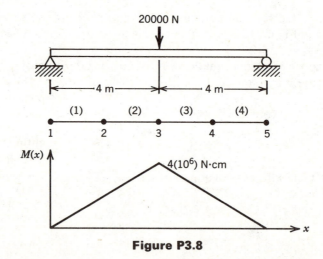

Figure P3.8

3.9 Use the residual equation in Problem 3.8 to obtain the nodal displacements for the beam shown in Figure P3.9. The governing differential equation is

$$D\frac{d^2\phi}{dx^2} - M(x) = 0$$

and $M(x)$ is given in the figure. Each element is 300 cm long; $EI = 2(10^{10})\text{N}\cdot\text{cm}^2$.

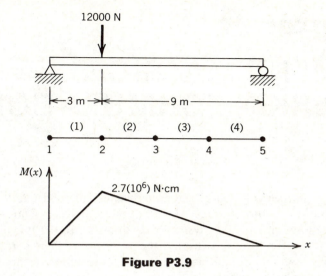

Figure P3.9

Chapter 4

ELEMENT MATRICES: GALERKIN FORMULATION

The computer implementation of the finite element method parallels the computer solution of structural frameworks. The matrix analysis of structures places the emphasis on the element. The system of equations is constructed by calculating the element's contribution and placing the values into their correct positions within the final system of equations. The final set of equations emerges after all of the elements have been considered.

The objective in this chapter is to determine the element contributions to the final system of equations and show where these contributions are located. Matrix notation is used, and an element stiffness matrix and an element force vector are defined for the differential equation analyzed in the previous chapter.

There are three points to keep in mind as the element matrices are developed. First, the residual equations are always arranged in numerical sequence, that is, $R_1, R_2, \ldots, R_{p-1}, R_p$ where there are p nodal values. Second, the nodal values $\Phi_1, \Phi_2, \ldots, \Phi_p$ are arranged sequentially within an equation. Third, an equation is developed for each node. The boundary conditions are incorporated after all of the equations have been developed.

4.1 ELEMENT MATRICES

The discussion starts by defining a column vector $\{R\}$. Each component of $\{R\}$ represents a residual equation. The vector is

$$\{R\} = \begin{Bmatrix} R_1 \\ R_2 \\ \vdots \\ R_{p-1} \\ R_p \end{Bmatrix} \tag{4.1}$$

where R_β is the residual equation for node β. The residual equation is further subdivided into element contributions. For example, $R_\beta^{(e)}$ is the contribution of element (e) to the residual equation for node β.

The weighting functions for a four-element grid are shown in Figure 4.1. Analyzing these functions from an element point of view rather than an equation point of view shows that element three contributes nonzero values to equations three and

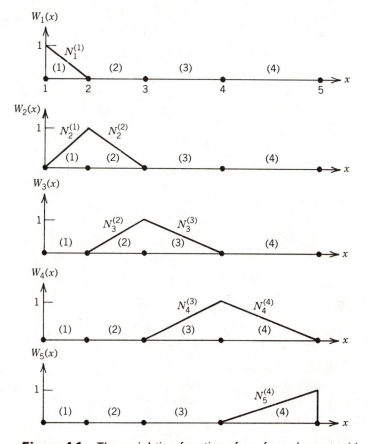

Figure 4.1 The weighting functions for a four-element grid.

four. Its contribution to equation three is

$$R_3^{(3)} = - \int_{X_3}^{X_4} N_3^{(3)} \left(D \frac{d^2\phi}{dx^2} + Q \right) dx \tag{4.2}$$

while its contribution to equation four is

$$R_4^{(3)} = - \int_{X_3}^{X_4} N_4^{(3)} \left(D \frac{d^2\phi}{dx^2} + Q \right) dx \tag{4.3}$$

Element three contributes nothing to the other equations because every weighting function except W_3 and W_4 are zero in element three.

The results for element three can be generalized. Given an arbitrary element with nodes i and j (Figure 4.2), we find that it contributes

$$R_i^{(e)} = - \int_{X_i}^{X_j} N_i(x) \left(D \frac{d^2\phi}{dx^2} + Q \right) dx \tag{4.4}$$

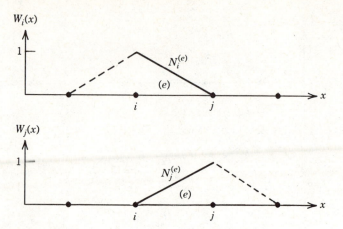

Figure 4.2.   The weighting functions for element (e).

to equation i and

$$R_j^{(e)} = -\int_{X_i}^{X_j} N_j(x)\left(D\frac{d^2\phi}{dx^2}+Q\right)dx \tag{4.5}$$

to equation j.

The integrals in (4.4) and (4.5) were evaluated in Chapter 3. The first integral, $R_i^{(e)}$, is equivalent to $R_s^{(e+1)}$, (3.24), with $s=i$ and $t=j$ whereas $R_j^{(e)}$ is the same as $R_s^{(e)}$, (3.18), with $r=i$ and $s=j$. Using (3.18) and (3.24) gives

$$R_i^{(e)} = D\frac{d\phi}{dx}\bigg|_{x=X_i} + \frac{D}{L}(\Phi_i-\Phi_j) - \frac{QL}{2} \tag{4.6}$$

$$R_j^{(e)} = -D\frac{d\phi}{dx}\bigg|_{x=X_j} + \frac{D}{L}(-\Phi_i+\Phi_j) - \frac{QL}{2} \tag{4.7}$$

Equations (4.6) and (4.7) can be written as

$$\begin{Bmatrix} R_i^{(e)} \\ R_j^{(e)} \end{Bmatrix} = \begin{Bmatrix} I_i^{(e)} \\ I_j^{(e)} \end{Bmatrix} + \frac{D}{L}\begin{bmatrix} 1 & -1 \\ -1 & 1 \end{bmatrix}\begin{Bmatrix} \Phi_i \\ \Phi_j \end{Bmatrix} - \frac{QL}{2}\begin{Bmatrix} 1 \\ 1 \end{Bmatrix} \tag{4.8}$$

or

$$\{R^{(e)}\} = \{I^{(e)}\} + [k^{(e)}]\{\Phi^{(e)}\} - \{f^{(e)}\} \tag{4.9}$$

where $\{R^{(e)}\}$ is the contribution of element (e) to the final system of equations. This contribution consists of an element stiffness matrix $[k^{(e)}]$ and an element force vector $\{f^{(e)}\}$. The other vectors are

$$\{\Phi^{(e)}\} = \begin{Bmatrix} \Phi_i \\ \Phi_j \end{Bmatrix}$$

which is the column vector of nodal values and

$$\{I^{(e)}\} = \begin{Bmatrix} I_i^{(e)} \\ I_j^{(e)} \end{Bmatrix} = \begin{Bmatrix} D\dfrac{d\phi}{dx}\Big|_{x=X_i} \\ -D\dfrac{d\phi}{dx}\Big|_{x=X_j} \end{Bmatrix} \tag{4.10}$$

which is the element contribution to the interelement requirement. Equation (4.10) is deleted from the discussion except when derivative boundary conditions are specified at node one or p for the reason given relative to (3.26) and (3.27).

The element matrices are the important results and they are

$$[k^{(e)}] = \frac{D}{L} \begin{bmatrix} 1 & -1 \\ -1 & 1 \end{bmatrix} \tag{4.11}$$

and

$$\{f^{(e)}\} = \frac{QL}{2} \begin{Bmatrix} 1 \\ 1 \end{Bmatrix} \tag{4.12}$$

The element stiffness matrix is the matrix that multiplies the column vector of nodal values, $\{\Phi^{(e)}\}$. Equations (4.11) and (4.12) are useful because they are easy to program for computer evaluation; also, it can be quickly determined where each coefficient is located in the final system of equations.

The vector $\{R\}$ represents a system of equations that symbolically is

$$\{R\} = [K]\{\Phi\} - \{F\} = \{0\} \tag{4.13}$$

Equation (4.8) states that the coefficients in the first row of $[k^{(e)}]$ and $\{f^{(e)}\}$ are located in row i of $[K]$ and $\{F\}$ because row i is the row associated with $R_i^{(e)}$. Similarly, the coefficients in the second row of $[k^{(e)}]$ and $\{f^{(e)}\}$ are located in row j of $[K]$ and $\{F\}$ because this row is associated with $R_j^{(e)}$. The coefficients of $[k^{(e)}]$ are located in columns i and j of $[K]$ because the coefficients in the first column multiply Φ_i and those in the second column multiply Φ_j.

4.2 DIRECT STIFFNESS METHOD

The direct stiffness method is the name given to the procedure for incorporating the element matrices into the final system of equations. The method is simple and straightforward. The numerical values of i and j for a specific element are written over the columns of $[k^{(e)}]$ and along the side of $[k^{(e)}]$ and $\{f^{(e)}\}$, that is,

$$[k^{(e)}] = \begin{matrix} & i & j \\ & \begin{bmatrix} k_{11} & k_{12} \\ k_{21} & k_{22} \end{bmatrix} & \begin{matrix} i \\ j \end{matrix} \end{matrix} \qquad \{f^{(e)}\} = \begin{Bmatrix} f_1 \\ f_2 \end{Bmatrix} \begin{matrix} i \\ j \end{matrix} \tag{4.14}$$

The direct stiffness procedure is illustrated by using the hypothetical set of matrices

$$[k^{(e)}] = \begin{bmatrix} 4 & 6 \\ 5 & 7 \end{bmatrix}, \qquad \{f^{(e)}\} = \begin{Bmatrix} 8 \\ 9 \end{Bmatrix}$$

for a linear element between nodes 2 and 3 ($i=2, j=3$). Using these values of i and j gives

$$[k^{(e)}] = \begin{matrix} 2 & 3 \\ \begin{bmatrix} 4 & 6 \\ 5 & 7 \end{bmatrix} & \begin{matrix} 2 \\ 3 \end{matrix} \end{matrix}, \qquad \{f^{(e)}\} = \begin{Bmatrix} 8 \\ 9 \end{Bmatrix} \begin{matrix} 2 \\ 3 \end{matrix}$$

and the location of the coefficients in $[K]$ and $\{F\}$ are

> 4 adds to K_{22}
> 6 adds to K_{23}
> 5 adds to K_{32}
> 7 adds to K_{33}
> 8 adds to F_2
> 9 adds to F_3

The word *add* is emphasized because there could be contributions to K_{22}, K_{23}, K_{32}, K_{33}, F_2, and F_3 from other elements which have not been considered.

The direct stiffness method is easily incorporated into a computer program. If we use the variable names

> GSM for $[K]$
> ESM for $[k^{(e)}]$
> GF for $\{F\}$
> EF for $\{f^{(e)}\}$

and assume that the numerical values of i and j are contained in the array NS, the computer implementation of the direct stiffness method is

```
          DO 10 I = 1,2
          II = NS(I)
          GF(II) = GF(II) + EF(I)
          DO 10 J = 1,2
          JJ = NS(J)
     10   GSM(II,JJ) = GSM(II,JJ) + ESM(I,J)
```

It is assumed that NS has a dimension of at least two and that the element has two nodes.

The direct stiffness procedure shows how to utilize the element matrices. The method is further illustrated in the following section where the beam deflection problem of Chapter 3 is reworked.

4.3 ANALYSIS OF A SIMPLY SUPPORTED BEAM

The direct stiffness method is now illustrated by reworking the beam deflection problem considered in Chapter 3. The beam and the element model in Figure 3.3 are presented in Figure 4.3 for convenience.

The element stiffness matrix $[k^{(e)}]$ is given by (4.11) and $\{f^{(e)}\}$ is given by (4.12).

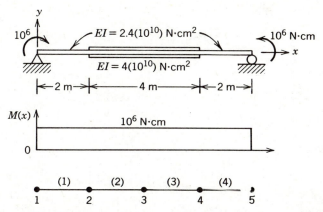

Figure 4.3 A simply supported beam subjected to concentrated end moments.

The element data in tabular form are

e	i	j	$\dfrac{D}{L}$	$\dfrac{QL}{2}$
1	1	2	$1.2(10^8)$	-10^8
2	2	3	$2.0(10^8)$	-10^8
3	3	4	$2.0(10^8)$	-10^8
4	4	5	$1.2(10^8)$	-10^8

The substitution of D/L into (4.11) and $QL/2$ into (4.12) gives the element matrices

$$[k^{(1)}]=10^8 \begin{matrix} 1 & 2 \\ \begin{bmatrix} 1.2 & -1.2 \\ -1.2 & 1.2 \end{bmatrix} & \begin{matrix} 1 \\ 2 \end{matrix} \end{matrix}, \quad \{f^{(1)}\}=-10^8 \begin{Bmatrix} 1 \\ 1 \end{Bmatrix} \begin{matrix} 1 \\ 2 \end{matrix}$$

$$[k^{(2)}]=10^8 \begin{matrix} 2 & 3 \\ \begin{bmatrix} 2 & -2 \\ -2 & 2 \end{bmatrix} & \begin{matrix} 2 \\ 3 \end{matrix} \end{matrix}, \quad \{f^{(2)}\}=-10^8 \begin{Bmatrix} 1 \\ 1 \end{Bmatrix} \begin{matrix} 2 \\ 3 \end{matrix}$$

$$[k^{(3)}]=10^8 \begin{matrix} 3 & 4 \\ \begin{bmatrix} 2 & -2 \\ -2 & 2 \end{bmatrix} & \begin{matrix} 3 \\ 4 \end{matrix} \end{matrix}, \quad \{f^{(3)}\}=-10^8 \begin{Bmatrix} 1 \\ 1 \end{Bmatrix} \begin{matrix} 3 \\ 4 \end{matrix}$$

$$[k^{(4)}]=10^8 \begin{matrix} 4 & 5 \\ \begin{bmatrix} 1.2 & -1.2 \\ -1.2 & 1.2 \end{bmatrix} & \begin{matrix} 4 \\ 5 \end{matrix} \end{matrix}, \quad \{f^{(4)}\}=-10^8 \begin{Bmatrix} 1 \\ 1 \end{Bmatrix} \begin{matrix} 4 \\ 5 \end{matrix}$$

The global matrices $[K]$ and $[F]$ are first initialized with zeros and then built by adding the coefficients of the element matrices an element at a time. Adding the values of element one gives

$$[K]=10^8 \begin{bmatrix} 1.2 & -1.2 & 0 & 0 & 0 \\ -1.2 & 1.2 & 0 & 0 & 0 \\ 0 & 0 & 0 & 0 & 0 \\ 0 & 0 & 0 & 0 & 0 \\ 0 & 0 & 0 & 0 & 0 \end{bmatrix}, \quad \{F\}=-10^8 \begin{Bmatrix} 1 \\ 1 \\ 0 \\ 0 \\ 0 \end{Bmatrix}$$

Adding element two gives

$$[K]=10^8 \begin{bmatrix} 1.2 & -1.2 & 0 & 0 & 0 \\ -1.2 & 3.2 & -2 & 0 & 0 \\ 0 & -2 & 2 & 0 & 0 \\ 0 & 0 & 0 & 0 & 0 \\ 0 & 0 & 0 & 0 & 0 \end{bmatrix}, \quad \{F\}=-10^8 \begin{Bmatrix} 1 \\ 2 \\ 1 \\ 0 \\ 0 \end{Bmatrix}$$

Adding element three gives

$$[K]=10^8 \begin{bmatrix} 1.2 & -1.2 & 0 & 0 & 0 \\ -1.2 & 3.2 & -2 & 0 & 0 \\ 0 & -2 & 4 & -2 & 0 \\ 0 & 0 & -2 & 2 & 0 \\ 0 & 0 & 0 & 0 & 0 \end{bmatrix}, \quad \{F\}=-10^8 \begin{Bmatrix} 1 \\ 2 \\ 2 \\ 1 \\ 0 \end{Bmatrix}$$

The addition of element four finishes the summation through the elements giving the system of equations

$$\begin{bmatrix} 1.2 & -1.2 & 0 & 0 & 0 \\ -1.2 & 3.2 & -2 & 0 & 0 \\ 0 & -2 & 4 & -2 & 0 \\ 0 & 0 & -2 & 3.2 & -1.2 \\ 0 & 0 & 0 & -1.2 & 1.2 \end{bmatrix} \begin{Bmatrix} Y_1 \\ Y_2 \\ Y_3 \\ Y_4 \\ Y_5 \end{Bmatrix} - \begin{Bmatrix} -1 \\ -2 \\ -2 \\ -2 \\ -1 \end{Bmatrix} = \begin{Bmatrix} 0 \\ 0 \\ 0 \\ 0 \\ 0 \end{Bmatrix} \quad (4.15)$$

after 10^8 has been canceled.

The final result is a 5×5 stiffness matrix $[K]$ and a 5×1 column vector $\{F\}$. The size of the system of equations differs from the three equations in (3.32) because the boundary conditions have not been incorporated.

To incorporate the boundary conditions, note that Y_1 and Y_5 are known quantities, $Y_1 = Y_5 = 0$. Equations one and five in (4.15) are eliminated because equations are not written for the nodes whose values are already known. Elimination of these two equations produces three equations containing the five nodal values. These equations are

$$R_2 = -1.2Y_1 + 3.2Y_2 - 2Y_3 \quad +2 = 0$$
$$R_3 = -2Y_2 \quad +4Y_3 \quad -2Y_4 \quad +2 = 0 \quad (4.16)$$
$$R_4 = -2Y_3 \quad +3.2Y_4 - 1.2Y_5 + 2 = 0$$

The nodal values Y_1 and Y_5 can be eliminated from the equations in (4.16) because they have known values; each is zero. The terms $-1.2Y_1$ and $-1.2Y_5$ vanish. If either or both Y_1 and Y_2 were nonzero values, then $-1.2Y_1$ and $-1.2Y_5$ are evaluated and added to the two coefficient in the respective row. Deleting the terms involving Y_1 and Y_5 yields the system of three equations

$$
\begin{aligned}
R_2 &= 3.2Y_2 - 2Y_3 && + 2 = 0 \\
R_3 &= -2Y_2 + 4Y_3 - 2Y_4 && + 2 = 0 \\
R_4 &= \qquad\; -2Y_3 + 3.2Y_4 + 2 = 0
\end{aligned}
\tag{4.17}
$$

These equations are identical to (3.32).

4.4 PROPERTIES OF THE GLOBAL STIFFNESS MATRIX

The stiffness matrix, $[K]$, is always symmetric and positive definite for structural problems and for governing differential equations that are self adjoint. The diagonal coefficients, K_{ii}, are always positive and relatively large when compared to the off-diagonal values in the same row.

Finite element equations are usually solved using Gaussian elimination or efficient modifications of Gaussian elimination (Conte and deBoor, 1980) because the system of equations is not always diagonally dominant; that is, K_{ii} can be less than the sum of the off-diagonal coefficients of row i. This occurs in some structural problems. The relatively large diagonal coefficient allows Gaussian elimination to be performed without interchanging rows (pivoting). This is an important fact because only the nonzero coefficients need to be stored within the computer. The symmetry property is also important because it eliminates the need to store the coefficients below the main diagonal.

The symmetry and positive definite properties are a result of the mathematical formulation. Another important property possessed by $[K]$ is related to the element grid and the node numbers. The global matrix $[K]$ is banded. A banded matrix has the characteristic that all of the nonzero coefficients are located relatively close to the main diagonal and all of the coefficients beyond the bandwidth are zero. This is illustrated schematically in Figure 4.4 where the bandwidth is shown by the dashed diagonal line. The C's denote nonzero terms. It is permissible to have zero coefficients within the bandwidth.

The bandwidth of $[K]$ is related to the numbering of the nodes. The bandwidth of a one-dimensional grid of the linear elements whose nodes are numbered in succession from left to right is two. The matrix consists of the main diagonal and a diagonal on each side of it. All of the other coefficients are zero. This fact can be observed by looking at (4.15).

The reason for the banded property is obtained by studying a system of equations as it is constructed. Each individual equation is associated with a node; that is, the third equation of a system is the residual equation for the third node. The nonzero coefficients in the third equation occur in the columns corresponding to the node numbers of the elements that touch node three. Consider the one-

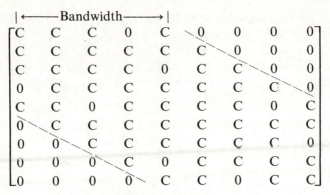

Figure 4.4 A banded matrix.

dimensional grid in Figure 4.5a. Elements two and three touch node three so that columns two, three, and four will contain nonzero coefficients. Columns one and five contain zero values because elements one and four do not touch node three. A more general situation occurs in two-dimensional grids. Consider the four triangles touching node 12, Figure 4.5b. Nonzero coefficients occur in columns 6, 10, 12, 14, and 21.

The bandwidth is one plus the greatest distance between the diagonal coefficient and the last nonzero coefficient in the row. All rows must be considered in this calculation. The general equation for calculating the bandwidth of $[K]$ for a finite element grid is

$$NBW = \max_{e} \left[BW^{(e)} \right] + 1 \tag{4.18}$$

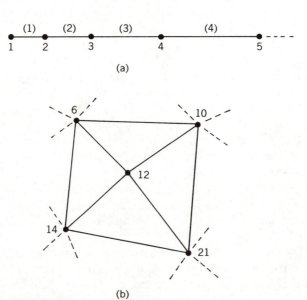

Figure 4.5. Node numbering in (a) a one-dimensional region, (b) a segment of a two-dimensional region.

where NBW is the bandwidth, and $BW^{(e)}$ is the difference between the largest and smallest node numbers for an element. The largest value of $BW^{(e)}$ is used in the calculation.

The way to minimize the bandwidth is to number the nodes such that $BW^{(e)}$ is as small as possible in each element. This is done by keeping the node numbers of each element as close as possible.

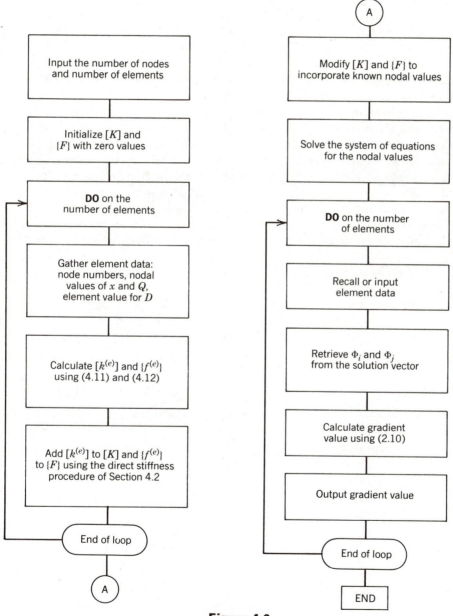

Figure 4.6

4.5   GENERAL FLOW OF THE COMPUTATIONS

The general flow of the computations using the direct stiffness procedure is shown in the flow diagram in Figure 4.6. The computations start with the information that defines the problem. These data include the number of nodes, the number of elements, the nodal coordinates, and the element values of D and Q. The global matrices $[K]$ and $\{F\}$ are initialized with zero values before starting to assemble the system of equations.

The direct stiffness construction of the global matrices is done within a DO loop over the number of elements. The first step is to collect the element data, which include the element node numbers and the values of D and Q. The length is calculated from the nodal coordinates. Once the element data are available, the element matrices $[k^{(e)}]$ and $\{f^{(e)}\}$ are calculated. This step is followed by the direct stiffness procedure during which the coefficients of $[k^{(e)}]$ and $\{f^{(e)}\}$ are added to the correct locations of $[K]$ and $\{F\}$.

The global matrices $[K]$ and $\{F\}$, which exist when the direct stiffness procedure is completed, must be modified to include the known nodal conditions. This modification produces a system of equations that is solved using a direct approach such as Gaussian elimination. Completion of the latter step produces a set of nodal values that is one of the desired results. In some situations, the nodal values are all that is desired and the problem is finished.

Whenever quantities related to the element gradient values are needed, another DO loop on the elements is necessary. The first step of this loop is to recall the element data from memory. Once the element data are available, the element nodal values of ϕ are selected from the solution vector and the gradient value is calculated.

The flow chart in Figure 4.6 gives the general flow of the computations. It is not, however, a flow diagram for an actual computer program. Good computer programs incorporate techniques that minimize memory requirements and make the program both computationally efficient and transferable.

PROBLEMS

4.1–4.4   Develop the system of equations for the following problems using the element matrix and direct stiffness concepts discussed in this chapter. Modify the system of equations to incorporate the boundary conditions and solve for the unknown nodal values.

4.1   Problem 3.1.

4.2   Problem 3.2.

4.3   Problem 3.3.

4.4(a–e)   Problem 3.4(a–e).

4.5   Solve the system of equations (4.15) when the left end of the beam has settled 1 cm, that is, $Y_1 = -1$ and $Y_5 = 0$.

Chapter 5

TWO-DIMENSIONAL ELEMENTS

A primary advantage of the finite element method is the ease with which it can be generalized to solve two-dimensional problems composed of several different materials and having irregular boundaries. Many general-purpose finite element programs are available for solving two-dimensional problems. All of these programs use triangular and rectangular elements or generalizations of these elements.

The discussion of two-dimensional problems begins by considering the linear triangular and bilinear rectangular elements. The shape functions and pertinent local coordinate systems are discussed in this chapter. Some convenient natural coordinate systems are discussed in the next chapter. These elements are then used to solve heat transfer, irrotational fluid flow, and solid mechanics problems. Generalizations of these elements are considered in Part Four of this book.

5.1 TWO-DIMENSIONAL GRIDS

The linear triangular element (Figure 5.1a) has straight sides and a node at each corner. The interpolation equation for a scalar quantity is

$$\phi = \alpha_1 + \alpha_2 x + \alpha_3 y \tag{5.1}$$

which is a complete linear polynomial because it contains a constant term and all of the possible linear terms, namely, x and y. As a result, the triangular element can take on any orientation and satisfy the continuity requirements involving adjacent elements.

The bilinear rectangular element (Figure 5.1b) has straight sides and a node at each corner. The interpolation equation for a scalar quantity is

$$\phi = C_1 + C_2 x + C_3 y + C_4 xy \tag{5.2}$$

This equation contains only one of the three possible second-order terms, xy. The rectangle cannot be arbitrarily oriented because the x^2 and y^2 terms are not present. The sides of the rectangle must remain parallel to the xy-coordinate system.

A grid of rectangular elements is easy to construct. All of the elements in a row parallel to the x-axis must be the same height. All of the elements in a column parallel to the y-axis must be the same width. It should be clear that the rectangular element is best suited for square or rectangular regions. Both triangular and

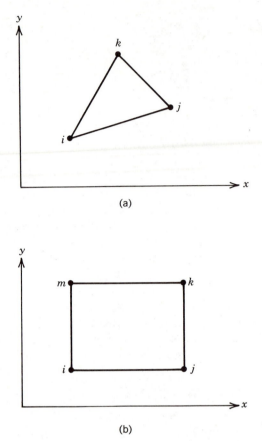

(a)

(b)

Figure 5.1 The linear triangular element and the bilinear rectangular element.

rectangular elements should be used in irregular regions. The triangular elements are used to model the irregular boundary.

The division of a region into triangular elements is most easily accomplished by first dividing it into large quadrilateral and triangular subregions. Each of these subregions is then divided into triangles. A triangular subregion is most easily divided into elements by specifying the same number of nodes along each side and then connecting the appropriate nodes by straight lines and placing nodes at the intersection points. The traiangular region shown in Figure 5.2a has been divided into nine elements after placing four nodes on a side. There is no reason why the nodes have to be equally spaced along a side. A variation in the spacing allows the size of the elements to be changed. There are $(n-1)^2$ triangular elements in a triangular region, where n is the number of nodes on a side.

When the triangular region has curved sides, the boundary elements model the curvature using straight-line segments. The division of a curved triangular region into linear triangular elements is shown in Figure 5.2b. The dashed line is the original shape and the solid lines denote the elements.

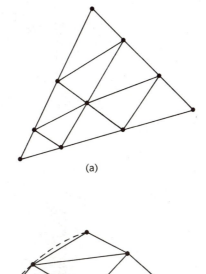

(a)

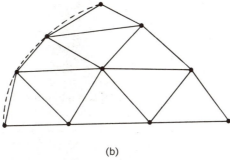

(b)

Figure 5.2   Regions divided into triangular elements.

The quadrilateral subregion is easily divided into triangular elements by connecting the nodes on opposite sides using line segments (Figure 5.3a). The interior nodes are located at the intersection points. The interior quadrilaterals are divided into triangular elements by inserting the shortest diagonal (Figure 5.3b). Division using the shortest diagonal is preferable because elements close to an equilateral shape produce more accurate results than long narrow triangles.

The number of nodes along adjacent sides of a quadrilateral subregion do not have to be the same, but the number of nodes on opposite sides should be equal unless the grid is to be refined (or enlarged). The spacing between boundary nodes can be varied to produce elements of difference sizes. There are $2(n-1)(m-1)$ triangular elements in a quadrilateral, where n and m are the number of nodes on a pair of adjacent sides.

The nodes on the boundary between subregions must be identical in number and must have the same relative position. This property is necessary to insure the continuity of ϕ across an element boundary.

The application of the concepts relative to the discretization of a region are illustrated in Figure 5.4. The node spacing has been varied along the edges of the quadrilateral in order to have smaller elements in the vicinity of the curved boundary.

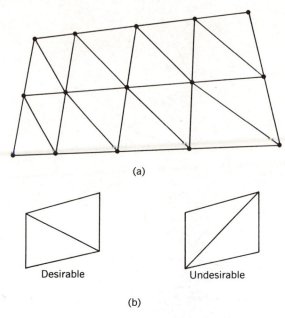

(a)

Desirable   Undesirable

(b)

Figure 5.3   Division of a quadrilateral subregion into triangular elements.

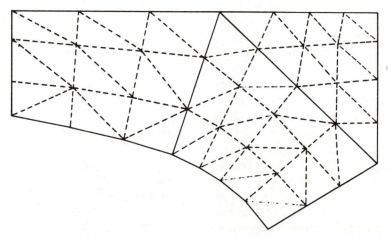

Figure 5.4   Division of a region into subregions and then into triangular elements.

A regular mesh with all elements the same size and shape is not necessary because there usually are regions in which the nodal variable is relatively constant. Larger elements can be used in these regions. The ability to vary the element size is an important advantage of the triangular element. The easiest way to make a transition in element size is to employ a quadrilateral region that has an unequal number of nodes on two opposite sides. A good combination is to place two nodes on one side for every three nodes on the opposite side. Such a region is shown in Figure 5.5.

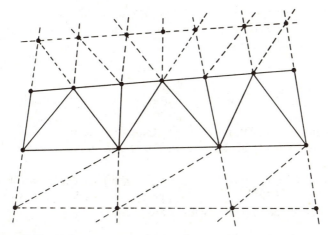

Figure 5.5 Using an expansion region to change the element size.

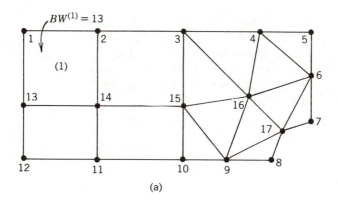

(a)

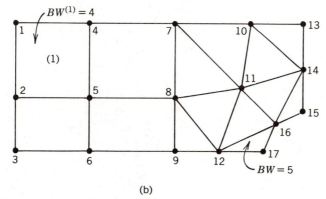

(b)

Figure 5.6 Two sets of node numbers resulting in different band widths.

Assigning the node numbers would be a trivial operation if the numbering did not influence the bandwidth of the system of equations. The bandwidth, NBW, as discussed in Chapter 4, is given by

$$NBW = \max_e \left[BW^{(e)} \right] + 1 \qquad (5.3)$$

where $BW^{(e)}$ is the difference between the largest and smallest node numbers in an element.

The minimization of NBW in simple regions can be accomplished by numbering across the dimension of the body with the fewest nodes when assigning node numbers.

Two different numbering schemes for a set of nodes are shown in Figure 5.6a and b. The values of $BW^{(1)}$ are 13 and 4, respectively. The largest values of $BW^{(e)}$ are 13 and 5. The respective bandwidths are 14 and 6.

5.2 LINEAR TRIANGULAR ELEMENT

The linear triangular element shown in Figure 5.7 has straight sides and three nodes, one at each corner. A consistent labeling of the nodes is a necessity and the labeling in this book proceeds counterclockwise from node i, which is arbitrarily specified. The nodal values of ϕ are Φ_i, Φ_j, and Φ_k whereas the nodal coordinates are (X_i, Y_i), (X_j, Y_j), and (X_k, Y_k).

The interpolation polynomial is

$$\phi = \alpha_1 + \alpha_2 x + \alpha_3 y \qquad (5.4)$$

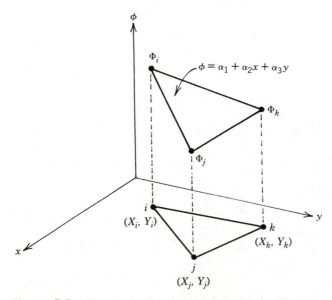

Figure 5.7 Parameters for the linear triangular element.

with the nodal conditions

$$\phi = \Phi_i \quad \text{at} \quad x = X_i, y = Y_i$$
$$\phi = \Phi_j \quad \text{at} \quad x = X_j, y = Y_j$$
$$\phi = \Phi_k \quad \text{at} \quad x = X_k, y = Y_k$$

Substitution of these conditions into (5.4) produces the system of equations

$$\Phi_i = \alpha_1 + \alpha_2 X_i + \alpha_3 Y_i$$
$$\Phi_j = \alpha_1 + \alpha_2 X_j + \alpha_3 Y_j$$
$$\Phi_k = \alpha_1 + \alpha_2 X_k + \alpha_3 Y_k \tag{5.5}$$

which yields

$$\alpha_1 = \frac{1}{2A} \left[(X_j Y_k - X_k Y_j)\Phi_i + (X_k Y_i - X_i Y_k)\Phi_j + (X_i Y_j - X_j Y_i)\Phi_k \right]$$

$$\alpha_2 = \frac{1}{2A} \left[(Y_j - Y_k)\Phi_i + (Y_k - Y_i)\Phi_j + (Y_i - Y_j)\Phi_k \right]$$

$$\alpha_3 = \frac{1}{2A} \left[(X_k - X_j)\Phi_i + (X_i - X_k)\Phi_j + (X_j - X_i)\Phi_k \right]$$

where the determinant

$$\begin{vmatrix} 1 & X_i & Y_i \\ 1 & X_j & Y_j \\ 1 & X_k & Y_k \end{vmatrix} = 2A \tag{5.6}$$

and A is the area of the triangle.

Substituting for α_1, α_2, and α_3 in (5.4) and rearranging produces an equation for ϕ in terms of three shape functions and Φ_i, Φ_j, and Φ_k that is

$$\phi = N_i \Phi_i + N_j \Phi_j + N_k \Phi_k \tag{5.7}$$

where

$$N_i = \frac{1}{2A} \left[a_i + b_i x + c_i y \right] \tag{5.8}$$

$$N_j = \frac{1}{2A} \left[a_j + b_j x + c_j y \right] \tag{5.9}$$

$$N_k = \frac{1}{2A} \left[a_k + b_k x + c_k y \right] \tag{5.10}$$

and

$$a_i = X_j Y_k - X_k Y_j, \quad b_i = Y_j - Y_k, \quad \text{and} \quad c_i = X_k - X_j$$
$$a_j = X_k Y_i - X_i Y_k, \quad b_j = Y_k - Y_i, \quad \text{and} \quad c_j = X_i - X_k$$
$$a_k = X_i Y_j - X_j Y_i, \quad b_k = Y_i - Y_j, \quad \text{and} \quad c_k = X_j - X_i$$

The scalar quantity ϕ is related to the nodal values by a set of shape functions that are linear in x and y. This means that the gradients $\partial\phi/\partial x$ and $\partial\phi/\partial y$ are constant within the element. For example,

$$\frac{\partial\phi}{\partial x}=\frac{\partial N_i}{\partial x}\Phi_i+\frac{\partial N_j}{\partial x}\Phi_j+\frac{\partial N_k}{\partial x}\Phi_k \qquad (5.11)$$

but

$$\frac{\partial N_\beta}{\partial x}=\frac{b_\beta}{2A} \qquad \beta=i,j,k$$

Therefore,

$$\frac{\partial\phi}{\partial x}=\frac{1}{2A}\left[b_i\Phi_i+b_j\Phi_j+b_k\Phi_k\right] \qquad (5.12)$$

Since b_i, b_j, and b_k are constants (they are fixed once the nodal coordinates are specified) and Φ_i, Φ_j, and Φ_k are independent of the space coordinates, the derivative has a constant value. A constant gradient within any element means that many small elements have to be used to accurately approximate a rapid change in ϕ.

ILLUSTRATIVE EXAMPLE

Evaluate the element shape functions and calculate the value of the pressure at point A in Figure 5.8 if the nodal values are $\Phi_i=40\ \text{N/cm}^2$, $\Phi_j=34\ \text{N/cm}^2$, and $\Phi_k=46\ \text{N/cm}^2$. Point A is located at $(2, 1.5)$.

The pressure ϕ is given by (5.7), and the shape functions are defined by (5.8),

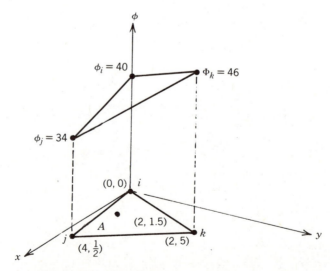

Figure 5.8   Parameters for the example problem.

(5.9), and (5.10). The coefficients for the shape function equations are

$$a_i = X_j Y_k - X_k Y_j = 4(5) - 2(0.5) = 19$$
$$a_j = X_k Y_i - X_i Y_k = 2(0) - 0(5) = 0$$
$$a_k = X_i Y_j - X_j Y_i = 0(0.5) - 4(0) = 0$$
$$b_i = Y_j - Y_k = 0.5 - 5 = -4.5$$
$$b_j = Y_k - Y_i = 5 - 0 = 5$$
$$b_k = Y_i - Y_j = 0 - 0.5 = -0.5$$
$$c_i = X_k - X_j = 2 - 4 = -2$$
$$c_j = X_i - X_k = 0 - 2 = -2$$
$$c_k = X_j - X_i = 4 - 0 = 4$$

whereas

$$2A = \begin{vmatrix} 1 & X_i & Y_i \\ 1 & X_j & Y_j \\ 1 & X_k & Y_k \end{vmatrix} = \begin{vmatrix} 1 & 0 & 0 \\ 1 & 4 & 0.5 \\ 1 & 2 & 5 \end{vmatrix}$$
$$= 20 - 1 = 19$$

Substitution of the coefficients into the shape function equations gives

$$N_i = \frac{19 - 4.5x - 2y}{19}$$

$$N_j = \frac{5x - 2y}{19}$$

$$N_k = \frac{-0.5x + 4y}{19}$$

Note that $N_i + N_j + N_k = 1$. The equation for the pressure is

$$\phi = \left(\frac{19 - 4.5x - 2y}{19}\right)\Phi_i + \left(\frac{5x - 2y}{19}\right)\Phi_j + \left(\frac{-0.5x + 4y}{19}\right)\Phi_k$$

The value of ϕ at point A, (2, 1.5), is

$$\phi = (\tfrac{7}{19})40 + (\tfrac{7}{19})34 + (\tfrac{5}{19})46$$
$$\phi = 39.4 \text{ N/cm}^2$$

The shape functions defined for the triangular element satisfy the properties discussed relative to the one-dimensional shape functions. Each shape function has a value of one at its own node and is zero at the other two. The three functions also sum to one. There are two other important properties whose proofs are left as exercises. First, a shape function varies linearly along the sides between its node and the other two nodes, that is, N_i varies linearly along sides ij and ik. A shape function is zero along the side opposite its node; that is, N_i is zero along side jk.

A result of the first property is that ϕ varies linearly along each of the three sides. Another important characteristic of ϕ is that any line of constant ϕ is a straight line intersecting two sides of the element (unless all nodes have the same value). These two properties make it easy to locate contour lines as illustrated in the following example.

ILLUSTRATIVE EXAMPLE

Determine the location of the 42 N/cm² contour line for the triangular element used in the previous illustrative example.

The pressure contour for 42 N/cm² intersects sides *ik* and *jk*. Simple ratios are used to obtain the coordinate values because the pressure varies linearly along each side. For side *jk*

$$\frac{46-42}{46-34}=\frac{2-x}{2-4} \quad \text{or} \quad \frac{4}{12}=\frac{2-x}{-2}$$

$$x=2.67 \text{ cm}$$

and

$$\frac{46-42}{46-34}=\frac{5-y}{5-0.5} \quad \text{or} \quad y=3.5 \text{ cm}$$

Similar ratios for side *ik* yield

$$x=\tfrac{2}{3} \text{ cm} \quad \text{and} \quad y=\tfrac{5}{3} \text{ cm}$$

The contour line is shown in Figure 5.9.

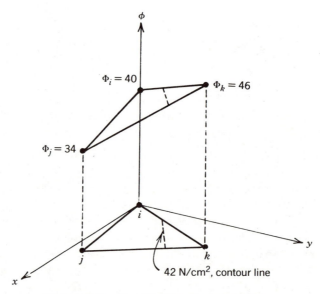

Figure 5.9   The 42 N/cm² contour line.

5.3 BILINEAR RECTANGULAR ELEMENT

The bilinear rectangular element has a length $2b$ and a height $2a$. The nodes are labeled i, j, k, and m with node i always at the lower left-hand corner. The element and the important coordinate systems are shown in Figure 5.10.

The interpolation equation (5.2) is written in terms of local coordinates s and t. There are at least three choices with

$$\phi = C_1 + C_2 s + C_3 t + C_4 st \tag{5.13}$$

being the most useful. The other choices would replace the st term by either s^2 or t^2. Equation (5.13) is used because ϕ is linear in s along any line of constant t and linear in t along any line of constant s. Because of these properties, the element is often said to be bilinear.

Equation (5.13) is written relative to a local coordinate system, whose origin is at node i because the shape functions are easier to evaluate in this reference frame. Another popular coordinate system is qr, which has its origin located at the center of the element (Figure 5.10).

The coefficients C_1, C_2, C_3, and C_4 in (5.13) are obtained by using the nodal values of ϕ and the nodal coordinates (in the st system) to generate four equations. These equations are

$$\Phi_i = C_1$$
$$\Phi_j = C_1 + (2b)C_2$$
$$\Phi_k = C_1 + (2b)C_2 + (2a)C_3 + (4ab)C_4$$
$$\Phi_m = C_1 + (2a)C_3$$

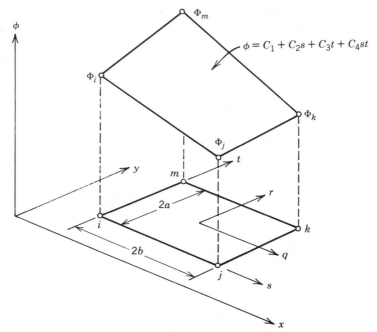

Figure 5.10. Parameters for the bilinear rectangular element.

Solving gives

$$C_1 = \Phi_i$$

$$C_2 = \frac{1}{2b}(\Phi_j - \Phi_i)$$

$$C_3 = \frac{1}{2a}(\Phi_m - \Phi_i) \tag{5.15}$$

$$C_4 = \frac{1}{4ab}(\Phi_i - \Phi_j + \Phi_k - \Phi_m)$$

Substitution of (5.15) into (5.13) and rearranging gives

$$\phi = N_i\Phi_i + N_j\Phi_j + N_k\Phi_k + N_m\Phi_m \tag{5.16}$$

where

$$N_i = \left(1 - \frac{s}{2b}\right)\left(1 - \frac{t}{2a}\right)$$

$$N_j = \frac{s}{2b}\left(1 - \frac{t}{2a}\right)$$

$$N_k = \frac{st}{4ab} \tag{5.17}$$

$$N_m = \frac{t}{2a}\left(1 - \frac{s}{2b}\right)$$

The shape functions for the bilinear rectangular element have properties similar to those possessed by the triangular element. Each shape function varies linearly along the edges between its node and the two adjacent nodes. For example, N_i varies linearly along sides ij and im. Each shape function is also zero along the sides its node does not touch, that is, N_i is zero along sides jk and km. The linear variation of ϕ along an edge of the rectangular element and an edge of the triangular element means that these two elements are compatible and can be used adjacent to one another.

The transformation equations between the qr and st coordinate systems are

$$s = b + q \qquad \text{and} \qquad t = a + r \tag{5.18}$$

Substitution of (5.18) into (5.17) gives the shape functions in terms of q and r

$$N_i = \frac{1}{4}\left(1 - \frac{q}{b}\right)\left(1 - \frac{r}{a}\right)$$

$$N_j = \frac{1}{4}\left(1 + \frac{q}{b}\right)\left(1 - \frac{r}{a}\right)$$

$$N_k = \frac{1}{4}\left(1 + \frac{q}{b}\right)\left(1 + \frac{r}{a}\right) \tag{5.19}$$

$$N_m = \frac{1}{4}\left(1 - \frac{q}{b}\right)\left(1 + \frac{r}{a}\right)$$

The shape functions defined by (5.19) are useful because they lead to a natural coordinate system that allows the rectangle to be deformed into a general quadrilateral.

A contour line in a rectangular element is generally curved. The intersection of the contour line with the edges can be obtained using linear interpolation. The easiest method for obtaining a third point is to set s or t to zero in the shape function equations and solve (5.16) for the other coordinate value. This procedure is illustrated in the following example.

ILLUSTRATIVE EXAMPLE

Determine three points on the 50°C contour line for the rectangular element shown in Figure 5.11. The nodal values are $\Phi_i = 42°C$, $\Phi_j = 54°C$, $\Phi_k = 56°C$, and $\Phi_m = 46°C$.

The lengths of the sides are

$$2b = X_j - X_i = 8 - 5 = 3$$
$$2a = Y_m - Y_i = 5 - 3 = 2$$

Substituting these values into (5.17) gives the shape functions

$$N_i = \left(1 - \frac{s}{3}\right)\left(1 - \frac{t}{2}\right), \qquad N_j = \frac{s}{3}\left(1 - \frac{t}{2}\right)$$

$$N_k = \frac{st}{6}, \qquad N_m = \frac{t}{2}\left(1 - \frac{s}{3}\right)$$

Inspection reveals that the 50°C contour line intersects the sides ij and km; therefore, we need to assume values of t and calculate values of s. Along side ij, $t = 0$ and

$$\phi = \left(1 - \frac{s}{3}\right)\Phi_i + \frac{s}{3}\Phi_j = 50$$

Substituting for Φ_i and Φ_j and solving gives $s = 2.0$. Along side km, $t = 2a = 2$ and

$$\phi = \frac{s}{3}\Phi_k + \left(1 - \frac{s}{3}\right)\Phi_m = 50$$

Substituting for Φ_k and Φ_m and solving gives $s = 1.2$.

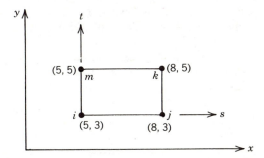

Figure 5.11. Nodal coordinates for the example problem.

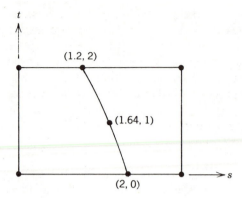

Figure 5.12.   The 50° contour line.

To obtain the third point, assume that $t=a=1$; then

$$\phi=\frac{1}{2}\left(1-\frac{s}{3}\right)\Phi_i+\frac{s}{6}\Phi_j+\frac{s}{6}\Phi_k+\frac{1}{2}\left(1-\frac{s}{3}\right)\Phi_m=50$$

Substituting the nodal values gives

$$\frac{s}{6}(-42+54+56-46)+\frac{1}{2}(42+46)=50$$

Solving yields $s=1.64$.

The st coordinates of the three points are (from top to bottom) (1.2, 2), (1.64, 1) and (2, 0). The xy coordinates of these points are (6.2, 5), (6.64, 4) and (7, 3). A straight line from (6.2, 5) to (7, 3) passes through the point (6.60, 4); therefore, the contour line is not straight (Figure 5.12).

5.4   A CONTINUOUS PIECEWISE SMOOTH EQUATION

The element equation for ϕ defined by either (5.7) or (5.16) can be used for any triangular or rectangular element by specifying the numerical values of i, j, and k or i, j, k, and m. Any node of a triangular element may be node i. An asterisk is used to distinguish it from the other nodes. Node i of the rectangular element is always at the origin of the st coordinate system.

The element nodal data for the four-element grid in Figure 5.13 is

e	i	j	k	m
1	1	4	5	2
2	2	5	6	3
3	3	6	7	
4	8	3	7	

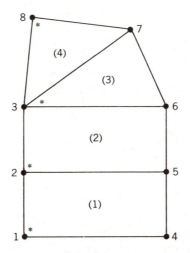

Figure 5.13. A four-element grid with node numbers.

The interpolation equation for element one is

$$\phi^{(1)} = N_1^{(1)}\Phi_1 + N_4^{(1)}\Phi_4 + N_5^{(1)}\Phi_5 + N_2^{(1)}\Phi_2 \qquad (5.20)$$

Note that the element node numbers are no longer consecutive. This is the usual case with two-dimensional elements. The shape functions in (5.17) are a function of the global coordinates only in the sense that

$$2b = X_j - X_i = X_4 - X_1$$

and

$$2a = Y_m - Y_i = Y_2 - Y_1$$

The interpolation equation for element four is

$$\phi^{(4)} = N_8^{(4)}\Phi_8 + N_3^{(4)}\Phi_3 + N_7^{(4)}\Phi_7 \qquad (5.21)$$

The shape functions in (5.21) are a function of the global coordinates and the specification of i, j, and k immediately indicates which coordinates to use. Consider, for example, $N_8^{(4)}$. Using (5.8) gives

$$N_8^{(4)} = \frac{1}{2A}(a_8^{(4)} + b_8^{(4)}x + c_8^{(4)}y)$$

where

$$a_8^{(4)} = X_3 Y_7 - X_7 Y_3$$
$$b_8^{(4)} = Y_3 - Y_7$$
$$c_8^{(4)} = X_7 - X_3$$

since $j = 3$ and $k = 7$. The area, A, is that of element four.

PROBLEMS

5.1 Verify that N_i for the triangular element is equal to one at node i and equal to zero at nodes j and k.

5.2 Verify that N_i for the triangular element in Figure 5.1a is zero everywhere along side jk. *Hint:* Write an equation of the form $y = d + mx$ for side jk and substitute for y in (5.8).

5.3 Verify that the shape functions for the triangular element sum to one, that is, $N_i + N_j + N_k = 1$. Comment on the behavior of the following summations:

(i) $a_i + a_j + a_k$.

(ii) $b_i + b_j + b_k$.

(iii) $c_i + c_j + c_k$.

5.4 Verify that the shape functions for the rectangular element given by (5.17) sum to one. Also check those given by (5.19).

5.5 Determine the requirement that the shape functions must satisfy in order to correctly model the condition that ϕ equals a constant within an element.

5.6 Verify that a line of constant ϕ in a rectangular element is in general not a straight line. Under what conditions will it be a straight line?

5.7–5.11 The nodal values for five different triangular elements are summarized below. Each column of values is associated with an element and a specific problem.

(a) Calculate the value of ϕ at the coordinates of point A.

(b) Determine the xy coordinates where the specified contour line intersects the element boundaries.

(c) Evaluate $\partial\phi/\partial x$ and $\partial\phi/\partial y$ within the element.

Element	Problem Number				
Quantity	5.7	5.8	5.9	5.10	5.11
X_i	0.13	0.31	0.13	0.13	0.44
Y_i	0.01	0.06	0.13	0.00	0.25
X_j	0.25	0.38	0.25	0.25	0.50
Y_j	0.06	0.09	0.13	0.00	0.25
X_k	0.13	0.31	0.19	0.25	0.50
Y_k	0.13	0.13	0.19	0.07	0.38
Φ_i	190	130	185	194	43
Φ_j	160	94	151	160	60
Φ_k	185	125	160	158	52
Point A					
x	0.20	0.36	0.18	0.20	0.47
y	0.06	0.09	0.13	0.03	0.30
Contour line	170	110	170	180	55

5.12–5.16 The nodal values for five different rectangular elements are summarized below. Each column of values is associated with an element and a specific problem.

(a) Calculate the value of ϕ at the coordinates of point B.

(b) Determine three sets of xy coordinates for the specified contour line.

(c) Evaluate $\partial\phi/\partial x$ and $\partial\phi/\partial y$ at point B.

Element	Problem Number				
Quantity	5.12	5.13	5.14	5.15	5.16
X_i	0.31	0.25	0.25	0.31	0.40
Y_i	0.18	0.18	0.06	0.13	0.00
X_j	0.38	0.31	0.31	0.38	0.44
Y_m	0.25	0.25	0.13	0.19	0.03
Φ_i	115	140	158	125	76
Φ_j	85	115	130	92	54
Φ_k	76	104	125	86	60
Φ_m	105	124	150	116	80
Point B					
x	0.35	0.26	0.28	0.34	0.42
y	0.22	0.22	0.10	0.15	0.01
Contour line	90	130	154	100	70

Chapter 6

COORDINATE SYSTEMS

All finite element solutions require the evaluation of integrals. Some of these are easily evaluated while others are very difficult. Many are impossible to evaluate analytically so that numerical techniques are employed.

The difficulties associated with evaluating an integral can often be decreased by changing the variables of integration. This involves writing the integral in a new coordinate system. The objective of this chapter is to discuss several coordinate systems that can be used to eliminate some of the difficulties associated with finite element integrals.

Local and natural coordinate systems are discussed in this chapter. These systems are discussed relative to the one-dimensional linear element, and then the two-dimensional triangular and rectangular elements.

6.1 LOCAL COORDINATE SYSTEMS

The linear shape functions developed in Chapter 2,

$$N_i(x) = \frac{X_j - x}{L} \quad \text{and} \quad N_j(x) = \frac{x - X_i}{L} \tag{6.1}$$

are for an element in which the origin of the coordinate system is to the left of node i. These are general equations valid for all linear elements regardless of their location. The disadvantage of these shape functions shows up when evaluating integrals involving products of the shape functions such as

$$\int_{X_i}^{X_j} N_i(x) N_j(x) \, dx \quad \text{or} \quad \int_{X_i}^{X_j} N_i^2(x) \, dx \tag{6.2}$$

Integrals similar to these occur in the consideration of both field problems and solid mechanics problems. The integrations in (6.2) are simplified by developing new shape functions defined relative to a coordinate system whose origin is located on the element. This type of system is called a local coordinate system.

The two most common local coordinate systems for the one-dimensional element have the origin located at node i or at the center of the element (Figure 6.1).

The shape functions for a coordinate system located at node i are obtained from (6.1) by replacing x by $x = X_i + s$. This substitution produces

$$N_i(s) = \frac{X_j - x}{L} = \frac{X_j - (X_i + s)}{L} = 1 - \frac{s}{L} \tag{6.3}$$

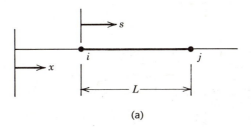

(a)

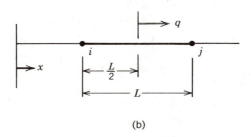

(b)

Figure 6.1 Local coordinate systems for the one-dimensional element.

and

$$N_j(s)=\frac{x-X_i}{L}=\frac{X_i+s-X_i}{L}=\frac{s}{L} \tag{6.4}$$

Note that each shape function equals one at its own node and is zero at the other node. The two sum to one as did the pair in (6.1).

The shape functions for a coordinate system located at the center of the element are obtained from (6.1) by replacing x by $x=X_i+(L/2)+q$. The shape functions relative to this origin are

$$N_i(q)=\left(\frac{1}{2}-\frac{q}{L}\right) \quad \text{and} \quad N_j(q)=\left(\frac{1}{2}+\frac{q}{L}\right) \tag{6.5}$$

where the coordinate variable q ranges from $-L/2$ to $L/2$.

The shape functions, (6.3) and (6.4), as well as the pair in (6.5), are useful only if a change in the integration variables is performed. The change of variable formula from integral calculus (Olmstead, 1961) is

$$\int_a^b f(x)\,dx=\int_{p_1}^{p_2} f(g(p))\left[\frac{d(g(p))}{dp}\right]dp \tag{6.6}$$

where p is the new coordinate variable and $g(p)$ is the equation relating x and p, that is, $x=g(p)$.

Interpretation of (6.6) relative to the coordinate systems in Figure 6.1 goes as follows. For the coordinate s, where $x=X_i+s$

$$\int_{X_i}^{X_j} f(x)\,dx = \int_{s_1}^{s_2} \frac{d(X_i+s)}{ds}\,ds = \int_0^L h(s)\,ds \tag{6.7}$$

where $h(s)$ is $f(x)$ written in terms of s. The limits of integration were obtained by substituting X_i and X_j for x in $x = X_i + s$ and solving for s.

For the coordinate q, where $x = X_i + L/2 + q$

$$\int_{X_i}^{X_j} f(x)\,dx = \int_{q_1}^{q_2} r(q) \frac{d(X_i + L/2 + q)}{dq}\,dq = \int_{-L/2}^{L/2} r(q)\,dq \tag{6.8}$$

where $r(q)$ is $f(x)$ written in terms of q.

The usefulness of (6.7) and (6.8) comes when integrals such as

$$\int_{X_i}^{X_j} N_i^2\,dx$$

are evaluated. Using the coordinate variable s, we obtain

$$\int_{X_i}^{X_j} N_i^2(x)\,dx = \int_0^L N_i^2(s)\,ds = \int_0^L \left(1 - \frac{s}{L}\right)^2 ds = \frac{L}{3}$$

Using the q coordinate, we obtain

$$\int_{X_i}^{X_j} N_i^2(x)\,dx = \int_{-L/2}^{L/2} N_i^2(q)\,dq = \int_{-L/2}^{L/2} \left(\frac{1}{2} - \frac{q}{L}\right)^2 dq = \frac{L}{3}$$

The result, $L/3$, is obtained from

$$\int_{X_i}^{X_j} N_i^2(x)\,dx$$

only after a rather complicated expression is recognized as being L^3.

6.2 NATURAL COORDINATE SYSTEMS

The local coordinate systems s and q can be converted to natural coordinate systems. A natural coordinate system is a local system that permits the specification of a point within the element by a dimensionless number whose absolute magnitude never exceeds unity.

Start with the q coordinate in Figure 6.1 and form the ratio $q/(L/2) = 2q/L = \xi$. The coordinate ξ varies from -1 to $+1$ (Figure 6.2a). The shape functions in (6.5) can be written in terms of ξ by replacing q by $q = \xi L/2$. The new shape functions are

$$N_i(\xi) = \tfrac{1}{2}(1 - \xi) \qquad \text{and} \qquad N_j(\xi) = \tfrac{1}{2}(1 + \xi) \tag{6.9}$$

The change of variables in the integration yields

$$\int_{-L/2}^{L/2} r(q)\,dq = \int_{\xi_1}^{\xi_2} g(\xi) \frac{d(\xi L/2)}{d\xi}\,d\xi = \frac{L}{2} \int_{-1}^{1} g(\xi)\,d\xi \tag{6.10}$$

where $g(\xi)$ is $r(q)$ written in terms of ξ.

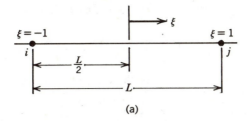

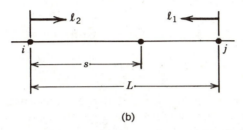

(b)

Figure 6.2 Natural coordinate systems for the one-dimensional element.

The advantage of the coordinate variable ξ is the -1 to $+1$ limits of integration. Most computer programs use numerical integration techniques to evaluate the element matrices. A numerical integration scheme used in finite element programs is the Gauss–Legendre method (Conte and deBoor, 1980), which has the sampling points and weighting coefficients defined on a $-1, +1$ interval.

Another interesting natural coordinate system consists of a pair of length ratios, Figure 6.2b. If s is the distance from node i, then ℓ_1 and ℓ_2 are defined as the ratios

$$\ell_1 = \frac{L-s}{L} \qquad \text{and} \qquad \ell_2 = \frac{s}{L} \tag{6.11}$$

This pair of coordinates is not independent because

$$\ell_1 + \ell_2 = 1 \tag{6.12}$$

The most important characteristic of (6.11) and (6.12) is that ℓ_1 and ℓ_2 are identical to the shape functions defined by (6.3) and (6.4). The usefulness of these coordinates is associated with the evaluation of integrals of the type

$$\int_0^L N_i^a(s) N_j^b(s) \, ds \tag{6.13}$$

which involve the product of shape functions. The length ratio coordinates result in a simple formula for evaluating an integral similar to (6.13).

The change of variable rule and the relationships $N_i(s) = \ell_1$, $N_j(s) = \ell_2$, $s = L\ell_2$, and $ds/d\ell_2 = L$ give

$$\int_0^L N_i^a(s) N_j^b(s) \, ds = \int_0^1 \ell_1^a \ell_2^b L \, d\ell_2 \tag{6.14}$$

The integral on the right-hand side of (6.14) can be changed to

$$L \int_0^1 (1-\ell_2)^a \ell_2^b \, d\ell_2 \tag{6.15}$$

using (6.12). The integral in (6.15) is of the same form as

$$\int_0^1 t^{z-1}(1-t)^{w-1} \, dt = \frac{\Gamma(z)\Gamma(w)}{\Gamma(z+w)} \tag{6.16}$$

where $\Gamma(n+1)=n!$ (Abramowitz and Stegun, 1964). Thus

$$L \int_0^1 \ell_1^a \ell_2^b \, d\ell_2 = L \frac{\Gamma(a+1)\Gamma(b+1)}{\Gamma(a+b+1+1)}$$

$$= L \frac{a!b!}{(a+b+1)!} \tag{6.17}$$

Equation (6.17) is useful because it states that a rather complicated integral can be evaluated using an equation which involves only the length of the element and the powers involved in the product.

Evaluation of a pair of integrals illustrates the usefulness of (6.17). Starting with

$$\int_{X_i}^{X_j} N_i^2(x) \, dx = \int_0^L N_i^2(s) \, ds$$

(6.12) gives

$$\int_0^L N_i^2(s) \, ds = L \int_0^1 \ell_1^2 \ell_2^0 \, d\ell_2 = L \frac{2!0!}{(2+0+1)!} = \frac{L}{3}$$

Table 6.1   Coordinate Systems and Limits of Integration for the One-Dimensional Element

Type of System	Coordinate Variable	Shape Functions		Limits of Integration
Global	x	$N_i = \dfrac{X_j - x}{L},$	$N_j = \dfrac{x - X_i}{L}$	$X_i, \quad X_j$
Local	s	$N_i = 1 - \dfrac{s}{L},$	$N_j = \dfrac{s}{L}$	$0, \quad L$
Local	q	$N_i = \left(\dfrac{1}{2} - \dfrac{q}{L}\right),$	$N_j = \left(\dfrac{1}{2} + \dfrac{q}{L}\right)$	$-\dfrac{L}{2}, \quad \dfrac{L}{2}$
Natural	ξ	$N_i = \dfrac{1}{2}(1-\xi),$	$N_j = \dfrac{1}{2}(1+\xi)$	$-1, \quad 1$
Natural	ℓ_2	$N_i = \ell_1,$	$N_j = \ell_2$	$0, \quad 1$

Another example is

$$\int_0^L N_i^3(s)N_j^2(s)\,ds = L\int_0^1 \ell_1^3\ell_2^2\,d\ell_2 = L\frac{3!2!}{(3+2+1)!} = \frac{L}{60}$$

The coordinate systems, shape functions, and limits of integration for the one-dimensional linear element are summarized in Table 6.1.

6.3 RECTANGULAR ELEMENT

Natural coordinate systems can be defined for two-dimensional elements; they have the same advantage as observed for the one-dimensional formulation. They are more convenient for both analytical and numerical integration.

The natural coordinate system for the rectangular element is shown in Figure 6.3. It is located at the center of the element and the coordinates are the length ratios

$$\xi = \frac{q}{b} \quad \text{and} \quad \eta = \frac{r}{a} \tag{6.18}$$

where q and r are the local coordinates. The shape functions in (5.19) are easily converted to the natural coordinate system. The results are

$$N_i = \tfrac{1}{4}(1-\xi)(1-\eta), \qquad N_j = \tfrac{1}{4}(1+\xi)(1-\eta)$$
$$N_k = \tfrac{1}{4}(1+\xi)(1+\eta), \qquad N_m = \tfrac{1}{4}(1-\xi)(1+\eta) \tag{6.19}$$

It should be clear that ξ and η range between plus and minus one, that is,

$$-1 \leqslant \xi \leqslant 1 \quad \text{and} \quad -1 \leqslant \eta \leqslant 1$$

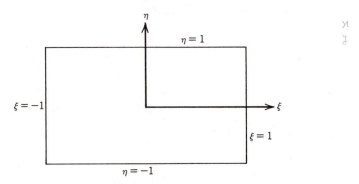

Figure 6.3. A natural coordinate system for the rectangular element.

6.4 TRIANGULAR ELEMENT: AREA COORDINATES

A natural coordinate system for the triangular element is obtained by defining the three length ratios L_1, L_2, and L_3 shown in Figure 6.4a. Each coordinate is the ratio of a perpendicular distance from one side, s, to the altitude, h, of that

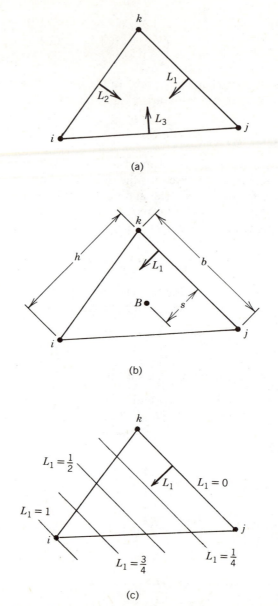

Figure 6.4. The three area coordinates for a triangular element.

same side. This is illustrated in Figure 6.4*b*. Each coordinate is a length ratio that varies between zero and one. The lines of constant L_1 are shown in Figure 6.4*c*. Each of these lines is parallel to the side from which L_1 is measured.

The coordinates L_1, L_2, and L_3 are called area coordinates because their values give the ratio of the area of a subtriangular region to the area of the complete triangle. Consider point B as shown in Figure 6.5. The area of the complete triangle

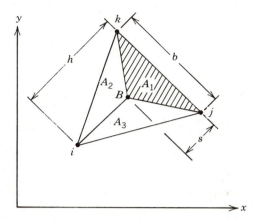

Figure 6.5. A triangle divided into the areas corresponding to the area coordinates.

is A and is given by

$$A = \frac{bh}{2}$$

whereas the area of the shaded triangle (B, j, k) is

$$A_1 = \frac{bs}{2} \tag{6.20}$$

Forming the ratio A_1/A yields

$$\frac{A_1}{A} = \frac{s}{h} = L_1 \tag{6.21}$$

The area coordinate L_1 is the ratio of the shaded area in Figure 6.5 to the total area. Similar equations can be written for L_2 and L_3 giving

$$L_2 = \frac{A_2}{A} \quad \text{and} \quad L_3 = \frac{A_3}{A} \tag{6.22}$$

Since $A_1 + A_2 + A_3 = A$,

$$L_1 + L_2 + L_3 = 1 \tag{6.23}$$

An equation relating the three coordinates was expected because the coordinates are not independent. The location of a point can be specified using two of the coordinates.

Equation (6.21) can be reworked into another form. Multiplying the top and bottom by two gives

$$L_1 = \frac{2A_1}{2A} \tag{6.24}$$

Using the determinant expansion for $2A_1$ produces

$$2A_1 = \begin{vmatrix} 1 & x & y \\ 1 & X_j & Y_j \\ 1 & X_k & Y_k \end{vmatrix}$$

or

$$2A_1 = (X_j Y_k - X_k Y_j) + (Y_j - Y_k)x + (X_k - X_j)y \tag{6.25}$$

where x and y are the coordinates of B in Figure 6.5. Substituting (6.25) into (6.24) yields

$$L_1 = \frac{1}{2A}[(X_j Y_k - X_k Y_j) + (Y_j - Y_k)x + (X_k - X_j)y] \tag{6.26}$$

Equation (6.26) is identical to (5.8); thus

$$L_1 = N_i \tag{6.27}$$

A similar analysis for L_2 and L_3 shows that

$$L_2 = N_j \quad \text{and} \quad L_3 = N_k \tag{6.28}$$

The area coordinates for the linear triangular element are identical to the shape functions, and the two sets of quantities can be interchanged.

The advantage of using the area coordinate system is the existence of an integration equation that simplifies the evaluation of area integrals (Eisenberg and Malvern, 1973). This integral equation is related to (6.17) and is

$$\int_A L_1^a L_2^b L_3^c \, dA = \frac{a!b!c!}{(a+b+c+2)!} 2A \tag{6.29}$$

The use of (6.29) can be illustrated by evaluating the shape function product

$$\int_A N_i(x, y)N_j(x, y) \, dA \tag{6.30}$$

over the area of a triangle. The area integral is

$$\int_A N_i N_j \, dA = \int_A L_1^1 L_2^1 L_3^0 \, dA$$

$$= \frac{1!1!0!}{(1+1+0+2)!} 2A = \frac{2A}{4!} = \frac{A}{12}$$

The area coordinates L_1 and L_2 can be substituted for N_i and N_j, respectively. Since N_k was not in the product, L_3 is included to the zero power. Zero factorial is defined as one.

The incorporation of derivative boundary conditions or surface loads into a finite element analysis requires the evaluation of an integral along the edge of an element. These integrals are easy to evaluate once it is known how the area coordinates behave on an edge. Consider point B on the side ij (Figure 6.6). The

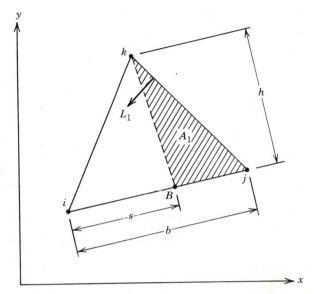

Figure 6.6. The area coordinates for a point on the edge of a triangle.

coordinate L_3 is zero and L_1 is the ratio of the shaded area to the total area. Define the coordinate variable s, which is parallel to side ij and measured from node i. If the coordinate of point B is s, and the length of the side is b, then

$$L_1=\frac{2A_1}{2A}=\frac{\dfrac{2h(b-s)}{2}}{\dfrac{2bh}{2}}=\frac{b-s}{b}=1-\frac{s}{b} \tag{6.31}$$

The area coordinate L_2 is

$$L_2=\frac{s}{b} \tag{6.32}$$

The area coordinates L_1 and L_2 reduce to the one-dimensional shape functions $N_i(s)$ and $N_j(s)$ defined by (6.3) and (6.4). Using the one-dimensional natural coordinates, ℓ_1 and ℓ_2, defined by (6.11), the area coordinates become

$$L_1=\ell_1 \quad\text{and}\quad L_2=\ell_2 \qquad \text{side } i\to j \tag{6.33}$$

The relationships for the other two sides are

$$L_2=\ell_1 \quad\text{and}\quad L_3=\ell_2 \qquad \text{side } j\to k \tag{6.34}$$

$$L_3=\ell_1 \quad\text{and}\quad L_1=\ell_2 \qquad \text{side } k\to i \tag{6.35}$$

The importance of the relationships in (6.33), (6.34), and (6.35) is that any integral over the edge of a triangular element can be replaced by a line integral written in terms of s or ℓ_2, that is,

$$\int_\Gamma f(L_1, L_2, L_3)\,d\Gamma = \int_0^L g(s)\,ds = L\int_0^1 h(\ell_2)\,d\ell_2 \qquad (6.36)$$

and evaluated using the factorial formula (6.17). The boundary of a two-dimensional element is denoted by Γ.

ILLUSTRATIVE EXAMPLE

Evaluate $\int_\Gamma [N]^T\,d\Gamma$ over side ik of a linear triangular element.
    The integral is

$$\int_\Gamma [N]^T\,d\Gamma = L_{ik}\int_0^1 \begin{Bmatrix} N_i \\ N_j \\ N_k \end{Bmatrix} d\ell_2 = L_{ik}\int_0^1 \begin{Bmatrix} L_1 \\ L_2 \\ L_3 \end{Bmatrix} d\ell_2$$

since the linear triangular shape functions and the area coordinates are equivalent. Along side ik, $L_1 = \ell_1$, $L_2 = 0$, and $L_3 = \ell_2$; thus

$$\int_\Gamma [N]^T\,d\Gamma = L_{ik}\int_0^1 \begin{Bmatrix} \ell_1 \\ 0 \\ \ell_2 \end{Bmatrix} d\ell_2 = \frac{L_{ik}}{2}\begin{Bmatrix} 1 \\ 0 \\ 1 \end{Bmatrix}$$

using (6.35) and then (6.17).

6.5 CONTINUITY

The function for approximating $\phi(x, y)$ consists of a set of continuous piecewise smooth equations, each defined over a single element. The need to integrate this piecewise smooth function places a requirement on the order of continuity between elements.
    The integral

$$\int_0^H \frac{d^n\phi}{dx^n}\,dx$$

is defined only if ϕ has continuity of order $(n-1)$ (Olmstead, 1961). This ensures that only finite jump discontinuities exist in the nth derivative. This requirement means that the first derivative of the approximating function must be continuous between elements if the integral contains second-derivative terms, $n=2$. All of the integrals in this book, except the beam element, contain first-derivative terms. Therefore, ϕ must be continuous between elements, but its derivatives do not have to be continuous. Continuity in the derivative is required for the beam element.
    Continuity of ϕ in the one-dimensional element is assured, since two adjacent elements have a common node. Continuity in ϕ along a common boundary between two rectangular elements is relatively easy to prove and is left as an

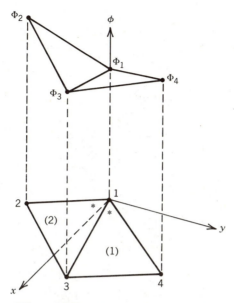

Figure 6.7.   A two-element grid.

exercise. Continuity of ϕ along a common boundary of two arbitrarily oriented triangular elements is more complicated and is considered here.

Consider two adjacent elements (Figure 6.7) with the coordinate system originating at node one. The nodal values are Φ_1, Φ_2, Φ_3, and Φ_4. The equations for ϕ are

$$\phi^{(1)} = N_1^{(1)}\Phi_1 + N_3^{(1)}\Phi_3 + N_4^{(1)}\Phi_4$$
$$\phi^{(2)} = N_1^{(2)}\Phi_1 + N_2^{(2)}\Phi_2 + N_3^{(2)}\Phi_3 \tag{6.37}$$

The properties of the shape functions indicate that $N_2^{(2)} = N_4^{(1)} = 0$ along the common boundary. Recalling the equality between the shape functions and the area coordinates, (6.27) and (6.28), allows (6.37) to be written as

$$\phi^{(1)} = L_1^{(1)}\Phi_1 + L_2^{(1)}\Phi_3$$
$$\phi^{(2)} = L_1^{(2)}\Phi_1 + L_3^{(2)}\Phi_3 \tag{6.38}$$

Remember that the subscripts on the area coordinates are not related to the node numbers.

Since $L_3^{(1)} = L_2^{(2)} = 0$, (6.38) can be reworked into

$$\phi^{(1)} = L_1^{(1)}\Phi_1 + (1 - L_1^{(1)})\Phi_3$$
$$\phi^{(2)} = L_1^{(2)}\Phi_1 + (1 - L_1^{(2)})\Phi_3 \tag{6.33}$$

using (6.23). The proof is completed when it is shown that $L_1^{(1)} = L_1^{(2)}$.

A point on the common boundary is shown in Figure 6.8 with the areas associated with $L_1^{(1)}$ and $L_1^{(2)}$ shaded. Defining the distance from point B to node three as c

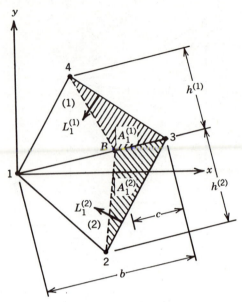

Figure 6.8. The area coordinates $L_1^{(1)}$ and $L_1^{(2)}$ along a common boundary.

and the length of side $1-3$ as b,

$$L_1^{(1)} = \frac{2A_1^{(1)}}{2A^{(1)}} = \frac{\dfrac{2ch^{(1)}}{2}}{\dfrac{2bh^{(1)}}{2}} = \frac{c}{b}$$

and

$$L_1^{(2)} = \frac{2A_1^{(2)}}{2A^{(2)}} = \frac{\dfrac{2ch^{(2)}}{2}}{\dfrac{2bh^{(2)}}{2}} = \frac{c}{b} = L_1^{(1)}$$

The proof is complete.

PROBLEMS

6.1 The shape functions for the three node one-dimensional quadratic element (Figure P6.1) relative to the local coordinate s follow. Write these shape functions in terms of the local coordinate q.

$$N_i = \left(1 - \frac{2s}{L}\right)\left(1 - \frac{s}{L}\right), \qquad N_j = \frac{4s}{L}\left(1 - \frac{s}{L}\right)$$

$$N_k = \frac{2s}{L^2}\left(s - \frac{L}{2}\right)$$

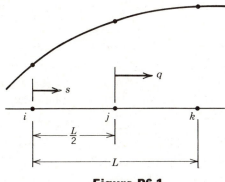

Figure P6.1

6.2 Do Problem 6.1 for the natural coordinate ξ (Figure P6.2).

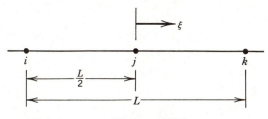

Figure P6.2

6.3 Do Problem 6.1 for the natural coordinates ℓ_1 and ℓ_2 (Figure P6.3).

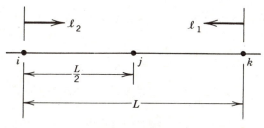

Figure P6.3

6.4 A local coordinate system r has its origin at the one-third point of a linear element (Figure P6.4). Develop the shape functions in terms of r:

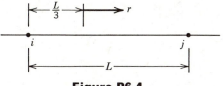

Figure P6.4

(a) Starting with the x-coordinate system.

(b) Starting with the s-coordinate system.

(c) Starting with the q-coordinate system.

6.5   The integration of $d\phi/dx$ occurs in many finite element applications. When N_i and N_j are written in terms of a new coordinate variable, $d\phi/dx$ must be evaluated using the chain rule, that is, for the local coordinate s,

$$\frac{d\phi}{ds}=\frac{d\phi}{dx}\frac{dx}{ds} \quad \text{and} \quad \frac{d\phi}{dx}=\frac{d\phi}{ds}\frac{1}{dx/ds}$$

Show that the following relationships hold for the one-dimensional coordinate systems studied in this chapter.

$$\frac{d\phi}{ds}=\frac{d\phi}{dx}, \quad \frac{d\phi}{dq}=\frac{d\phi}{dx} \quad \text{and} \quad \frac{d\phi}{d\xi}=\frac{L}{2}\frac{d\phi}{dx}$$

6.6   The quadratic shape functions given in Problem 6.1 are

$$N_i=\ell_1-2\ell_1\ell_2, \qquad N_j=4\ell_1\ell_2 \qquad N_k=2\ell_2^2-\ell_2$$

when written in terms of the natural coordinates ℓ_1 and ℓ_2. Using these equations for the shape functions and (6.17), evaluate

(a) $\displaystyle\int_0^L N_iN_j\,ds$ (b) $\displaystyle\int_0^L N_jN_k\,ds$ (c) $\displaystyle\int_0^L N_k^2\,ds$

6.7   Verify that $L_2=N_j$ for the linear triangular element.

6.8   Verify that any line of constant ϕ in a triangular element is a straight line. *Hint:* Investigate the value of ϕ along the side of a triangle when two nodes have the same value.

6.9   Evaluate the following integrals using (6.29).

(a) $\displaystyle\int_A N_i^2N_k\,dA$ (b) $\displaystyle\int_A N_iN_jN_k\,dA$

(c) $\displaystyle\int_A (N_i^2+N_j)\,dA$ (d) $\displaystyle\int_A (N_j^2N_k+N_i)\,dA$

6.10   Verify that $\phi^{(1)}=\phi^{(2)}$ along the common boundary between the two rectangular elements shown in Figure P6.10.

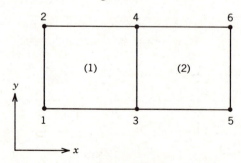

Figure P6.10

6.11 Verify that $\phi^{(1)} = \phi^{(2)}$ along the common boundary between the triangular and rectangular elements in Figure P6.11. *Hint:* Use the st-coordinate system for the rectangular element.

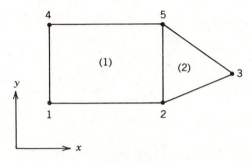

Figure P6.11

6.12 Show that the shape functions for the rectangular element, (6.19), reduce to the linear shape functions, (6.9), along side ij.

6.13 The three shape functions along one edge of a quadratic triangular element are given in Figure P6.13. Show that these equations reduce to the shape functions for the one-dimensional quadratic element that are given in Problem 6.6.

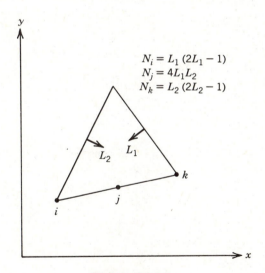

$$N_i = L_1(2L_1 - 1)$$
$$N_j = 4L_1L_2$$
$$N_k = L_2(2L_2 - 1)$$

Figure P6.13

PART TWO
FIELD PROBLEMS

Chapters 7 through 16 cover the finite element solution of steady state and time-dependent field problems. Specific application areas include heat transfer, irrotational flow, and acoustical vibrations. The reader whose interest is in structural applications can go directly to Part Three, which covers structural and solid mechanics applications.

Chapter 7

TWO-DIMENSIONAL FIELD EQUATION

The implementation of the finite element method can be subdivided into three broad steps: (1) establishing the element interpolation properties, (2) evaluating the element matrices, and (3) solving an actual problem. We have discussed the properties of a pair of two-dimensional elements in Chapters 5 and 6. Specific application areas are considered in the next several chapters. Our immediate objective is centered on the discussion of the element matrices associated with the two-dimensional field equation:

$$D_x \frac{\partial^2 \phi}{\partial x^2} + D_y \frac{\partial^2 \phi}{\partial y^2} - G\phi + Q = 0 \tag{7.1}$$

We start by briefly discussing the governing differential equations for several physical problems embedded within (7.1). The emphasis in the remainder of the chapter is on the derivation of the integral equations for the element matrices and the evaluation of these equations for the linear triangular and bilinear rectangular elements.

7.1 GOVERNING DIFFERENTIAL EQUATIONS

The general field equation, (7.1), has many important applications in the physical sciences. A few of these applications are discussed in this section. Since our objective is to establish the usefulness of (7.1), a discussion of the boundary conditions is deferred to the application chapters that follow.

The first application area is the torsion of noncircular sections. The governing differential equation is

$$\frac{1}{g} \frac{\partial^2 \phi}{\partial x^2} + \frac{1}{g} \frac{\partial^2 \phi}{\partial y^2} + 2\theta = 0 \tag{7.2}$$

where g is the shear modulus of the material and θ is the angle of twist. Equation (7.2) is obtained from (7.1) by noting that $D_x = D_y = 1/g$, $G = 0$ and $Q = 2\theta$. The variable ϕ is a stress function, and the shear stresses within the shaft are related the derivatives of ϕ with respect to x and y.

Several fluid mechanics problems are embedded within (7.1). The streamline and potential formulations for an ideal irrotational fluid are governed by

$$\frac{\partial^2 \phi}{\partial x^2} + \frac{\partial^2 \phi}{\partial y^2} = 0 \tag{7.3}$$

and

$$\frac{\partial^2 \psi}{\partial x^2} + \frac{\partial^2 \psi}{\partial y^2} = 0 \tag{7.4}$$

respectively. The streamlines, ψ, are perpendicular to the constant potential lines ϕ, and the velocity components are related to the derivatives of either ϕ or ψ with respect to x and y. Each of (7.3) and (7.4) are obtained from (7.1) using $D_x = D_y = 1$, and $G = Q = 0$.

The flow of water within the earth is governed by equations that are embedded within (7.1). The seepage of water under a dam or retaining wall and within a confined aquifer is governed by

$$D_x \frac{\partial^2 \phi}{\partial x^2} + D_y \frac{\partial^2 \phi}{\partial y^2} = 0 \tag{7.5}$$

where D_x and D_y are the permeabilities of the earth material and ϕ represents the piezometric head. The water level around a well during the pumping process is governed by

$$D_x \frac{\partial^2 \phi}{\partial x^2} + D_y \frac{\partial^2 \phi}{\partial y^2} + Q = 0 \tag{7.6}$$

where Q is a point sink term. The other coefficients are the same as defined for (7.5), and the aquifer is assumed to be confined.

There are two heat transfer equations embedded within (7.1). The heat transfer from a two-dimensional fin to the surrounding fluid by convection is governed by

$$D_x \frac{\partial^2 T}{\partial x^2} + D_y \frac{\partial^2 T}{\partial y^2} - \frac{2h}{t} T + \frac{2hT_f}{t} = 0 \tag{7.7}$$

The coefficients D_x and D_y represent the thermal conductivities in the x and y directions, respectively; h is the convection coefficient; t is the thickness of the fin; T_f is the ambient temperature of the surrounding fluid; and T is the temperature of the fin. Relating (7.7) to (7.1), we find that $G = 2h/t$ and $Q = 2hT_f/t$.

The fin of (7.7) is assumed to be thin and the heat loss from the edges is neglected. When the body is very long in the z-direction and the temperature is a function of only the x- and y-coordinate directions, the heat transfer is governed by

$$D_x \frac{\partial^2 T}{\partial x^2} + D_y \frac{\partial^2 T}{\partial y_2} = 0 \tag{7.8}$$

where D_x, D_y, and T represent the same variables as given for (7.7). Heat transfer by convection is related to (7.8) through the boundary conditions.

When G in (7.1) is a negative coefficient and Q equals zero, the differential equation is called the Helmholtz equation. A negative G leads to the solution of an eigenvalue problem. A couple of physical problems governed by the Helmholtz equation are the Seiche motion of water and acoustic vibrations.

Seiche motion, which describes the standing waves on a bounded shallow body of water, is governed by

$$h\frac{\partial^2 w}{\partial x^2}+h\frac{\partial^2 w}{\partial y^2}+\frac{4\pi^2}{gT^2}w=0 \tag{7.9}$$

where h is the water depth at the quiescent state, w is the wave height above the quiescent level, g is the gravitational constant, and T is the period of oscillation.

A fluid vibrating within closed volume is governed by

$$\frac{\partial^2 P}{\partial x^2}+\frac{\partial^2 P}{\partial y^2}+\frac{w^2}{c^2}P=0 \tag{7.10}$$

where P is the pressure excess above ambient pressure, w is the wave frequency, and c is the wave velocity in the medium. The waves described by (7.10) would not be a function of the z-direction.

Equations (7.2) through (7.10) describe nine different physical problems that are contained within the general differential equation (7.1). It should be clear that a discussion of applying the finite element method to solve (7.1) is worthy of our time.

7.2 INTEGRAL EQUATIONS FOR THE ELEMENT MATRICES

Our immediate objective is to derive the integral equations that define the element matrices for the group of problems embedded in (7.1). The element contribution to the system of equations is given by

$$\{R^{(e)}\}=-\int_A [N]^T\left(D_x\frac{\partial^2\phi}{\partial x^2}+D_y\frac{\partial^2\phi}{\partial y^2}-G\phi+Q\right)dA \tag{7.11}$$

where $[N]$ is the row vector containing the element shape functions. Since the interpolation function, $\phi(x,y)$, does not have continuous derivatives between the elements, the second-derivative terms in (7.11) must be replaced by first-derivative terms.

The second-derivative terms in (7.11) can be replaced by applying the product rule for differentiation. Consider the quantity

$$\frac{\partial}{\partial x}\left([N]^T\frac{\partial\phi}{\partial x}\right) \tag{7.12}$$

Differentiation gives

$$\frac{\partial}{\partial x}\left([N]^T\frac{\partial\phi}{\partial x}\right)=[N]^T\frac{\partial^2\phi}{\partial x^2}+\frac{\partial[N]^T}{\partial x}\frac{\partial\phi}{\partial x} \tag{7.13}$$

Rearranging and substituting for $[N]^T\partial^2\phi/\partial x^2$ in (7.11) produces

$$-\int_A [N]^T D_x\frac{\partial^2\phi}{\partial x^2}dA=-\int_A D_x\frac{\partial}{\partial x}\left([N]^T\frac{\partial\phi}{\partial x}\right)dA+\int_A D_x\frac{\partial[N]^T}{\partial x}\frac{\partial\phi}{\partial x}dA \tag{7.14}$$

The first integral on the right-hand side of (7.14) can be replaced by an integral around the boundary using Green's theorem (Olmstead, 1961). Application of the theorem yields

$$\int_A \frac{\partial}{\partial x}\left([N]^T \frac{\partial \phi}{\partial x}\right) dA = \int_\Gamma [N]^T \frac{\partial \phi}{\partial x} \cos \theta \, d\Gamma \tag{7.15}$$

where θ is the angle to the outward normal and Γ is the element boundary. Substituting (7.15) into (7.14) gives the final relationship for the second-derivative term as

$$-\int_A D_x [N]^T \frac{\partial^2 \phi}{\partial x^2} dA = -\int_\Gamma D_x [N]^T \frac{\partial \phi}{\partial x} \cos \theta \, d\Gamma + \int_A D_x \frac{\partial [N]^T}{\partial x} \frac{\partial \phi}{\partial x} dA \tag{7.16}$$

A similar set of operations starting with

$$\frac{\partial}{\partial y}\left([N]^T \frac{\partial \phi}{\partial y}\right)$$

produces

$$-\int_A D_y [N]^T \frac{\partial^2 \phi}{\partial y^2} dA = -\int_\Gamma D_y [N]^T \frac{\partial \phi}{\partial y} \sin \theta \, d\Gamma + \int_A D_y \frac{\partial [N]^T}{\partial y} \frac{\partial \phi}{\partial y} dA \tag{7.17}$$

Substitution of (7.16) and (7.17) into (7.11) gives

$$\{R^{(e)}\} = -\int_\Gamma \left([N]^T \left(D_x \frac{\partial \phi}{\partial x} \cos \theta + D_y \frac{\partial \phi}{\partial y} \sin \theta\right)\right) d\Gamma$$

$$+ \int_A \left(D_x \frac{\partial [N]^T}{\partial x} \frac{\partial \phi}{\partial x} + D_y \frac{\partial [N]^T}{\partial y} \frac{\partial \phi}{\partial y}\right) dA$$

$$+ \int_A G[N]^T \phi \, da - \int_A Q[N]^T dA \tag{7.18}$$

Equation (7.18) is close to the desired form. It can be put in a final form by substituting for ϕ using the relationship

$$\phi^{(e)} = [N]\{\Phi^{(e)}\} \tag{7.19}$$

Substitution and rearranging gives

$$\{R^{(e)}\} = -\int_\Gamma [N]^T \left(D_x \frac{\partial \phi}{\partial x} \cos \theta + D_y \frac{\partial \phi}{\partial y} \sin \theta\right) d\Gamma$$

$$+ \left(\int_A \left(D_x \frac{\partial [N]^T}{\partial x} \frac{\partial [N]}{\partial x} + D_y \frac{\partial [N]^T}{\partial y} \frac{\partial [N]}{\partial y}\right)\right) dA \{\Phi^{(e)}\}$$

$$+ \left(\int_A G[N]^T [N] \, dA\right) \{\Phi^{(e)}\} - \int_A Q[N]^T dA \tag{7.20}$$

which has the general form

$$\{R^{(e)}\} = \{I^{(e)}\} + [k^{(e)}]\{\Phi^{(e)}\} - \{f^{(e)}\} \tag{7.21}$$

where

$$\{I^{(e)}\} = -\int_\Gamma [N]^T \left(D_x \frac{\partial \phi}{\partial x} \cos \theta + D_y \frac{\partial \phi}{\partial y} \sin \theta\right) d\Gamma \tag{7.22}$$

$$[k^{(e)}] = \int_A \left(D_x \frac{\partial [N]^T}{\partial x} \frac{\partial [N]}{\partial x} + D_y \frac{\partial [N]^T}{\partial y} \frac{\partial [N]}{\partial y} \right) dA + \int_A G[N]^T[N]\, dA \qquad (7.23)$$

and

$$\{f^{(e)}\} = \int_A Q[N]^T\, dA \qquad (7.24)$$

The variable ϕ in (7.22) was not replaced because the quantity

$$D_x \frac{\partial \phi}{\partial x} \cos\theta + D_y \frac{\partial \phi}{\partial y} \sin\theta \qquad (7.25)$$

occurs in the derivative boundary conditions and is considered in detail in Chapter 9.

The first integral in (7.23) can be written more compactly by defining

$$[D] = \begin{bmatrix} D_x & 0 \\ 0 & D_y \end{bmatrix} \qquad (7.26)$$

and the gradient vector

$$\{gv\} = \begin{Bmatrix} \dfrac{\partial \phi}{\partial x} \\ \dfrac{\partial \phi}{\partial y} \end{Bmatrix} = \begin{bmatrix} \dfrac{\partial [N]}{\partial x} \\ \dfrac{\partial [N]}{\partial y} \end{bmatrix} \{\Phi^{(e)}\} = [B]\{\Phi^{(e)}\} \qquad (7.27)$$

The first row of $\{gv\}$ is the derivative of $[N]$ with respect to x; the second row is the derivative of $[N]$ with respect to y. The transpose of $[B]$ contains two columns and is given by

$$[B]^T = \begin{bmatrix} \dfrac{\partial [N]^T}{\partial x} & \dfrac{\partial [N]^T}{\partial y} \end{bmatrix} \qquad (7.28)$$

If we use (7.26), (7.27), and (7.28), it is easy to verify that

$$\int_A [B]^T[D][B]\, dA = \int_A \left(D_x \frac{\partial [N]^T}{\partial x} \frac{\partial [N]}{\partial x} + D_y \frac{\partial [N]^T}{\partial y} \frac{\partial [N]}{\partial y} \right) dA \qquad (7.29)$$

The stiffness matrix for field problems is usually written as

$$[k^{(e)}] = \int_A [B]^T[D][B]\, dA + \int_A G[N]^T[N]\, dA \qquad (7.30)$$

and the individual integrals are denoted in this book by $[k_D^{(e)}]$ and $[k_G^{(e)}]$, where

$$[k^{(e)}] = [k_D^{(e)}] + [k_G^{(e)}] \qquad (7.31)$$

7.3 ELEMENT MATRICES: TRIANGULAR ELEMENTS

Our objective for the rest of this chapter is to evaluate the element matrices for the two-dimensional elements discussed in Chapter 5.

The scalar quantity ϕ is defined over a triangular region by

$$\phi^{(e)} = [N_i \quad N_j \quad N_k]\{\Phi^{(e)}\} \qquad (7.32)$$

where

$$N_i = \frac{1}{2A}(a_i + b_i x + c_i y)$$

$$N_j = \frac{1}{2A}(a_j + b_j x + c_j y)$$

$$N_k = \frac{1}{2A}(a_k + b_k x + c_k y)$$

and the a, b, and c coefficients were defined in Chapter 5. The gradient vector for this element is

$$\{gv\} = \begin{bmatrix} \dfrac{\partial N_i}{\partial x} & \dfrac{\partial N_j}{\partial x} & \dfrac{\partial N_k}{\partial x} \\[2mm] \dfrac{\partial N_i}{\partial y} & \dfrac{\partial N_j}{\partial y} & \dfrac{\partial N_k}{\partial y} \end{bmatrix} \{\Phi^{(e)}\} \qquad (7.33)$$

or

$$\{gv\} = \frac{1}{2A}\begin{bmatrix} b_i & b_j & b_k \\ c_i & c_j & c_k \end{bmatrix} \{\Phi^{(e)}\} = [B]\{\Phi^{(e)}\} \qquad (7.34)$$

The two matrices, $[B]$, as given by (7.34) and $[D]$, as given by (7.26), consist entirely of constants, since b_β, c_β, $\beta = i, j, k$ are constants and D_x and D_y are material coefficients. The first integral of (7.30), therefore, is easily evaluated. This integral becomes

$$[k_D^{(e)}] = \int_A [B]^T[D][B]\,dA = [B]^T[D][B]\int_A dA$$

or

$$[k_D^{(e)}] = [B]^T[D][B]A \qquad (7.35)$$

Expanding the matrix product yields

$$[k_D^{(e)}] = \frac{D_x}{4A}\begin{bmatrix} b_i^2 & b_i b_j & b_i b_k \\ b_i b_j & b_j^2 & b_j b_k \\ b_i b_k & b_j b_k & b_k^2 \end{bmatrix} + \frac{D_y}{4A}\begin{bmatrix} c_i^2 & c_i c_j & c_i c_k \\ c_i c_j & c_j^2 & c_j c_k \\ c_i c_k & c_j c_k & c_k^2 \end{bmatrix} \qquad (7.36)$$

The second integral of (7.30) involves the shape functions. If we assume that G is constant within the element, this integral becomes

$$[k_G^{(e)}] = \int_A G[N]^T[N]\, dA = G \int_A \begin{Bmatrix} N_i \\ N_j \\ N_k \end{Bmatrix} \begin{bmatrix} N_i & N_j & N_k \end{bmatrix}\, dA$$

$$= G \int_A \begin{bmatrix} N_i^2 & N_iN_j & N_iN_k \\ N_iN_j & N_j^2 & N_jN_k \\ N_iN_k & N_jN_k & N_k^2 \end{bmatrix} dA$$

$$= G \int_A \begin{bmatrix} L_1^2 & L_1L_2 & L_1L_3 \\ L_1L_2 & L_2^2 & L_2L_3 \\ L_1L_3 & L_2L_3 & L_3^2 \end{bmatrix} dA \tag{7.37}$$

since $N_i = L_1$, $N_j = L_2$, and $N_k = L_3$ for the linear triangle. Using the factorial formula (6.29) to evaluate each integral yields

$$[k_G^{(e)}] = \frac{GA}{12} \begin{bmatrix} 2 & 1 & 1 \\ 1 & 2 & 1 \\ 1 & 1 & 2 \end{bmatrix} \tag{7.38}$$

The element stiffness matrix for the triangular element is the sum of (7.36) and (7.38) when G is nonzero.

The element force vector also involves the shape functions, and the evaluation of (7.24) is quite similar to the evaluation of $[k_G^{(e)}]$. Substitution gives

$$\int_A Q[N]^T\, dA = Q \int_A \begin{Bmatrix} N_i \\ N_j \\ N_k \end{Bmatrix} dA = Q \int_A \begin{Bmatrix} L_1 \\ L_2 \\ L_3 \end{Bmatrix} dA \tag{7.39}$$

assuming that Q is constant within the element. Using the integration equation (6.29) produces

$$\{f^{(e)}\} = \frac{QA}{3} \begin{Bmatrix} 1 \\ 1 \\ 1 \end{Bmatrix} \tag{7.40}$$

ILLUSTRATIVE EXAMPLE

Heat transfer from a two-dimensional fin is governed by (7.7). Calculate the element matrices for the element in Figure 5.8 when $D_x = D_y = 0.5$ W/cm-C, $h = 0.01$ W/cm^2-C, $t = 0.5$ cm, and $T_f = 10$C.

The element stiffness matrix $[k^{(e)}]$ is given by (7.30), where $[k_D^{(e)}]$ and $[k_G^{(e)}]$ are defined by (7.36) and (7.38), respectively. The b and c coefficients for (7.36) were calculated in the example associated with Figure 5.8. These values are

$$b_i = -4.5, \quad b_j = 5, \quad b_k = -0.5$$
$$c_i = -2, \quad c_j = -2, \quad c_k = 4$$

The area was also calculated and is $9.5\,\text{cm}^2$. The parameters multiplying the matrices in (7.36) and (7.38) are

$$\frac{D_x}{4A}=\frac{D_y}{4A}=\frac{0.5}{4(9.5)}=0.0132$$

$$\frac{GA}{12}=\frac{2hA}{12t}=\frac{2(0.01)9.5}{12(0.5)}=0.0317$$

Substitution of the b and c coefficients and the calculated coefficients into (7.36) and (7.38) gives $[k^{(e)}]$ as

$$0.0132\begin{bmatrix} 20.3 & -22.5 & 2.25 \\ -22.5 & 25 & -2.50 \\ 2.25 & -2.50 & 0.250 \end{bmatrix}+0.0132\begin{bmatrix} 4 & 4 & -8 \\ 4 & 4 & -8 \\ -8 & -8 & 16 \end{bmatrix}$$

$$+0.0317\begin{bmatrix} 2 & 1 & 1 \\ 1 & 2 & 1 \\ 1 & 1 & 2 \end{bmatrix}$$

or

$$[k^{(e)}]=\begin{bmatrix} 0.384 & -0.213 & -0.0442 \\ -0.213 & 0.446 & -0.107 \\ -0.0442 & -0.107 & 0.278 \end{bmatrix}$$

The element force vector $\{f^{(e)}\}$ is given by (7.40). In this example, $Q=2h\phi_f/t$; thus

$$\frac{QA}{3}=\frac{2h\phi_f A}{3t}=\frac{2(0.01)(9.5)(10)}{3(0.5)}=1.27$$

and

$$\{f^{(e)}\}=\begin{Bmatrix} 1.27 \\ 1.27 \\ 1.27 \end{Bmatrix}$$

7.4 ELEMENT MATRICES: RECTANGULAR ELEMENT

Evaluation of the element matrices for the rectangular element cannot be performed as quickly as the integrations in the previous section. Each coefficient involves integrating a polynomial over an area. The integrals can be evaluated using the shape functions given by either of (5.17) or (5.19). We will use (5.17) because of the similarity between the st- and xy-coordinate systems.

The shape functions in (5.17) were developed relative to the st-coordinate system; this presents a minor problem. All of the integrals are defined relative to the xy system. In particular, the gradient matrix $[B]$ has coefficients related to the derivative of the shape functions with respect to x or y.

The change of variables equation for double integrals is discussed in detail in Chapter 27 with regard to quadratic elements. The application of this equation to a rectangular element defined relative to an st-coordinate system can be summarized as follows. Since the st-coordinate system is parallel to the xy-coordinate system and a unit length in either s or t is the same as a unit length of x or y,

$$\int_A f(x, y) \, dx \, dy = \int_A f(s, t) \, ds \, dt \tag{7.41}$$

Equally important is the relationship between the derivatives. The chain rule gives

$$\frac{\partial N_\beta}{\partial x} = \frac{\partial N_\beta}{\partial s} \quad \text{and} \quad \frac{\partial N_\beta}{\partial y} = \frac{\partial N_\beta}{\partial t} \tag{7.42}$$

The shape functions (5.17) are

$$N_i = 1 - \frac{s}{2b} - \frac{t}{2a} + \frac{st}{4ab}, \qquad N_j = \frac{s}{2b} - \frac{st}{4ab}$$

$$N_k = \frac{st}{4ab}, \qquad N_m = \frac{t}{2a} - \frac{st}{4ab}$$

The evaluation of $[k^{(e)}]$ and $\{f^{(e)}\}$ is illustrated by considering a specific integral in each case. The easiest integrations are associated with $\{f^{(e)}\}$, which is

$$\{f^{(e)}\} = \int_A Q[N]^T \, dA = \int_0^{2b} \int_0^{2a} Q \begin{Bmatrix} N_i \\ N_j \\ N_k \\ N_m \end{Bmatrix} dt \, ds \tag{7.44}$$

Considering the third coefficient gives

$$\int_0^{2b} \int_0^{2b} N_k \, dt \, ds = \int_0^{2b} \int_0^{2a} \frac{st}{4ab} \, dt \, ds$$

$$= \int_0^{2b} \frac{st^2}{8ab} \Big|_0^{2a} \, ds = \int_0^{2b} \frac{as}{2b} \, ds = \frac{A}{4} \tag{7.45}$$

The other three integrals yield the same result and

$$\{f^{(e)}\} = \frac{QA}{4} \begin{Bmatrix} 1 \\ 1 \\ 1 \\ 1 \end{Bmatrix} \tag{7.46}$$

The integral associated with $[k_G^{(e)}]$ is

$$[k_G^{(e)}] = \int_A G[N]^T[N] \, dA$$

$$\int_A G \begin{bmatrix} N_i^2 & N_i N_j & N_i N_k & N_i N_m \\ N_i N_j & N_j^2 & N_j N_k & N_j N_m \\ N_i N_k & N_j N_k & N_k^2 & N_k N_m \\ N_i N_m & N_j N_m & N_k N_m & N_m^2 \end{bmatrix} dA \tag{7.47}$$

Selecting the N_k^2 term, we have

$$\int_0^{2b} \int_0^{2a} \left(\frac{st}{4ab}\right)^2 dt\, ds = \int_0^{2b} \int_0^{2a} \frac{s^2 t^2}{16a^2 b^2} dt\, ds = \frac{4ab}{9} = \frac{A}{9} \tag{7.48}$$

The complete set of coefficients is

$$[k_G^{(e)}] = \frac{GA}{36} \begin{bmatrix} 4 & 2 & 1 & 2 \\ 2 & 4 & 2 & 1 \\ 1 & 2 & 4 & 2 \\ 2 & 1 & 2 & 4 \end{bmatrix} \tag{7.49}$$

The evaluation of $[k_D^{(e)}]$ involves the derivatives of the shape functions. The gradient matrix $[B]$ is

$$[B] = \begin{bmatrix} \dfrac{\partial N_i}{\partial x} & \dfrac{\partial N_j}{\partial x} & \dfrac{\partial N_k}{\partial x} & \dfrac{\partial N_m}{\partial x} \\[2mm] \dfrac{\partial N_i}{\partial y} & \dfrac{\partial N_j}{\partial y} & \dfrac{\partial N_k}{\partial y} & \dfrac{\partial N_m}{\partial y} \end{bmatrix} \tag{7.50}$$

Using the relationships given in (7.42) allows $[B]$ to be written in terms of s and t. Differentiation of the shape functions gives

$$[B] = \frac{1}{4ab} \begin{bmatrix} -(2a-t) & (2a-t) & t & -t \\ -(2b-s) & -s & s & (2b-s) \end{bmatrix} \tag{7.51}$$

The coefficient in the first row and first column of $[k_D^{(e)}]$ is available after the multiplications $[B]^T[D][B]$ have been performed. This coefficient is

$$\frac{D_x}{16a^2 b^2}(2a-t)^2 + \frac{D_y}{16a^2 b^2}(2b-s)^2 \tag{7.52}$$

and the associated integral is

$$\int_0^{2b} \int_0^{2a} \frac{D_x}{16a^2 b^2}(2a-t)^2\, dt\, ds + \int_0^{2b} \int_0^{2a} \frac{D_y}{16a^2 b^2}(2b-s)^2\, dt\, ds \tag{7.53}$$

which integrates to

$$\frac{D_x a}{3b} + \frac{D_y b}{3a} \tag{7.54}$$

The complete result for $[k_D^{(e)}]$ is

$$[k_D^{(e)}] = \frac{D_x a}{6b} \begin{bmatrix} 2 & -2 & -1 & 1 \\ -2 & 2 & 1 & -1 \\ -1 & 1 & 2 & -2 \\ 1 & -1 & -2 & 2 \end{bmatrix} + \frac{D_y b}{6a} \begin{bmatrix} 2 & 1 & -1 & -2 \\ 1 & 2 & -2 & -1 \\ -1 & -2 & 2 & 1 \\ -2 & -1 & 1 & 2 \end{bmatrix} \tag{7.55}$$

The element stiffness matrix $[k^{(e)}]$ for the rectangular element is the sum of (7.49) and (7.55). The element force vector is given by (7.46).

PROBLEMS

7.1 Evaluate one of the following integrals related to the force vector for the rectangular element.

(a) $\int_A QN_i \, dA$ (b) $\int_A QN_j \, dA$ (c) $\int_A QN_m \, dA$

7.2 Evaluate one of the following integrals related to $[k_G^{(e)}]$ for the rectangular element.

(a) $\int_A GN_iN_j \, dA$ (b) $\int_A GN_kN_m \, dA$ (c) $\int_A GN_m^2 \, dA$

7.3 Evaluate the integrals for one of the following coefficients in $[k_D^{(e)}]$ for the rectangular element.

(a) Row 1, column 3.

(b) Row 2, column 4.

(c) Row 4, column 4.

7.4 Evaluate the coefficients in (7.24) when Q varies linearly over a triangular element, that is, $Q(x, y)$ is given by

$$Q(x, y) = N_iQ_i + N_jQ_j + N_kQ_k$$

7.5 Evaluate the coefficients in (7.24) when Q varies linearly over a rectangular element, that is, $Q(s, t)$ is given by

$$Q(s, t) = N_iQ_i + N_jQ_j + N_kQ_k + N_mQ_m$$

7.6 Write the transformation equations that apply between the xy- and st-coordinate systems of the rectangular element. Use these equations and the chain rule equations

$$\frac{\partial N_\beta}{\partial s} = \frac{\partial N_\beta}{\partial x}\frac{\partial x}{\partial s} + \frac{\partial N_\beta}{\partial y}\frac{\partial y}{\partial s}$$

$$\frac{\partial N_\beta}{\partial t} = \frac{\partial N_\beta}{\partial x}\frac{\partial x}{\partial t} + \frac{\partial N_\beta}{\partial y}\frac{\partial y}{\partial t}$$

to verify the relationships given in (7.42).

7.7 Write the transformation equations that apply between the xy- and qr-coordinate systems. Use these equations and chain rule equations similar to those in Problem 7.6 to show that

$$\frac{\partial N_\beta}{\partial x} = \frac{\partial N_\beta}{\partial q} \quad \text{and} \quad \frac{\partial N_\beta}{\partial y} = \frac{\partial N_\beta}{\partial r}$$

7.8   Develop a relationship for $[B]$ similar to (7.51) that applies in the qr-coordinate system.

7.9   Calculate $[k^{(e)}]$ and $\{f^{(e)}\}$ for a linear triangular element when

(a) $D_x = D_y = 1$, $G = 4$, $Q = 5$, and the coordinates are those given in Problem 5.7.

(b) $D_x = D_y = 0.25$, $G = 6$, $Q = 3$, and the coordinates are those given in Problem 5.8.

(c) $D_x = D_y = 2$, $G = 6$, $Q = 6$, and the coordinates are those given in Problem 5.9.

7.10   Calculate $[k^{(e)}]$ and $\{f^{(e)}\}$ for a bilinear rectangular element when

(a) $D_x = D_y = 1$, $G = 12$, $Q = 5$, and the coordinates are those given in Problem 5.12.

(b) $D_x = D_y = 0.5$, $G = 10$, $Q = 40$, and the coordinates are those given in Problem 5.13.

(c) $D_x = D_y = 2$, $G = 2$, $Q = 6$, and the coordinates are those given in Problem 5.14.

7.11   Given the definitions $\phi^{(e)} = [N]\{\Phi^{(e)}\}$ and

$$\frac{d\phi^{(e)}}{dx} = \frac{d[N]}{dx}\{\Phi^{(e)}\} = [B]\{\Phi^{(e)}\}$$

start with

$$\{R^{(e)}\} = -\int_{X_i}^{X_j} [N]^T \left(D\frac{d^2\phi}{dx^2} + Q \right) dx$$

and obtain the general form

$$\{R^{(e)}\} = \{I^{(e)}\} + [k^{(e)}]\{\Phi^{(e)}\} - \{f^{(e)}\}$$

where

$$\{I^{(e)}\} = \begin{Bmatrix} I_i^{(e)} \\ I_j^{(e)} \end{Bmatrix} = \begin{Bmatrix} D\dfrac{d\phi}{dx}\Big|_{x=X_i} \\[2ex] -D\dfrac{d\phi}{dx}\Big|_{x=X_j} \end{Bmatrix}$$

$$[k^{(e)}] = \int_{X_i}^{X_j} D[B]^T[B]\,dx$$

and

$$\{f^{(e)}\} = \int_{X_i}^{X_j} Q[N]^T\,dx$$

for the one-dimensional linear element.

7.12   The mixed-derivative term $\partial^2\phi/\partial x\,\partial y$ occurs in some two-dimensional differential equations. Show that

$$-\int_A [N]^T \frac{\partial^2 \phi}{\partial x \, \partial y} \, dA = \frac{1}{2} \int_A \left(\frac{\partial [N]^T}{\partial x} \frac{\partial \phi}{\partial y} + \frac{\partial [N]^T}{\partial y} \frac{\partial \phi}{\partial x} \right) dA$$

$$-\int_\Gamma [N]^T \left(\frac{\partial \phi}{\partial y} \cos \theta + \frac{\partial \phi}{\partial x} \sin \theta \right) d\Gamma$$

7.13 Evaluate

$$\frac{1}{2} \int_A \left(\frac{\partial [N]^T}{\partial x} \frac{\partial \phi}{\partial y} + \frac{\partial [N]^T}{\partial y} \frac{\partial \phi}{\partial x} \right) dA$$

given that $\phi = [N]\{\Phi^{(e)}\}$ and $[N]$ consists of the shape functions for the linear triangular element.

Chapter 8

TORSION OF NONCIRCULAR SECTIONS

The element matrices for the linear triangular and bilinear rectangular elements were evaluated in the previous chapter. Application of this information to obtain a numerical solution of a realistic problem is discussed in this chapter. We will calculate the shear stresses in a square steel bar subjected to a twisting torque. The torsion of noncircular sections has been selected as the initial application area because it has the simplest of the possible boundary conditions; ϕ is zero on the boundary.

8.1 GENERAL THEORY

There are two theories for calculating the shear stresses in a solid noncircular shaft subjected to torsion. St. Venant developed one theory, and Prandtl proposed the other. Both theories are discussed by Fung (1965). Prandtl's theory is used in this chapter.

The shear stress components in a noncircular shaft subjected to a twisting moment T about the z axis (Figure 8.1a) can be calculated using

$$\tau_{zx} = \frac{\partial \phi}{\partial y} \qquad \text{and} \qquad \tau_{zy} = -\frac{\partial \phi}{\partial x} \tag{8.1}$$

where $\phi(x, y)$ is a stress function. The governing differential equation is

$$\frac{1}{g}\frac{\partial^2 \phi}{\partial x^2} + \frac{1}{g}\frac{\partial^2 \phi}{\partial y^2} + 2\theta = 0 \tag{8.2}$$

with

$$\phi = 0 \tag{8.3}$$

on the boundary. The physical parameters in (8.2) are the shear modulus, g, (N/cm^2), and the angle of twist per unit length, θ, (rad/cm).

Prandtl's formulation does not have the applied torque, T, (N · cm) in the governing equation. Instead, T is calculated using

$$T = 2\int_A \phi \, dA \tag{8.4}$$

once $\phi(x, y)$ is known.

100

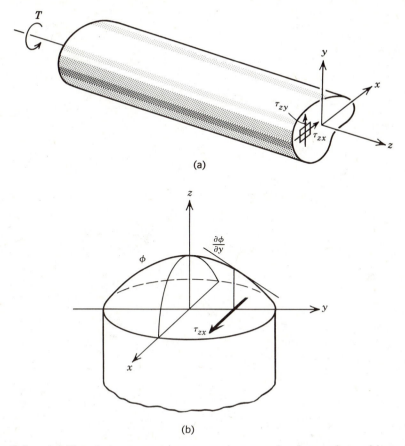

Figure 8.1. (a) The shear stress components in a noncircular section subjected to a torque loading. (b) The ϕ surface and related shear stress components.

The stress function represents a surface covering the cross section of the shaft (Figure 8.1b). The torque is proportional to the volume under the surface while the shear stresses are related to the surface gradients in the x and y coordinate directions.

Equation (8.2) is usually written as

$$\frac{\partial^2 \phi}{\partial x^2} + \frac{\partial^2 \phi}{\partial y^2} + 2g\theta = 0 \tag{8.5}$$

when the shaft is composed of a single material. This equation is obtained from (7.1) by noting that $D_x = D_y = 1$, $G = 0$, and $Q = 2g\theta$.

8.2   TWISTING OF A SQUARE BAR

The square shaft (Figure 8.2a) is used to illustrate the evaluation and the assemblage of the element matrices into a set of linear equations. This shaft has four axes of

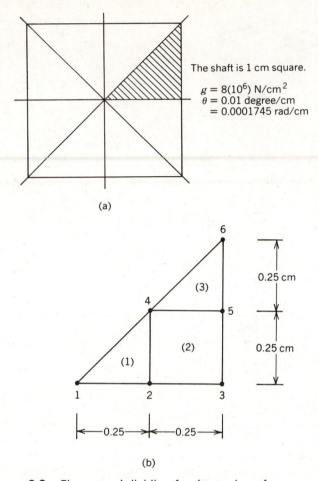

(a)

(b)

Figure 8.2 Element subdividion for the torsion of a square shaft.

symmetry; therefore, only one-eighth of the cross section needs to be analyzed. This fractional portion is divided into three elements (Figure 8.2*b*). Three elements are not sufficient to obtain an accurate answer, but they are enough to illustrate the calculations. The calculations have three significant digits of accuracy.

The element node numbers are

e	i	j	k	m
1	1	2	4	
2	2	3	5	4
3	4	5	6	

Elements one and three have the same orientation and the same dimensions; therefore, their matrices are identical.

The matrices for the triangular element are given by (7.36) and (7.40) while those for the rectangle are given by (7.46) and (7.55). The matrices become

$$[k_D^{(e)}] = \frac{1}{4A} \begin{bmatrix} b_i^2 & b_i b_j & b_i b_k \\ b_i b_j & b_j^2 & b_j b_k \\ b_i b_k & b_j b_k & b_k^2 \end{bmatrix} + \frac{1}{4A} \begin{bmatrix} c_i^2 & c_i c_j & c_i c_k \\ c_i c_j & c_j^2 & c_j c_k \\ c_i c_k & c_j c_k & c_k^2 \end{bmatrix} \qquad (8.6)$$

and

$$\{f^{(e)}\} = \frac{2g\theta A}{3} \begin{Bmatrix} 1 \\ 1 \\ 1 \end{Bmatrix} \qquad (8.7)$$

for the triangular element and

$$[k^{(e)}] = \frac{1}{6} \begin{bmatrix} 4 & -1 & -2 & -1 \\ -1 & 4 & -1 & -2 \\ -2 & -1 & 4 & -1 \\ -1 & -2 & -1 & 4 \end{bmatrix} \qquad (8.8)$$

and

$$\{f^{(e)}\} = \frac{2g\theta A}{4} \begin{Bmatrix} 1 \\ 1 \\ 1 \\ 1 \end{Bmatrix} \qquad (8.9)$$

for the rectangular element. Equation (8.8) also incorporates the fact that element two is square, $2a = 2b$.

We shall now evaluate (8.6) for elements one and three. The element area is $\frac{1}{32}$ and $4A^{(1)} = \frac{1}{8}$. The b and c coefficients are

$$\begin{array}{ll} b_1^{(1)} = Y_2 - Y_4 = -0.25, & c_1^{(1)} = X_4 - X_2 = 0 \\ b_2^{(1)} = Y_4 - Y_1 = 0.25, & c_2^{(1)} = X_1 - X_4 = -0.25 \\ b_4^{(1)} = Y_1 - Y_2 = 0, & c_4^{(1)} = X_2 - X_1 = 0.25 \end{array}$$

Substituting these values into (8.6) gives

$$[k^{(1)}] = \frac{8}{16} \begin{bmatrix} 1 & -1 & 0 \\ -1 & 1 & 0 \\ 0 & 0 & 0 \end{bmatrix} + \frac{8}{16} \begin{bmatrix} 0 & 0 & 0 \\ 0 & 1 & -1 \\ 0 & -1 & 1 \end{bmatrix} \qquad (8.10)$$

after noting that $1/4A^{(1)} = 8$ and the $b_i b_j$, $c_i c_j$-type products are either zero or $\pm \frac{1}{16}$. Adding the two matrices yields

$$[k^{(1)}] = \frac{1}{2} \begin{bmatrix} 1 & -1 & 0 \\ -1 & 2 & -1 \\ 0 & -1 & 1 \end{bmatrix} = [k^{(3)}] \qquad (8.11)$$

The element force vector, $\{f^{(1)}\}$, is readily obtained once the $2g\theta$ parameter has been evaluated. Substituting the values given in Figure 8.2 produces

$$2g\theta = 2(8)(10^6)(0.01)\left(\frac{\pi}{180}\right) = 2790$$

Substitution of 2790 and $A^{(1)} = \frac{1}{32}$ into (8.7) gives

$$\{f^{(1)}\}^T = [29.1 \quad 29.1 \quad 29.1] = \{f^{(3)}\}^T. \tag{8.12}$$

The stiffness matrix for element two is given by (8.8). The force vector, (8.9), is

$$\{f^{(2)}\}^T = [43.6 \quad 43.6 \quad 43.6 \quad 43.6] \tag{8.13}$$

since $A^{(2)} = \frac{1}{16}$ and $2g\theta$ is 2790.

The element matrices are summarized below. The node numbers indicate the rows and columns of $[K]$ and $\{F\}$ to which the individual coefficients add.

$$[k^{(1)}] = \frac{1}{2}\begin{matrix} & 1 & 2 & 4 \\ 1 & \begin{bmatrix} 1 & -1 & 0 \\ -1 & 2 & -1 \\ 0 & -1 & 1 \end{bmatrix} \end{matrix}, \quad \{f^{(1)}\} = \begin{Bmatrix} 29.1 \\ 29.1 \\ 29.1 \end{Bmatrix} \begin{matrix} 1 \\ 2 \\ 4 \end{matrix} \tag{8.14}$$

$$[k^{(2)}] = \frac{1}{6}\begin{matrix} & 2 & 3 & 5 & 4 \\ & \begin{bmatrix} 4 & -1 & -2 & -1 \\ -1 & 4 & -1 & -2 \\ -2 & -1 & 4 & -1 \\ -1 & -2 & -1 & 4 \end{bmatrix} \end{matrix}, \quad \{f^{(2)}\} = \begin{Bmatrix} 43.6 \\ 43.6 \\ 43.6 \\ 43.6 \end{Bmatrix} \begin{matrix} 2 \\ 3 \\ 5 \\ 4 \end{matrix} \tag{8.15}$$

$$[k^{(3)}] = \frac{1}{2}\begin{matrix} & 4 & 5 & 6 \\ & \begin{bmatrix} 1 & -1 & 0 \\ -1 & 2 & -1 \\ 0 & -1 & 1 \end{bmatrix} \end{matrix}, \quad \{f^{(3)}\} = \begin{Bmatrix} 29.1 \\ 29.1 \\ 29.1 \end{Bmatrix} \begin{matrix} 4 \\ 5 \\ 6 \end{matrix} \tag{8.16}$$

Adding the element contributions using the direct stiffness procedure and multiplying through by six gives the system of equations

$$\begin{bmatrix} 3 & -3 & 0 & 0 & 0 & 0 \\ -3 & 10 & -1 & -4 & -2 & 0 \\ 0 & -1 & 4 & -2 & -1 & 0 \\ 0 & -4 & -2 & 10 & -4 & 0 \\ 0 & -2 & -1 & -4 & 10 & -3 \\ 0 & 0 & 0 & 0 & -3 & 3 \end{bmatrix} \begin{Bmatrix} \Phi_1 \\ \Phi_2 \\ \Phi_3 \\ \Phi_4 \\ \Phi_5 \\ \Phi_6 \end{Bmatrix} - \begin{Bmatrix} 175 \\ 436 \\ 262 \\ 611 \\ 436 \\ 175 \end{Bmatrix} = \begin{Bmatrix} 0 \\ 0 \\ 0 \\ 0 \\ 0 \\ 0 \end{Bmatrix} \tag{8.17}$$

The nodal values Φ_3, Φ_5, and Φ_6 are on the external boundary and each is zero; therefore, equations three, five, and six are eliminated. Columns three, five, and six must be incorporated into $\{F\}$. Since $\Phi_3 = \Phi_5 = \Phi_6 = 0$, they contribute nothing to $\{F\}$ and the modified system of equations is

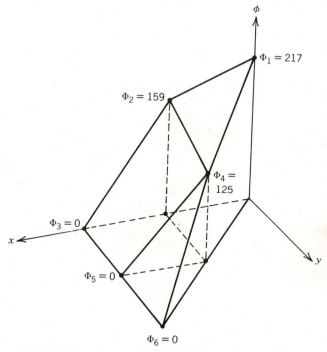

Figure 8.3. The ϕ surface for the torsion problem.

$$\begin{bmatrix} 3 & -3 & 0 \\ -3 & 10 & -4 \\ 0 & -4 & 10 \end{bmatrix} \begin{Bmatrix} \Phi_1 \\ \Phi_2 \\ \Phi_4 \end{Bmatrix} - \begin{Bmatrix} 175 \\ 436 \\ 611 \end{Bmatrix} = \begin{Bmatrix} 0 \\ 0 \\ 0 \end{Bmatrix}$$

Solution yields

$$\Phi_1 = 217, \qquad \Phi_2 = 159, \qquad \text{and} \qquad \Phi_4 = 125$$

The ϕ surface for this set of nodal values is shown in Figure 8.3.

The determination of the nodal values is a major part of the solution, but there is usually a set of element quantities that must be calculated once the nodal values are known. The shear stress components and the twisting torque are of interest in the present example. The evaluation of these quantities is discussed in the next two sections.

8.3 SHEAR STRESS COMPONENTS

The gradients of the nodal parameter, ϕ, are important because the shear stress components are related to these gradients by

$$\tau_{zx} = \frac{\partial \phi}{\partial y} \qquad \text{and} \qquad \tau_{zy} = -\frac{\partial \phi}{\partial x} \tag{8.18}$$

The gradient vector for the triangular element is given by (7.34) as

$$\{gv\} = \frac{1}{2A}\begin{bmatrix} b_i & b_j & b_k \\ c_i & c_j & c_k \end{bmatrix}\begin{Bmatrix} \Phi_i \\ \Phi_j \\ \Phi_k \end{Bmatrix} \tag{8.19}$$

The gradient vector for the rectangular element is given by (7.51) as

$$\{gv\} = \frac{1}{4ab}\begin{bmatrix} -(2a-t) & (2a-t) & t & -t \\ -(2b-s) & -s & s & (2b-s) \end{bmatrix}\begin{Bmatrix} \Phi_i \\ \Phi_j \\ \Phi_k \\ \Phi_m \end{Bmatrix} \tag{8.20}$$

The area as well as the b and c coefficients are the same for both triangular elements and they were evaluated prior to (8.10). Using these values and the calculated values for Φ_1, Φ_2, and Φ_4 gives

$$\{gv^{(1)}\} = \frac{16}{4}\begin{bmatrix} -1 & 1 & 0 \\ 0 & -1 & 1 \end{bmatrix}\begin{Bmatrix} 217 \\ 159 \\ 125 \end{Bmatrix} = \begin{Bmatrix} -232 \\ -136 \end{Bmatrix}$$

and

$$\tau_{zx}^{(1)} = \frac{\partial \phi}{\partial y} = -136 \text{ N/cm}^2$$

$$\tau_{zy}^{(1)} = -\frac{\partial \phi}{\partial x} = 232 \text{ N/cm}^2$$

The stress components for element three are calculated in a similar manner giving

$$\tau_{zx}^{(3)} = 0 \quad \text{and} \quad \tau_{zy}^{(3)} = 500 \text{ N/cm}^2$$

The gradient values are not constant within the rectangular element. Thus we can calculate the shear stress components at node three, which is where the largest value of τ_{zy} occurs. The local coordinates of node three are $s=2b$ and $t=0$; also, $2a=2b=0.25$. The gradient vector is

$$\{gv^{(2)}\} = \frac{16}{4}\begin{bmatrix} -1 & 1 & 0 & 0 \\ 0 & -1 & 1 & 0 \end{bmatrix}\begin{Bmatrix} \Phi_2 \\ \Phi_3 \\ \Phi_5 \\ \Phi_4 \end{Bmatrix}$$

or

$$\{gv^{(2)}\} = \begin{bmatrix} -4 & 4 & 0 & 0 \\ 0 & -4 & 4 & 0 \end{bmatrix}\begin{Bmatrix} 159 \\ 0 \\ 0 \\ 125 \end{Bmatrix} = \begin{Bmatrix} -636 \\ 0 \end{Bmatrix}$$

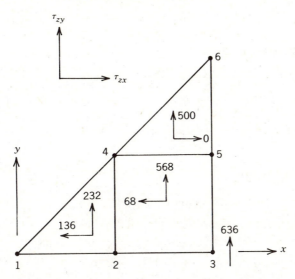

Figure 8.4   The shear stress values for the square shaft.

and

$$\tau_{zx}^{(2)} = 0 \quad \text{and} \quad \tau_{zy}^{(2)} = 636 \text{ N/cm}^2$$

The shear stress values calculated above are shown in Figure 8.4 along with the values of τ_{zx} and τ_{zy} at the center of the rectangular element. The shear stress values calculated for each triangular element are constant within the element and are usually assumed to be the values at the center of the element.

There are at least two ways to improve the stress values obtained for this example. First, a larger number of elements can be used. As the element size decreases, the existence of a constant value within the element becomes more realistic. An alternative approach is to use elements with more nodes and a quadratic or cubic interpolating polynomial. Differentiation will then yield gradients that are a function of the location within the element.

8.4   EVALUATION OF THE TWISTING TORQUE

Another quantity of interest in the analysis of a torsion member is the twisting torque T defined in (8.4). This integral is equivalent to

$$T = \sum_{e=1}^{n} 2 \int_{A^{(e)}} \phi^{(e)} \, dA \tag{8.21}$$

The integral

$$\int_{A^{(e)}} \phi^{(e)} \, dA$$

is

$$\int_{A(e)} \phi^{(e)} \, dA = \frac{A}{3} (\Phi_i + \Phi_j + \Phi_k) \tag{8.22}$$

for the triangular element and

$$\int_{A(e)} \phi^{(e)} \, dA = \frac{A}{4} (\Phi_i + \Phi_j + \Phi_k + \Phi_m) \tag{8.23}$$

for the rectangular element

The torque, T, is the sum of the element contributions or

$$T = T^{(1)} + T^{(2)} + T^{(3)}$$

The element contributions are

$$T^{(1)} = 2(\tfrac{1}{32})(\tfrac{1}{3})(\Phi_1 + \Phi_2 + \Phi_4)$$
$$= \tfrac{1}{48}(217 + 159 + 125) = 10.4$$
$$T^{(2)} = 2(\tfrac{1}{16})(\tfrac{1}{4})(\Phi_2 + \Phi_3 + \Phi_5 + \Phi_4) = 8.88$$
$$T^{(3)} = \tfrac{1}{48}(\Phi_4 + \Phi_5 + \Phi_6) = 2.60$$

The torque is

$$T = 10.4 + 8.88 + 2.60 = 21.9 \text{ N} \cdot \text{cm} \tag{8.24}$$

This torque acts on one-eighth of the cross section. Thus the torque acting on the square bar is $8(21.9) = 175$ N $\cdot$ cm.

It takes a torque of 175 N $\cdot$ cm to produce a twist of $1°$ in a 1-cm square steel shaft that is 100 cm long. The accuracy of this result is questionable, however, because of the coarseness of the grid. In fact, our answer is 11 percent below the theoretical value* of 196 N $\cdot$ cm.

The calculated value of the torque seldom corresponds to the actual applied value because we have guessed at the angle of twist, θ. The correct values of τ_{zx}, τ_{zy}, and θ are obtained by scaling the calculated value by the ratio $T_{\text{actual}}/T_{\text{calculated}}$. For example, suppose that the torque applied to the shaft in Figure 8.2 was 250 N $\cdot$ cm. The true angle of twist is

$$\theta_{\text{true}} = \frac{T_{\text{act}}}{T_{\text{cal}}} \theta_{\text{assumed}}$$

or

$$\theta_{\text{true}} = \left(\frac{250}{175}\right)(0.01) = 0.0143 \text{ rad/cm} \tag{8.25}$$

The shear stress values are scaled in the same way. The largest value of τ_{zy} for

---

*The relationship between the applied torque and the angle of twist for a square of dimension $2a$ is given by $T = 0.1406g\theta(2a)^4$ (Timoshenko and Goodier, 1970, equation 170, p. 313). For our example, $2a = 1$ and $T = 0.1406g\theta = 196$ N $\cdot$ cm.

the actual applied torque is

$$\tau_{zy} = \left(\frac{T_{\text{act}}}{T_{\text{cal}}}\right) \tau_{zy}$$

$$= \left(\frac{250}{175}\right)(636) = 909 \text{ N/cm}^2 \tag{8.26}$$

8.5 COMPUTER SOLUTIONS FOR THE SQUARE BAR

The three-element grid discussed in the previous sections is the smallest number of elements that can be used to solve the torsion problem involving a square bar. More accurate values for θ and the shear stresses are obtained using grids with a larger number of linear elements or by using elements with a higher level of interpolation. Three computer solutions for the square bar are summarized in this section. The first grid consists of 50 elements and is a mixture of the linear triangle and the bilinear rectangular elements (Figure 8.5). The other two (Figure 8.6) are mixtures of the quadratic triangular and quadrilateral elements. One grid has three elements, the other seven.

The maximum shear stress and the maximum value of ϕ for the various grids are summarized in Table 8.1. The 50-element grid is a definite improvement over

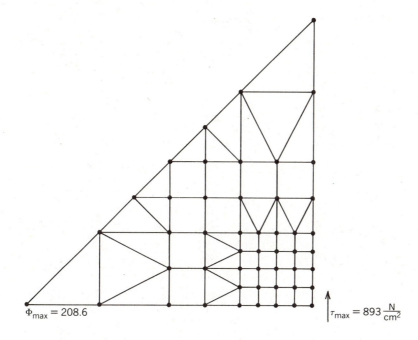

49 nodes, 50 elements

Figure 8.5. A 50-element grid for the square shaft.

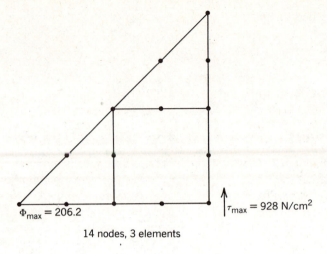

$\Phi_{max} = 206.2$ $\tau_{max} = 928 \text{ N/cm}^2$

14 nodes, 3 elements

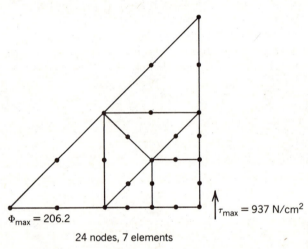

$\Phi_{max} = 206.2$ $\tau_{max} = 937 \text{ N/cm}^2$

24 nodes, 7 elements

Figure 8.6.   Two grids using quadratic elements to solve the square shaft problem.

the 3-element grid of Figure 8.2*b*. The grids with the quadratic elements, however, give more accurate results than the grid with 50 linear elements.

Table 8.1 indicates why the higher-order elements are included in many commercial computer programs: Greater accuracy can be obtained with fewer elements. Fewer elements means less work preparing the element data. A comparison of the results for the quadratic grids indicates that additional elements are not needed.

Table 8.1 also emphasizes a fact that should be kept in mind when constructing the grids for a finite element problem. A rather coarse grid gives good results if we are only interested in the nodal values. Note that Φ_{max} for the three-element

Table 8.1 Computer Solutions of a Square Bar

Grid	Torque	Φ_{max}	τ_{max}
Figure 8.2*b*	175	217	640
Figure 8.5	192	212	893
Figure 8.6*a*	196	206	928
Figure 8.6*b*	196	206	937
Theoretical	196	207	945

grid is within 6 percent of its theoretical value. If we are interested in derivative-related quantities, then we need a fine grid of linear elements or several quadratic elements. The maximum shear stress for the 50-element grid is 893 N/cm², which is 5.5 percent below the theoretical maximum. It took 50 elements to obtain the same accuracy in a derivative quantity that three elements produced in the nodal values. This ratio will not be the same for all problems, but it should be kept in mind.

PROBLEMS

8.1 Solve the torsion problem for a square shaft using the four triangular elements shown in Figure P8.1. Set up the equations, solve for the nodal values of ϕ, and calculate the maximum shear stress in element three. Use $2g\theta = 2790$.

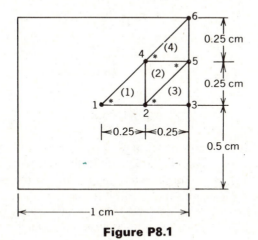

Figure P8.1

8.2 Solve the torsion problem for a square shaft using the four rectangular elements shown in Figure P8.2. Set up the equations, solve for the nodal values of ϕ, and calculate the maximum shear stress at node three. Use $2g\theta = 2790$.

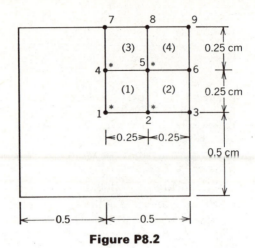

Figure P8.2

8.3   Do Problem 8.2 using the four-element grid shown in Figure P8.3.

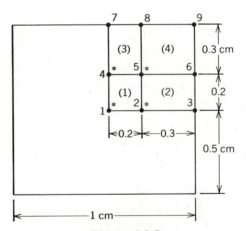

Figure P8.3

8.4–8.8   Use the nodal coordinates and ϕ values for the triangular elements given in Problems 5.7–5.11 and calculate the values of τ_{zx} and τ_{zy} for the element. Also calculate the element contribution to the twisting torque $T^{(e)}$.

8.9–8.13   Use the nodal coordinates and ϕ values for the rectangular elements given in Problems 5.12 to 5.16 and calculate the values of τ_{zx} and τ_{zy} at the center of the element. Also calculate the element contribution to the twisting torque $T^{(e)}$.

8.14–8.15   The twisting of the shapes shown in Figure P8.14 and P8.15 have analytical solutions. Obtain finite element solutions for these problems using the computer program **TDFIELD** (Chapter 16) and a reasonable number of elements (20 to 50). Compare the maximum values of ϕ and the shear stress for the finite element solution with the analytical values. Use $2g\theta = 2790$.

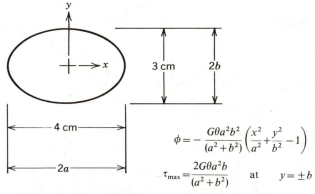

$$\phi = -\frac{G\theta a^2 b^2}{(a^2+b^2)}\left(\frac{x^2}{a^2}+\frac{y^2}{b^2}-1\right)$$

$$\tau_{max} = \frac{2G\theta a^2 b}{(a^2+b^2)} \quad \text{at} \quad y = \pm b$$

P8.14

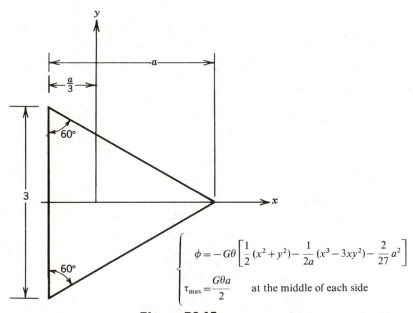

$$\phi = -G\theta\left[\frac{1}{2}(x^2+y^2)-\frac{1}{2a}(x^3-3xy^2)-\frac{2}{27}a^2\right]$$

$$\tau_{max} = \frac{G\theta a}{2} \quad \text{at the middle of each side}$$

Figure P8.15

8.16–8.19 Obtain finite element solutions to the torsion problem for the cross sections shown in Figures P8.16 to P8.19 using the computer program **TDFIELD** (Chapter 16). Use a minimum of 30 elements. Use $2g\theta = 2790$.

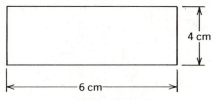

Figure P8.16

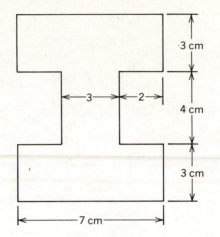

Figure P8.17

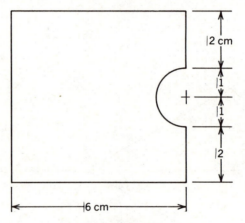

Figure P8.18

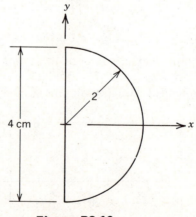

Figure P8.19

Chapter 9

DERIVATIVE BOUNDARY CONDITIONS: POINT SOURCES AND SINKS

The torsion of noncircular sections is a unique two-dimensional problem because the value of $\phi(x, y)$ is specified around the entire boundary. Most physical problems have a mixture of boundary conditions. The values of ϕ are specified on part of the boundary, and values related to the derivatives $\partial\phi/\partial x$ and $\partial\phi/\partial y$ are specified on other parts of the boundary. This latter situation occurs in both fluid flow and heat transfer problems. We cannot discuss either of these topics until we have studied how to handle derivative boundary conditions.

Another phenomenon that occurs in fluid flow and heat transfer problems is the point source or sink. In this situation, Q is concentrated at a point. Physical examples of a sink and a source include the pumping of water from an aquifer and heat generated by electrical lines embedded within a material.

The objective in this chapter is to discuss the concepts related to derivative boundary conditions and sources and sinks before we begin the study of irrotational flow and heat transfer.

9.1 DERIVATIVE BOUNDARY CONDITIONS

The two types of boundary conditions for two-dimensional field problems are shown schematically in Figure 9.1. Over part of the boundary, call it Γ_1, ϕ is specified. A boundary condition of the type

$$D_x \frac{\partial\phi}{\partial x} \cos\theta + D_y \frac{\partial\phi}{\partial y} \sin\theta = -M\phi_b + S \tag{9.1}$$

is specified on the rest of the boundary, Γ_2. When $D_x = D_y$, (9.1) reduces to

$$D_x \frac{\partial\phi}{\partial n} = -M\phi_b + S \tag{9.2}$$

where $\partial\phi/\partial n$ is the derivative normal to the boundary. In each case, ϕ_b represents the value of ϕ on Γ_2 and is unknown. Equations (9.1) and (9.2) simplify to

$$D_x \frac{\partial\phi}{\partial x} \cos\theta + D_y \frac{\partial\phi}{\partial y} \sin\theta = 0 \tag{9.3}$$

and

$$\frac{\partial\phi}{\partial n} = 0 \tag{9.4}$$

115

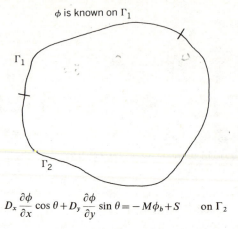

ϕ is known on Γ_1

$$D_x \frac{\partial \phi}{\partial x} \cos \theta + D_y \frac{\partial \phi}{\partial y} \sin \theta = -M\phi_b + S \qquad \text{on } \Gamma_2$$

Figure 9.1. The two types of boundary conditions for field problems.

when $M=S=0$. This situation occurs on insulated or impermeable boundaries or on axes of symmetry. The situation where either M or S, but not both, are zero is also possible.

The inclusion of the derivative boundary condition into the finite element analysis of field problems is done using the interelement vector $\{I^{(e)}\}$ given by (7.22). This vector is

$$\{I^{(e)}\} = -\int_{\Gamma} [N]^T \left(D_x \frac{\partial \phi}{\partial x} \cos \theta + D_y \frac{\partial \phi}{\partial y} \sin \theta \right) d\Gamma \tag{9.5}$$

where the integral is around the boundary of the element in a counterclockwise direction.

The integral in (9.5) is the sum of three integrals (one for each side) when integrating around a triangular element and the sum of four integrals for the rectangular element. We shall separate $\{I^{(e)}\}$ into two components

$$\{I^{(e)}\} = \{I_{bc}^{(e)}\} + \{I_i^{(e)}\} \tag{9.6}$$

where

$$\{I_{bc}^{(e)}\} = -\int_{\Gamma_{bc}} [N]^T \left(D_x \frac{\partial \phi}{\partial x} \cos \theta + D_y \frac{\partial \phi}{\partial y} \sin \theta \right) d\Gamma \tag{9.7}$$

and Γ_{bc} is the side of the element over which the boundary condition is specified. The vector $\{I_i^{(e)}\}$ contains the integrals of (9.5), which occur on the element sides that do not have a boundary condition specified on them. These integrals lead to the interelement requirements that must be satisfied before Galerkin's residual is zero.

Using our definition for $\{I_{bc}^{(e)}\}$ and substituting the relationship in (9.1) gives

$$\{I_{bc}^{(e)}\} = \int_{\Gamma_{bc}} [N]^T (M\phi_b - S) d\Gamma \tag{9.8}$$

where ϕ_b is given by the element equation

$$\phi^{(e)} = [N]\{\Phi^{(e)}\} \tag{9.9}$$

Substituting (9.9) into (9.8) produces

$$\{I_{bc}^{(e)}\} = \int_{\Gamma_{bc}} [N]^T (M[N]\{\Phi^{(e)}\} - S) \, d\Gamma \tag{9.10}$$

which can be separated into

$$\{I_{bc}^{(e)}\} = \left(\int_{\Gamma_{bc}} M[N]^T[N] \, d\Gamma \right) \{\Phi^{(e)}\} - \int_{\Gamma_{bc}} S[N]^T \, d\Gamma \tag{9.11}$$

The boundary condition produces two components. One component adds to $[k^{(e)}]$ because it multiplies $\{\Phi^{(e)}\}$. The other adds to $\{f^{(e)}\}$. The integrals, however, are different from those previously encountered because they are line integrals instead of area integrals.

The two components in $\{I_{bc}^{(e)}\}$ in (9.11) can be defined as

$$\{I_{bc}^{(e)}\} = [k_M^{(e)}]\{\Phi^{(e)}\} - \{f_S^{(e)}\} \tag{9.12}$$

where

$$[k_M^{(e)}] = \int_{\Gamma_{bc}} M[N]^T[N] \, d\Gamma \tag{9.13}$$

and

$$\{f_S^{(e)}\} = \int_{\Gamma_{bc}} S[N]^T \, d\Gamma \tag{9.14}$$

Before we proceed with the evaluation of the element integrals, it should be

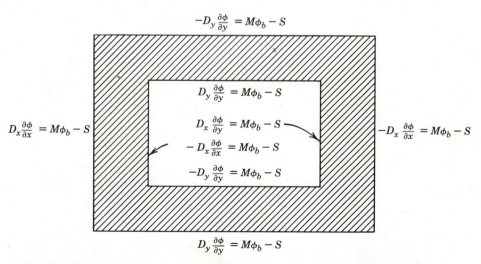

Figure 9.2. The derivative boundary conditions for a hollow rectangular region.

noted that the boundary condition (9.1) takes on different forms as we traverse the outside and inside of a rectangular region with a hole. The various forms for

$$-\left(D_x \frac{\partial \phi}{\partial x} \cos \theta + D_y \frac{\partial \phi}{\partial y} \sin \theta\right) \tag{9.15}$$

are given in Figure 9.2. The sign associated with M and S must be determined relative to the information contained in Figure 9.2. This determination is illustrated in the application chapters that follow. Remember that θ is the angle from the x-axis to the outward normal.

9.2 EVALUATION OF THE ELEMENT INTEGRALS

The integrals in (9.13) and (9.14) are valid for any two-dimensional element and can be evaluated once the element shape functions are known. We shall evaluate these for the linear triangular and bilinear rectangular elements that were studied in Chapter 5. We shall start with (9.14) and the rectangular element because this combination is the easiest to evaluate.

Assuming that S is specified over side ij and that the element has a unit thickness gives

$$\int_{\Gamma_{bc}} S[N]^T \, d\Gamma = \int_{-b}^{b} S \begin{Bmatrix} N_i \\ N_j \\ N_k \\ N_m \end{Bmatrix} dq \tag{9.16}$$

where the shape functions are those defined by (5.19) for the qr-coordinate system. Note, however, that $N_k = N_m = 0$ along side ij. Substituting for the nonzero shape functions and noting that $r = -a$,

$$\{f_S^{(e)}\} = \int_{-b}^{b} \frac{S}{2b} \begin{Bmatrix} (b-q) \\ (b+q) \\ 0 \\ 0 \end{Bmatrix} dq = \frac{SL_{ij}}{2} \begin{Bmatrix} 1 \\ 1 \\ 0 \\ 0 \end{Bmatrix} \tag{9.17}$$

The quantity S is multiplied by the length of side ij, L_{ij}, which is $2b$ and divided equally between the two nodes on side ij.

There are three other evaluations for the surface integral, one for each of the remaining three sides. It is left as an exercise to show that the other results for $\{f_S^{(e)}\}$ are

$$\{f_S^{(e)}\} = \frac{SL_{jk}}{2} \begin{Bmatrix} 0 \\ 1 \\ 1 \\ 0 \end{Bmatrix}, \qquad \frac{SL_{km}}{2} \begin{Bmatrix} 0 \\ 0 \\ 1 \\ 1 \end{Bmatrix}, \qquad \text{and} \qquad \frac{SL_{im}}{2} \begin{Bmatrix} 1 \\ 0 \\ 0 \\ 1 \end{Bmatrix} \tag{9.18}$$

for sides jk, km, and im, respectively. If S is specified on more than one side of an element, the values for $\{f_S^{(e)}\}$ for the appropriate sides are added together.

The evaluation of (9.14) for the triangular element gives results that are very similar to those in (9.17) and (9.18). The results are

$$\{f_S^{(e)}\} = \frac{SL_{ij}}{2} \begin{Bmatrix} 1 \\ 1 \\ 0 \end{Bmatrix}, \qquad \frac{SL_{jk}}{2} \begin{Bmatrix} 0 \\ 1 \\ 1 \end{Bmatrix}, \qquad \text{and} \qquad \frac{SL_{ik}}{2} \begin{Bmatrix} 1 \\ 0 \\ 1 \end{Bmatrix} \qquad (9.19)$$

for sides ij, jk, and ik, respectively. The quantities L_{ij}, L_{jk}, and L_{ik} are the lengths of the respective sides. They are not the area coordinates. The area coordinates have numerical subscripts.

The first result of (9.19) is obtained as follows. Given side ij,

$$\{f_S^{(e)}\} = \int_{\Gamma_{bc}} S[N]^T d\Gamma = L_{ij} \int_0^1 S \begin{Bmatrix} N_i \\ N_j \\ N_k \end{Bmatrix} d\ell_2 \qquad (9.20)$$

Since N_k is zero along side ij,

$$\{f_S^{(e)}\} = L_{ij} \int_0^1 S \begin{Bmatrix} N_i \\ N_j \\ 0 \end{Bmatrix} d\ell_2 = L_{ij} \int_0^1 S \begin{Bmatrix} \ell_1 \\ \ell_2 \\ 0 \end{Bmatrix} d\ell_2 \qquad (9.21)$$

because the shape functions N_i and N_j reduce to

$$N_i = L_1 = \ell_1 \qquad \text{and} \qquad N_j = L_2 = \ell_2 \qquad (9.22)$$

along side ij. This fact was discussed in Chapter 6. The integration is along a line; thus we can use the factorial formula (6.17), and the result in (9.19) follows immediately.

The integrals associated with $[k_M^{(e)}]$ are evaluated in a manner identical to those just discussed. The major difference is that there are more terms to consider. The integral in (9.13) expands into

$$[k_M^{(e)}] = \int_{\Gamma_{bc}} M \begin{bmatrix} N_i^2 & N_i N_j & N_i N_k & N_i N_m \\ N_i N_j & N_j^2 & N_j N_k & N_j N_m \\ N_i N_k & N_j N_k & N_k^2 & N_k N_m \\ N_i N_m & N_j N_m & N_k N_m & N_m^2 \end{bmatrix} d\Gamma \qquad (9.23)$$

for the rectangular element. If we assume that M is specified over side ij, then $N_k = N_m = 0$ and (9.23) becomes

$$[k_M^{(e)}] = \int_{-b}^b M \begin{bmatrix} N_i^2 & N_i N_j & 0 & 0 \\ N_i N_j & N_j^2 & 0 & 0 \\ 0 & 0 & 0 & 0 \\ 0 & 0 & 0 & 0 \end{bmatrix} dq \qquad (9.24)$$

Evaluation of the individual coefficients after noting $r = -a$ gives

$$\int_{-b}^b N_i^2 \, dq = \int_{-b}^b \frac{(b-q)^2}{4b^2} \, dq = \frac{2b}{3} = \frac{L_{ij}}{3} \qquad (9.25)$$

$$\int_{-b}^{b} N_i N_j \, dq = \int_{-b}^{b} \frac{(b-q)(b+q)}{4b^2} \, dq = \frac{2b}{6} = \frac{L_{ij}}{6} \tag{9.26}$$

and

$$\int_{-b}^{b} N_j^2 \, dq = \int_{-b}^{b} \frac{(b+q)^2}{4b^2} \, dq = \frac{2b}{3} = \frac{L_{ij}}{3} \tag{9.27}$$

Using these results, we have

$$[k_M^{(e)}] = \frac{ML_{ij}}{6} \begin{bmatrix} 2 & 1 & 0 & 0 \\ 1 & 2 & 0 & 0 \\ 0 & 0 & 0 & 0 \\ 0 & 0 & 0 & 0 \end{bmatrix} \tag{9.28}$$

There are three other results for $[k_M^{(e)}]$, one for each of the other sides. These results are

$$[k_M^{(e)}] = \frac{ML_{jk}}{6} \begin{bmatrix} 0 & 0 & 0 & 0 \\ 0 & 2 & 1 & 0 \\ 0 & 1 & 2 & 0 \\ 0 & 0 & 0 & 0 \end{bmatrix} \tag{9.29}$$

$$[k_M^{(e)}] = \frac{ML_{km}}{6} \begin{bmatrix} 0 & 0 & 0 & 0 \\ 0 & 0 & 0 & 0 \\ 0 & 0 & 2 & 1 \\ 0 & 0 & 1 & 2 \end{bmatrix} \tag{9.30}$$

and

$$[k_M^{(e)}] = \frac{ML_{im}}{6} \begin{bmatrix} 2 & 0 & 0 & 1 \\ 0 & 0 & 0 & 0 \\ 0 & 0 & 0 & 0 \\ 1 & 0 & 0 & 2 \end{bmatrix} \tag{9.31}$$

where L_{jk}, L_{km}, and L_{im} are the lengths of the respective sides.

The evaluation of (9.13) for the triangular element leads to

$$[k_M^{(e)}] = \frac{ML_{ij}}{6} \begin{bmatrix} 2 & 1 & 0 \\ 1 & 2 & 0 \\ 0 & 0 & 0 \end{bmatrix} \tag{9.32}$$

$$[k_M^{(e)}] = \frac{ML_{jk}}{6} \begin{bmatrix} 0 & 0 & 0 \\ 0 & 2 & 1 \\ 0 & 1 & 2 \end{bmatrix} \tag{9.33}$$

and

$$[k_M^{(e)}] = \frac{ML_{ik}}{6} \begin{bmatrix} 2 & 0 & 1 \\ 0 & 0 & 0 \\ 1 & 0 & 2 \end{bmatrix} \tag{9.34}$$

9.3 POINT SOURCES AND SINKS

An important physical situation is the concept of a point source or sink. A source or sink is said to exist whenever Q occurs over a very small area. Examples of line sources include steam and/or hot water pipes within the earth and conducting electrical wires embedded within a product. In each case, the cross-sectional area of the pipe or conductor is very small compared with the surrounding media. Point sinks occur in groundwater problems: They are pumps removing water from an aquifer.

Sources and sinks occur often enough in the real world to warrant our attention. Our discussion is structured around the two-dimensional element, but the procedure can be quickly modified to handle the one- or three-dimensional element.

Consider the triangular element in Figure 9.3 with a source Q^* located at (X_0, Y_0). Since the source is located at a point, Q is no longer constant throughout the element but is a function of x and y. Using unit impulse functions, $\delta(x - X_0)$ and $\delta(y - Y_0)$ (Kaplan, 1962), we can write

$$Q = Q^* \, \delta(x - X_0) \, \delta(y - Y_0) \tag{9.35}$$

The integral

$$\{f_Q^{(e)}\} = \int_A Q[N]^T \, dA \tag{9.36}$$

becomes

$$\{f_Q^{(e)}\} = Q^* \int_A \begin{Bmatrix} N_i \\ N_j \\ N_k \end{Bmatrix} \delta(x - X_0) \, \delta(y - Y_0) \, dx \, dy \tag{9.37}$$

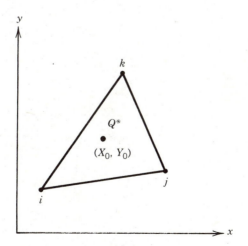

Figure 9.3. An element with a point source or sink.

The integral of a quantity multiplied by an impulse function, however, is equal to the quantity evaluated at X_0 and Y_0. Therefore,

$$\{f_Q^{(e)}\} = Q^* \begin{Bmatrix} N_i(X_0, Y_0) \\ N_j(X_0, Y_0) \\ N_k(X_0, Y_0) \end{Bmatrix} \tag{9.38}$$

The proportion of Q^* allocated to each node is based on the relative values of N_i, N_j, and N_k evaluated using the coordinates of the point source. Since the shape functions sum to one at every point within the element, we are not allocating more than Q^*.

ILLUSTRATIVE EXAMPLE

A line source $Q^* = 52$ W/cm is located at (5, 2) in the element shown in Figure 9.4. Determine the amount of Q^* allocated to each node.

The values of the a, b, and c constants are

$$\begin{array}{lll} a_i = 28, & a_j = 6, & a_k = -21 \\ b_i = -4, & b_j = 1, & b_k = 3 \\ c_i = -1, & c_j = -3, & c_k = 4 \end{array}$$

The shape function equations can be written after recalling that

$$a_i + a_j + a_k = 2A = 13$$

The equations are

$$N_i = \tfrac{1}{13}[28 - 4x - y]$$
$$N_j = \tfrac{1}{13}[6 + x - 3y]$$
$$N_k = \tfrac{1}{13}[-21 + 3x + 4y]$$

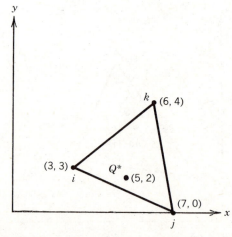

Figure 9.4.   A point source in a triangular element.

Substituting $x = X_0 = 5$ and $y = Y_0 = 2$ produces

$$N_i = \tfrac{1}{13}[28 - 4(5) - 2] = \tfrac{6}{13}$$
$$N_j = \tfrac{1}{13}[6 + 5 - 3(2)] = \tfrac{5}{13}$$
$$N_k = \tfrac{1}{13}[-21 + 3(5) + 4(2)] = \tfrac{2}{13}$$

The value of Q^* is allotted to nodes i, j, and k by the fractions $\tfrac{6}{13}$, $\tfrac{5}{13}$ and $\tfrac{2}{13}$, respectively. Therefore,

$$\{f_Q^{(e)}\} = \frac{52}{13} \begin{Bmatrix} 6 \\ 5 \\ 2 \end{Bmatrix} = \begin{Bmatrix} 24 \\ 20 \\ 8 \end{Bmatrix}$$

The best location for a source or sink is at a node. This location changes the result given in (9.38). If we assume that the source is at node j (Figure 9.5), then $N_i = N_k = 0$ and

$$\{f_Q^{(e)}\} = Q^* \begin{Bmatrix} 0 \\ 1 \\ 0 \end{Bmatrix} \qquad (9.39)$$

The magnitude of Q^*, however, must be modified when the source (sink) is shared by more than one element. The magnitude of the source is divided among the elements joining at the node. The source is allocated according to the ratio of the angle in the element to 360. The correct equation for element (e) in Figure 9.5 is

$$\{f_Q^{(e)}\} = \frac{\alpha Q^*}{360} \begin{Bmatrix} 0 \\ 1 \\ 0 \end{Bmatrix} \qquad (9.40)$$

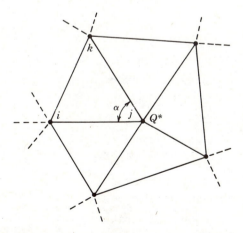

Figure 9.5. A point source at a node.

There is no need to evaluate α for the various elements. When the equations are assembled using the direct stiffness method, the element contributions at this node add to Q^*. An easier procedure for implementing a node source is to add the value of Q^* to the row of $\{F\}$ corresponding to the node number. A source is positive whereas the sink has a negative sign.

PROBLEMS

9.1 The boundary condition around the outside of a rectangular region follows. Determine the magnitude and sign for M and S on each of the sides; the sides are labeled as shown in Figure P9.1.

(a) $D_x \dfrac{\partial \phi}{\partial x} = 6\phi_b - 3,$ side 2

(b) $D_y \dfrac{\partial \phi}{\partial y} = -4\phi_b + 6,$ side 1

(c) $D_x \dfrac{\partial \phi}{\partial x} = 5\phi_b + 2,$ side 4

(d) $D_y \dfrac{\partial \phi}{\partial y} = 8\phi_b - 4,$ side 3

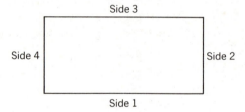

Figure P9.1

9.2 Evaluate the integral in (9.14) for:

(a) Side jk of a rectangular element.

(b) Side km of a rectangular element.

(c) Side im of a rectangular element.

(d) Side jk of a triangular element.

(e) Side ik of a triangular element.

9.3 Evaluate the integral in (9.13) for

(a) Side jk of a rectangular element.

(b) Side km of a rectangular element.

(c) Side im of a rectangular element.

(d) Side jk of a triangular element.

(e) Side ik of a triangular element.

9.4–9.8 Evaluate $\{f_Q^{(e)}\}$ for a point source, $Q^* = 40$ W/cm, located at point A for the corresponding triangular element in Problems 5.7–5.11.

9.9–9.13 Evaluate $\{f_Q^{(e)}\}$ for a point source, $Q^* = 40$ W/cm, located at point B for the corresponding rectangular element in Problems 5.12 to 5.16.

Chapter **10**

IRROTATIONAL FLOW

The irrotational flow of an ideal fluid has been studied extensively because of the information that can be obtained about flow around corners, over weirs, through constructions, and about airfoils. Ideal irrotational flow is an approximation. It assumes that no friction occurs between the fluid and a surface (ideal) and that no rotation or distortion of the fluid particles occurs during the movement (irrotational).

The flow of water through the earth can also be closely approximated by assuming irrotational flow. The analysis of groundwater flow is an important aspect of regional planning because many regions within the country depend wholly or in part on groundwater for their water supply. The flow of water through and under dams as well as into underground drains are some other important areas that can be studied using the theory.

10.1 FLOW OF AN IDEAL FLUID

The two-dimensional flow of an ideal fluid can be formulated in terms of a stream function ψ or a velocity potential function ϕ. Lines of constant ψ are perpendicular to lines of constant ϕ and the governing differential equations are identical. These equations are

$$\frac{\partial^2 \psi}{\partial x^2} + \frac{\partial^2 \psi}{\partial y^2} = 0 \quad \text{and} \quad \frac{\partial^2 \phi}{\partial x^2} + \frac{\partial^2 \phi}{\partial y^2} = 0 \tag{10.1}$$

respectively. The boundary conditions for ψ and ϕ, however, are not the same, and this difference leads to different calculated values. Both of these formulations are discussed in this section.

10.1.1 Streamline Formulation

Lines of constant ψ are called streamlines. The volume flow rate, Q_{ij}, between any pair of streamlines is equal to the difference in their values,

$$Q_{ij} = \psi_i - \psi_j \tag{10.2}$$

There is no flow perpendicular to a streamline. The velocity components are obtained from the calculated values of ψ using

$$V_x = \frac{\partial \psi}{\partial y} \quad \text{and} \quad V_y = -\frac{\partial \psi}{\partial x} \tag{10.3}$$

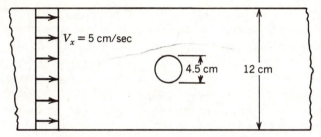

Figure 10.1. Flow around a cylinder

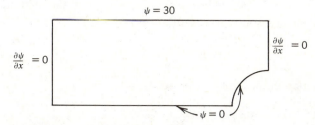

Figure 10.2. Boundary conditions for streamline flow around a cylinder.

The assumption of an ideal fluid implies that the motion of the fluid does not penetrate into the surrounding body or separate from the surface of the body and leave empty spaces. These conditions imply that the component of the fluid velocity normal to the surface is equal to the velocity of the surface in the same direction (Duncan et al., 1970). The above implies no flow perpendicular to a fixed boundary and, therefore, no velocity perpendicular to the boundary. Fixed boundaries as well as an axis of symmetry parallel to the flow are streamlines because there is no fluid velocity perpendicular to them.

The boundary conditions for streamline flow are discussed relative to the problem illustrated in Figure 10.1. On the left boundary, V_x is a uniform 5 cm/sec. Since $V_y=0$, it is concluded from (10.3) that $\partial\psi/\partial x=0$ on this edge. The same boundary condition applies to the right-hand vertical side because it is an axis of symmetry and the streamlines must be symmetrical about this edge.

The horizontal axis of symmetry and the cylinder boundary form a streamline as well as the upper boundary. A zero value is assigned to the lower streamline. The upper streamline can have any nonzero value; an appropriate value is 30 because the flow rate is 30 cm^3/sec for a unit thickness (one-half of the total flow). The four boundary conditions are shown in Figure 10.2.

10.1.2 Potential Formulation

The velocity components in the potential formulation are related to ϕ by

$$V_x=\frac{\partial\phi}{\partial x} \quad \text{and} \quad V_y=\frac{\partial\phi}{\partial y} \tag{10.4}$$

Relating these to the flow around the cylinder (Figure 10.1) gives $\partial\phi/\partial x = 5$ on the left-hand edge and $\partial\phi/\partial y = 0$ along the horizontal axis of symmetry and the upper boundary. Since the velocity normal to the cylinder is zero, $\partial\phi/\partial n = 0$ along the cylinder. We draw upon our knowledge that potential lines are perpendicular to streamlines to establish the last boundary condition. The right-hand edge must be a potential line because it is an axis of symmetry and all the streamlines are perpendicular to it. It is assigned an arbitrary value of 50. The value does not influence the results, since the velocity components are related to the gradient values. The boundary conditions are shown in Figure 10.3.

The numerical value of the flux boundary condition along the left edge is determined by comparing the actual boundary condition with the theoretical boundary condition. The actual boundary condition is

$$\frac{\partial\phi}{\partial x} = 5 \tag{10.5}$$

and the theoretical equation (see Figure 9.2) is

$$-D_x\frac{\partial\phi}{\partial x} = S \quad\text{or}\quad \frac{\partial\phi}{\partial x} = -S \tag{10.6}$$

when $D_x = 1$. Equating the second of (10.6) with (10.5) gives $S = -5$.

The boundary condition $\phi = 50$ on the right edge of Figure 10.3 is a special case. When the body is irregular in shape, the grid must include the downstream

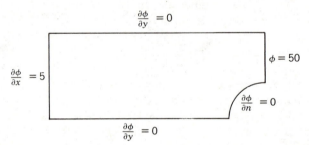

Figure 10.3. Boundary conditions for potential flow around a cylinder.

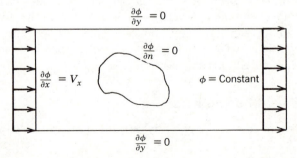

Figure 10.4. Boundary conditions for potential flow around an irregular shape.

side of the shape until a uniform velocity is again attained (Figure 10.4). At this point, constant potential lines are perpendicular to the flow and ϕ can be given an arbitrary value.

10.2 GROUNDWATER FLOW

There are two important groundwater problems governed by the field equation. The first is the seepage of groundwater under dams. The governing equation is

$$D_x \frac{\partial^2 \phi}{\partial x^2} + D_y \frac{\partial^2 \phi}{\partial y^2} = 0 \tag{10.7}$$

where D_x and D_y are the coefficients of permeability (m/day) and ϕ is the piezometric head, in meters, measured from the bottom of a confined aquifer. The boundary conditions generally consist of known values of ϕ beneath the water, and a zero seepage condition on the other boundaries (Figure 10.5). A impermeable vertical wall beneath the dam is modeled by using a narrow gap for the wall. The finite element method automatically enforces the impermeable boundary condition, $\partial \phi / \partial n = 0$, on each side of this gap, when no other conditions are specified.

The second groundwater problem is the calculation of the drawdown at a well that is removing water from an aquifer. The governing equation for a confined aquifer is

$$D_x \frac{\partial^2 \phi}{\partial x^2} + D_y \frac{\partial^2 \phi}{\partial y^2} + Q = 0 \tag{10.8}$$

where D_x, D_y, and ϕ are the same as defined for (10.7). The Q term in (10.8) represents a point sink, the well, and should be evaluated using the concepts discussed in Chapter 9. The best results occur when the well is located at a node.

The boundary conditions associated with (10.8) consist of known values on all

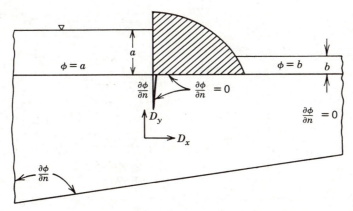

Figure 10.5. Boundary conditions for groundwater seepage under a dam.

or a part of the boundary and/or the seepage of water into the aquifer along the boundary. Seepage is described by the derivative boundary condition

$$-\left(D_x \frac{\partial \phi}{\partial x} \cos \theta + D_y \frac{\partial \phi}{\partial y} \sin \theta\right) = S \tag{10.9}$$

The fluid velocity components are calculated using Darcy's law

$$V_x = -D_x \frac{\partial \phi}{\partial x}$$

$$V_y = -D_y \frac{\partial \phi}{\partial y} \tag{10.10}$$

10.3 COMPUTER EXAMPLES

The computer solutions for two problems using the steady-state field program **TDFIELD** given in Chapter 16 are discussed here. The first problem involves ideal flow around a cylinder as shown in Figure 10.1. The second involves the pumping of water from a confined aquifer.

10.3.1 Flow Around A Cylinder

A grid consisting of rectangular and triangular elements for the configuration in Figure 10.2 is shown in Figure 10.6. The grid consists of 38 elements and 37 nodes.

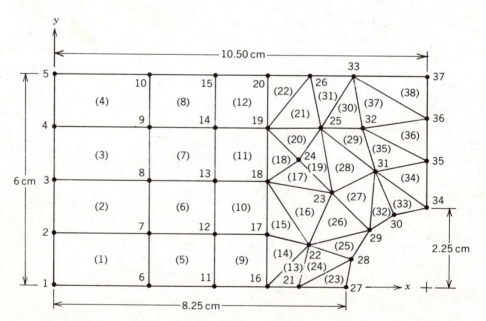

Figure 10.6.   Finite element grid for irrotational flow around a cylinder.

The length of 8.25 cm from the cylinder represents a guess as to where a uniform velocity exists on the upstream side of the cylinder.

The nodal coordinates and calculated values of ϕ and ψ are given in Table 10.1. Contour lines for various values of ϕ and ψ are shown in Figure 10.7. The $\phi=0$ contour line is not vertical so that a uniform velocity, V_x, does not exist along the left edge.

Table 10.1 Coordinate Data and Results for Irrotational Flow Around a Cylinder

Node	X	Y	ϕ	ψ
1	0.0	0.0	−10.33	0.0
2	0.0	1.5	−10.30	7.43
3	0.0	3.0	−10.24	14.9
4	0.0	4.5	−10.19	22.5
5	0.0	6.0	−10.16	30.0
6	2.7	0.0	4.04	0.0
7	2.7	1.5	4.10	7.32
8	2.7	3.0	4.26	110.8
9	2.7	4.5	4.41	22.3
10	2.7	6.0	4.47	30.0
11	4.5	0.0	10.6	0.0
12	4.5	1.5	10.8	7.01
13	4.5	3.0	10.3	14.3
14	4.5	4.5	10.7	22.0
15	4.5	6.0	10.8	30.0
16	6.0	0.0	110.3	0.0
17	6.0	1.5	110.8	6.25
18	6.0	3.0	21.0	10.5
19	6.0	4.5	21.7	21.5
20	6.0	6.0	22.0	30.0
21	6.9	0.0	22.9	0.0
22	7.2	1.2	24.8	0.0
23	7.8	2.7	30.1	10.66
24	6.9	3.6	26.1	15.9
25	7.5	4.5	30.2	20.8
26	7.2	6.0	28.8	30.0
27	8.25	0.0	26.5	0.0
28	8.42	0.86	28.3	0.0
29	8.91	1.59	33.3	0.0
30	10.64	2.08	41.1	0.0
31	10.0	3.3	38.5	10.3
32	8.7	4.5	37.6	110.9
33	8.4	6.0	36.1	30.0
34	10.5	2.25	50.0	0.0
35	10.5	3.6	50.0	10.3
36	10.5	4.8	50.0	21.5
37	10.5	6.0	50.0	30.0

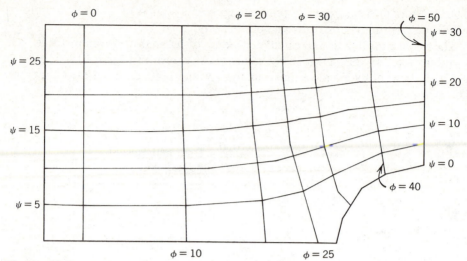

Figure 10.7. Streamlined and constant potential lines for irrotational flow around a cylinder.

10.3.2 Regional Aquifer

A small regional aquifer (Figure 10.8) has a single pump removing water. The upper and lower edges are impermeable, and the left and right edges are far enough from the pump so that a constant head of 200 m is maintained. The pump has a capacity of 1500 m^3/day. The permeabilities, D_x and D_y are 15 m/day.

The grid used to solve the problem is shown in Figure 10.9. The nodal values and contour lines for ϕ are given in Figure 10.10. The nodal values of X and Y as well as the calculated values of ϕ are given in Table 10.2. The maximum drawdown

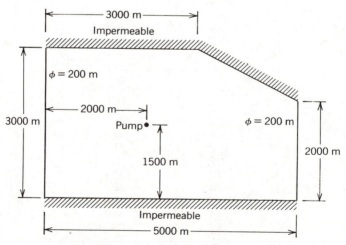

Figure 10.8. A pump in a regional aquifer.

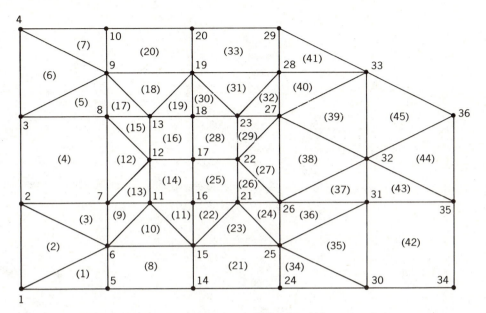

Figure 10.9. Finite element grid for the regional aquifer.

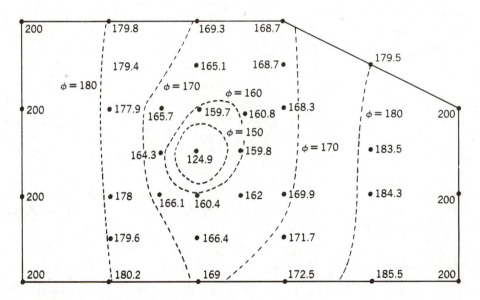

Figure 10.10. Nodal values and contour lines for the regional aquifer.

133

Table 10.2 Coordinate Data and Results for the Regional Aquifer Problem

Node	X	Y	ϕ
1	0	0	200.0
2	0	1000	200.0
3	0	2000	200.0
4	0	3000	200.0
5	1000	0	180.2
6	1000	500	179.6
7	1000	1000	178.0
8	1000	2000	177.9
9	1000	2500	179.4
10	1000	3000	179.8
11	1500	1000	166.1
12	1500	1500	164.3
13	1500	2000	165.7
14	2000	0	169.0
15	2000	500	166.4
16	2000	1000	160.4
17	2000	1500	124.9
18	2000	2000	159.7
19	2000	2500	165.1
20	2000	3000	169.3
21	2500	1000	162.0
22	2500	1500	159.8
23	2500	2000	160.8
24	3000	0	172.5
25	3000	500	171.7
26	3000	1000	169.9
27	3000	2000	168.3
28	3000	2500	168.7
29	3000	3000	168.7
30	4000	0	185.5
31	4000	1000	184.3
32	4000	1500	183.5
33	4000	2500	179.5
34	5000	0	200.0
35	5000	1000	200.0
36	5000	2000	200.0

occurs at the pump as expected, but this occurs only because the pump is at a node. The greatest drawdown occurs at the node closest to the pump when the pump is not located at a node.

PROBLEMS

Use the computer program **TDFIELD** discussed in Chapter 16 to analyze the following problems.

10.1   Determine and plot the location of several streamlines and equal potential lines for the flow of an ideal fluid about the elliptical shape shown in Figure P10.1.

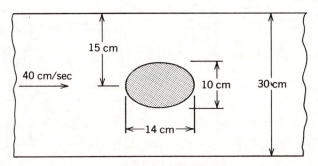

Figure P10.1

10.2   Do Problem 10.1 for the rectangular shape shown in Figure P10.2.

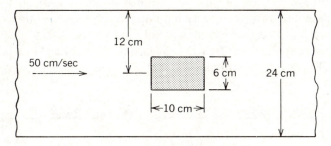

Figure P10.2

10.3   Do Problem 10.1 for the junction region shown in Figure P10.3.

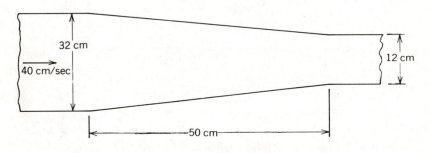

Figure P10.3

10.4 Do Problem 10.1 for the triangular region shown in Figure P10.4.

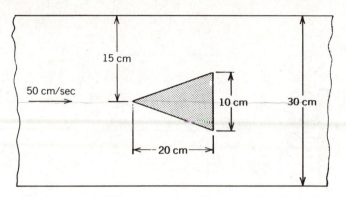

Figure P10.4

10.5 Determine the drawdown at the pump in an infinite media when the pump is removing 2500 m³/day, $D_x = D_y = 20$ m/day and $\phi = 300$ m at a distance far removed from the pump.

10.6 Two pumps are located in the confined aquifer shown in Figure P10.6. Calculate the drawdown at each pump and plot some of the contour lines. The pumping rates are 600 m³/day for pump 1 and 400 m³/day for pump 2. The permeabilities are $D_x = D_y = 30$ m/day. The region is impermeable on two sides, and ϕ is maintained at 100 m on the other two sides.

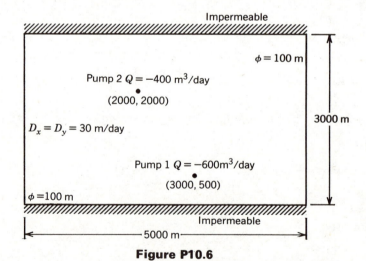

Figure P10.6

10.7 Determine the location and plot some of the equal potential lines for the groundwater seepage problem shown in Figure P10.7. The bottom layer is impermeable. The permeabilities are $D_x = D_y = 20$ m/day.

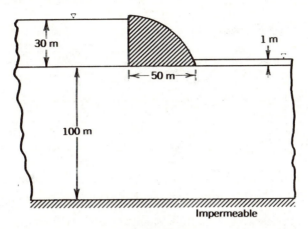

Figure P10.7

10.8 Repeat Problem 10.7 for the region shown in Figure P10.8. The permeabilities are $D_x = 20$ m/day and $D_y = 15$ m/day. The lower layer is impermeable. An impermeable wall 30 m deep has been placed on the upstream side of the dam.

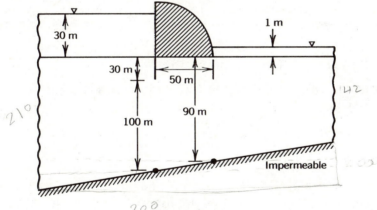

Figure P10.8

Chapter **11**

HEAT TRANSFER BY CONDUCTION AND CONVECTION

One of the first applications of the finite element method to nonstructural problems was in the area of heat transfer by conduction and convection (Visser, 1965). The solution of heat transfer problems using the finite element method is particularly popular with individuals responsible for analyzing thermal stress problems. The solution of the heat transfer problem becomes input to the stress analysis problem and the same grid can be used to solve both problems.

We shall discuss the solution of four different heat transfer problems in this chapter. Two of these problems involve heat transfer from a fin. The third is the analysis of a composite wall, and the fourth is the classical two-dimensional problem with convection boundary conditions.

11.1 THE ONE-DIMENSIONAL FIN

The governing differential equation for steady-state heat transfer from a one-dimensional fin is

$$kA\frac{d^2\phi}{dx^2} - hP\phi + hP\phi_f = 0 \qquad (11.1)$$

where k is the thermal conductivity, h is the convection coefficient, A is the cross-sectional area, P is the distance around the fin, and ϕ is the temperature. The temperature has a single value for all points on the cross section for a particular value of x. The boundary conditions associated with (11.1) are usually a specified temperature at $x=0$

$$\phi(0) = \phi_0 \qquad (11.2)$$

and convection heat loss at the free end

$$-kA\frac{d\phi}{dx} = hA(\phi_b - \phi_f) \qquad \text{at } x = H \qquad (11.3)$$

where ϕ_b is the temperature at the end of the fin and is not known prior to the solution of the problem. The convection coefficient in (11.3) may or may not be the same as the one in (11.1).

138

The governing differential equation (11.1) has the general form

$$D\frac{d^2\phi}{dx^2} - G\phi + Q = 0 \tag{11.4}$$

where $D=kA$, $G=hP$, and $Q=hP\phi_f$. The element contribution to the Galerkin residual equation $\{R^{(e)}\}$ is

$$\{R^{(e)}\} = -\int_{X_i}^{X_j} [N]^T\left(D\frac{d^2\phi}{dx^2} - G\phi + Q\right) dx$$

$$= -\int_{X_i}^{X_j} [N]^T\left(D\frac{d^2\phi}{dx^2} + Q\right) dx + \int_{X_i}^{X_j} G[N]^T\phi\, dx \tag{11.5}$$

The first integral in (11.5) is the one we studied in Chapters 3 and 4 and is equivalent to $\{I^{(e)}\} + [k^{(e)}]\{\Phi\} - \{f^{(e)}\}$. The second integral in (11.5) is a new quantity that needs to be evaluated. Recalling that $\phi^{(e)} = [N]\{\Phi^{(e)}\}$ and substituting this into the integral, we find that

$$\int_{X_i}^{X_j} G[N]^T\, dx = \left(\int_{X_i}^{X_j} G[N]^T[N]\, dx\right)\{\Phi^{(e)}\} \tag{11.6}$$

Since the integral multiplies $\{\Phi^{(e)}\}$, it is a part of the element stiffness matrix. If we define

$$[k_G^{(e)}] = \int_{X_i}^{X_j} G[N]^T[N]\, dx \tag{11.7}$$

then

$$\{R^{(e)}\} = \{I^{(e)}\} + ([k_D^{(e)}] + [k_G^{(e)}])\{\Phi^{(e)}\} - \{f_Q^{(e)}\} \tag{11.8}$$

where $[k_D^{(e)}]$ is given by (4.11) and $\{f_Q^{(e)}\}$ by (4.12).

The integral in (11.7) is most easily evaluated in either the s or ℓ_1, ℓ_2 coordinate systems discussed in Chapter 6. It is left as an exercise for the reader to show that

$$[k_G^{(e)}] = \frac{GL}{6}\begin{bmatrix} 2 & 1 \\ 1 & 2 \end{bmatrix} \tag{11.9}$$

The derivative boundary condition defined by (11.3) is incorporated into the formulation using the interelement vector, $\{I^{(e)}\}$. This vector is, (4.10),

$$\{I^{(e)}\} = \left\{\begin{array}{c} D\dfrac{d\phi}{dx}\Big|_{x=X_i} \\[2mm] -D\dfrac{d\phi}{dx}\Big|_{x=X_j} \end{array}\right\} \tag{11.10}$$

and can be split into

$$\{I^{(e)}\} = \left\{\begin{array}{c} D\dfrac{d\phi}{dx}\Big|_{x=X_i} \\[2mm] 0 \end{array}\right\} + \left\{\begin{array}{c} 0 \\[2mm] -D\dfrac{d\phi}{dx}\Big|_{x=X_j} \end{array}\right\} \tag{11.11}$$

which is equivalent to

$$\{I^{(e)}\} = \{I_i^{(e)}\} + \{I_b^{(e)}\}$$

where $\{I_i^{(e)}\}$ is the interelement requirement and $\{I_b^{(e)}\}$ is associated with the boundary condition. The nonzero term in $\{I_b^{(e)}\}$, however, is the left-hand side of (11.3). Thus

$$\{I_b^{(e)}\} = \left\{ \begin{matrix} 0 \\ hA(\phi_b - \phi_f) \end{matrix} \right\} = \left\{ \begin{matrix} 0 \\ hA\phi_j \end{matrix} \right\} - \left\{ \begin{matrix} 0 \\ hA\phi_f \end{matrix} \right\} \qquad (11.12)$$

since ϕ_b is the same as Φ_j. Equation (11.12) is equivalent to

$$\{I_b^{(e)}\} = \begin{bmatrix} 0 & 0 \\ 0 & hA \end{bmatrix} \left\{ \begin{matrix} \Phi_i \\ \Phi_j \end{matrix} \right\} - \left\{ \begin{matrix} 0 \\ hA\phi_f \end{matrix} \right\}$$

$$= [k_M^{(e)}]\{\Phi^{(e)}\} - \{f_S^{(e)}\} \qquad (11.13)$$

where

$$[k_M^{(e)}] = \begin{bmatrix} 0 & 0 \\ 0 & hA \end{bmatrix}, \qquad \{f_S^{(e)}\} = \left\{ \begin{matrix} 0 \\ hA\phi_f \end{matrix} \right\} \qquad (11.14)$$

The complete residual equation is obtained by substituting for $\{I^{(e)}\}$ in (11.8) and is

$$\{R^{(e)}\} = \{I_i^{(e)}\} + ([k_D^{(e)}] + [k_G^{(e)}] + [k_M^{(e)}])\{\Phi^{(e)}\} - \{f_Q^{(e)}\} - \{f_S^{(e)}\} \qquad (11.15)$$

Neglecting the interelement requirement $\{I_i^{(e)}\}$ gives

$$\{R^{(e)}\} = [k^{(e)}]\{\Phi^{(e)}\} - \{f^{(e)}\} \qquad (11.16)$$

The contribution of $[k_M^{(e)}]$ to $[k^{(e)}]$ occurs only for the last element of the fin and only when h is nonzero for the end of the fin. For example, $[k_M^{(e)}]$ is zero if the end of the fin is insulated.

ILLUSTRATIVE EXAMPLE

Calculate the temperature distribution in a one-dimensional fin with the physical properties given in Figure 11.1. The fin is rectangular in shape, and is 8 cm long, 4 cm wide, and 1 cm thick. Assume that convection heat loss occurs from the end of the fin.

The fin is modeled by four elements each with a length of 2 cm. The element matrices are

$$[k^{(e)}] = \frac{kA}{L} \begin{bmatrix} 1 & -1 \\ -1 & 1 \end{bmatrix} + \frac{hPL}{6} \begin{bmatrix} 2 & 1 \\ 1 & 2 \end{bmatrix} + \begin{bmatrix} 0 & 0 \\ 0 & hA \end{bmatrix}$$

and

$$\{f^{(e)}\} = \frac{hPL\phi_f}{2} \left\{ \begin{matrix} 1 \\ 1 \end{matrix} \right\} + \left\{ \begin{matrix} 0 \\ hA\phi_f \end{matrix} \right\}$$

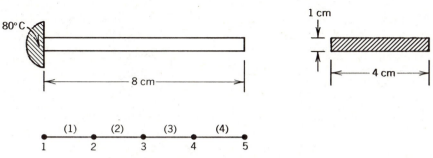

$$k = 3 \frac{W}{cm\text{-}°C}, \qquad h = 0.1 \frac{W}{cm^2 \cdot °C}, \qquad \phi_f = 20°C$$

Figure 11.1.   A rectangular fin.

where the third matrix in $[k^{(e)}]$ and the second vector in $\{f^{(e)}\}$ are applicable only for element four. The values of the various parameters are

$$\frac{kA}{L} = \frac{3(4)}{2} = 6 \frac{W}{°C}$$

$$\frac{hPL}{6} = \frac{0.1(10)2}{6} \doteq 0.333 \frac{W}{°C}$$

$$hA = 0.1(4) = 0.400 \frac{W}{°C}$$

$$\frac{hPL\phi_f}{2} = \frac{0.1(10)(20)(2)}{2} = 20 \ W$$

$$hA\phi_f = 0.1(4)(20) = 8 \ W$$

The element quantities are

$$[k^{(e)}] = \begin{bmatrix} 6.666 & -5.667 \\ -5.667 & 6.666 \end{bmatrix}, \qquad \{f^{(e)}\} = \begin{Bmatrix} 20 \\ 20 \end{Bmatrix}$$

for elements one, two, and three and

$$[k^{(e)}] = \begin{bmatrix} 6.666 & -5.667 \\ -5.667 & 7.066 \end{bmatrix}, \qquad \{f^{(e)}\} = \begin{Bmatrix} 20 \\ 28 \end{Bmatrix}$$

for element four. The assembly of the element matrices using the direct stiffness procedure produces the system of equations

$$\begin{bmatrix} 6.666 & -5.667 & 0 & 0 & 0 \\ -5.667 & 13.33 & -5.667 & 0 & 0 \\ 0 & -5.667 & 13.33 & -5.667 & 0 \\ 0 & 0 & -5.667 & 13.33 & -5.667 \\ 0 & 0 & 0 & -5.667 & 7.006 \end{bmatrix} \begin{Bmatrix} \Phi_1 \\ \Phi_2 \\ \Phi_3 \\ \Phi_4 \\ \Phi_5 \end{Bmatrix} = \begin{Bmatrix} 20 \\ 40 \\ 40 \\ 40 \\ 28 \end{Bmatrix}$$

The temperature at node one is known, $\Phi_1 = 80°C$; thus the first equation must be deleted and the others modified. The new system of equations is

$$
\begin{bmatrix}
13.33 & -5.667 & 0 & 0 \\
-5.667 & 13.33 & -5.667 & 0 \\
0 & -5.667 & 13.33 & -5.667 \\
0 & 0 & -5.667 & 7.066
\end{bmatrix}
\begin{Bmatrix}
\Phi_2 \\ \Phi_3 \\ \Phi_4 \\ \Phi_5
\end{Bmatrix}
=
\begin{Bmatrix}
493 \\ 40 \\ 40 \\ 28
\end{Bmatrix}
$$

The nodal values are

$$\{\Phi\}^T = [80 \quad 53.9 \quad 39.9 \quad 32.8 \quad 30.3]$$

which compares very favorably with the theoretical values of

$$\{\Phi_{th}\}^T = [80 \quad 54.3 \quad 40.2 \quad 33.2 \quad 30.6]$$

obtained using equation 2-47 in (Kreith, 1973).

11.2 THE COMPOSITE WALL

The governing differential equation for heat transfer through a composite wall is

$$kA\frac{d^2\phi}{dx^2} = 0 \tag{11.17}$$

where either ϕ is known at one or both surfaces or convection heat loss occurs from one or both surfaces. The convection boundary conditions are

$$kA\frac{d\phi}{dx} = hA(\phi_b - \phi_f) \qquad \text{at } x=0 \tag{11.18}$$

and

$$-kA\frac{d\phi}{dx} = hA(\phi_b - \phi_f) \qquad \text{at } x=H \tag{11.19}$$

The element stiffness matrix is given by

$$
[k^{(e)}] = \frac{kA}{L}\begin{bmatrix} 1 & -1 \\ -1 & 1 \end{bmatrix} + \begin{bmatrix} hA_i & 0 \\ 0 & 0 \end{bmatrix} + \begin{bmatrix} 0 & 0 \\ 0 & hA_j \end{bmatrix}
$$

$$
= [k_D^{(e)}] + [k_{M_i}^{(e)}] + [k_{M_j}^{(e)}] \tag{11.20}
$$

where the second matrix, $[k_{M_i}^{(e)}]$, results from the convection boundary condition at node i and the third matrix, $[k_{M_j}^{(e)}]$, results from the convection boundary condition at node j and is the same as given in (11.14). The element force vector is

$$\{f^{(e)}\} = \begin{Bmatrix} hA_i\phi_f \\ 0 \end{Bmatrix} + \begin{Bmatrix} 0 \\ hA_j\phi_f \end{Bmatrix} \tag{11.21}$$

where the first vector comes from (11.18) and the second vector from (11.19). The quantities, $\{f^{(e)}\}$, $[K_{M_i}^{(e)}]$, and $[K_{M_j}^{(e)}]$ are neglected when temperature is specified at both surfaces.

ILLUSTRATIVE EXAMPLE

Determine the temperature distribution through the composite wall shown in Figure 11.2 when convection heat loss occurs on the left surface. Assume a unit area.

The parameters for element one are

$$\frac{kA}{L} = \frac{0.2(1)}{2} = 0.1 \ \frac{W}{°C}$$

$$hA = 0.1(1) = 0.1 \ \frac{W}{°C}$$

$$hA\phi_f = 0.1(1)(-5) = -0.5 \ W$$

and

$$[k^{(1)}] = \begin{bmatrix} 0.2 & -0.1 \\ -0.1 & 0.1 \end{bmatrix}, \qquad \{f^{(e)}\} = \begin{Bmatrix} -0.5 \\ 0 \end{Bmatrix}$$

The parameter for element two is

$$\frac{kA}{L} = \frac{0.06(1)}{6} = 0.01 \ \frac{W}{°C}$$

and

$$[k^{(2)}] = \begin{bmatrix} 0.01 & -0.01 \\ -0.01 & 0.01 \end{bmatrix}$$

Assembly of the element matrices using the direct stiffness procedure gives

$$\begin{bmatrix} 0.20 & -0.10 & 0 \\ -0.10 & 0.21 & -0.01 \\ 0 & -0.01 & 0.01 \end{bmatrix} \begin{Bmatrix} \Phi_1 \\ \Phi_2 \\ \Phi_3 \end{Bmatrix} = \begin{Bmatrix} -0.5 \\ 0 \\ 0 \end{Bmatrix}$$

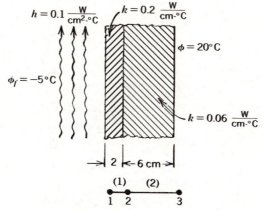

Figure 11.2. A composite wall.

Since $\Phi_3 = 20°C$, equation three is deleted and the second equation is modified. The final system is

$$\begin{bmatrix} 0.20 & -0.10 \\ -0.10 & 0.21 \end{bmatrix} \begin{Bmatrix} \Phi_1 \\ \Phi_2 \end{Bmatrix} = \begin{Bmatrix} -0.50 \\ 0.20 \end{Bmatrix}$$

and the nodal values are

$$\{\Phi\} = \begin{bmatrix} -2.66 & -0.31 & 20 \end{bmatrix}$$

The calculated values are also the theoretical values because the solution of the differential equation consists of straight-line segments.

11.3 THE TWO-DIMENSIONAL FIN

A two-dimensional fin is a thin piece of metal attached to a hot water or steam pipe as shown in Figure 11.3. Heat is transferred from the pipe to the fin by conduction and from the fin to the surrounding media (usually air) by convection. Convection heat loss occurs from both faces of the fin and the edge. The surface area of the edge, however, is small compared to the area of the two faces and the convection losses from the edge can be neglected.

The governing differential equation for the two-dimensional fin is

$$k_x t \frac{\partial^2 \phi}{\partial x^2} + k_y t \frac{\partial^2 \phi}{\partial y^2} - 2h\phi + 2h\phi_f = 0 \tag{11.22}$$

where k_x and k_y are the thermal conductivities in the x- and y-directions, respectively, t is the thickness of the fin, h is the convection coefficient, and ϕ_f is the temperature of the surrounding fluid. The boundary conditions are

$$\phi(\Gamma) = \phi_s \qquad \text{along the pipe boundary} \tag{11.23}$$

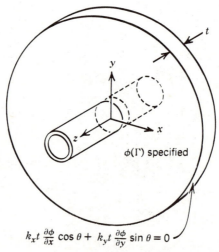

$$k_x t \frac{\partial \phi}{\partial x} \cos\theta + k_y t \frac{\partial \phi}{\partial y} \sin\theta = 0$$

Figure 11.3. A two-dimensional fin and boundary conditions.

and

$$k_x t \frac{\partial \phi}{\partial x} \cos \theta + k_y t \frac{\partial \phi}{\partial y} \sin \theta = 0 \qquad (11.24)$$

along the outer edge. The latter condition is the insulated boundary condition. The heat transfer in the fin is a two-dimensional problem because the fin is too thin to develop a temperature gradient in the z-direction.

The governing equation (11.22) has a form identical to (7.1) where

$$D_x = tk_x, \qquad D_y = tk_y, \qquad G = 2h, \qquad Q = 2h\phi_f \qquad (11.25)$$

The element matrices developed in Chapter 7 are applicable without modification. The element matrices for an element in a fin were calculated in the Illustrative Example in Section 7.3 and should be reviewed at this time.

11.4   LONG TWO-DIMENSIONAL BODIES

Another form of two-dimensional heat transfer is that which occurs in long bodies of constant cross section subjected to the same boundary conditions along the entire length of the body. In this situation, the temperature gradient in the z-direction is zero and the governing differential equation is

$$k_x \frac{\partial^2 \phi}{\partial x^2} + k_y \frac{\partial^2 \phi}{\partial y^2} + Q = 0 \qquad (11.26)$$

where k_x and k_y are the thermal conductivities and Q is an internal heat source or sink. The internal source or sink must exist along the entire z-direction for the heat transfer to be two-dimensional. A two-dimensional problem exists because the temperature can vary in the xy plane.

The differential equation (11.26) is embedded within (7.1). The parameters for (7.1) are

$$D_x = k_x, \qquad D_y = k_y, \qquad G = 0, \qquad Q = Q \qquad (11.27)$$

and the element matrices developed in Chapter 7 are applicable.

The item that makes heat transfer in a long body different from that in a fin is the boundary conditions. The possible boundary conditions are shown in Figure 11.4 and consist of prescribed temperature values, convection heat transfer, and surface heat fluxes. The latter two conditions were treated in general terms in Chapter 9. We shall now look at these conditions relative to heat transfer and assume that $k_x = k_y = k$, which changes the derivative boundary condition (9.1) to

$$k \frac{\partial \phi}{\partial n} = -M\phi_b + S \qquad (11.28)$$

11.4.1   Convection Boundary Condition

The objective here is to determine the parameters that comprise M and S in (11.28) as well as the signs on these two quantities for the convection boundary

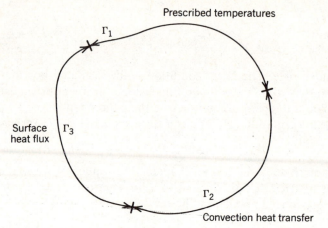

Figure 11.4. Types of boundary conditions for the classical heat transfer problem.

condition. We shall consider two situations: heat leaving the body and heat entering the body.

Consider the convection condition shown in Figure 11.5a, where heat is leaving the body. If heat is leaving, then the gradient $\partial\phi/\partial x$ must be negative and the heat conducted to the surface is

$$q_c = -kA\frac{\partial\phi}{\partial n} \tag{11.29}$$

The heat conducted to the surface must equal that leaving the surface; therefore,

$$-kA\frac{\partial\phi}{\partial n} = hA(\phi_b - \phi_f) \tag{11.30}$$

Rewriting (11.30) in the form of (11.28) gives

$$k\frac{\partial\phi}{\partial n} = -h\phi_b + h\phi_f \tag{11.31}$$

and we conclude that

$$M = h \quad \text{and} \quad S = h\phi_f \tag{11.32}$$

We now consider the convection situation when heat is entering the body (Figure 11.5b). The gradient $\partial\phi/\partial n$ is positive; therefore,

$$q_c = kA\frac{\partial\phi}{\partial n} \tag{11.33}$$

A negative sign is not needed because the gradient is positive. Equating (11.33) to the heat coming to the surface gives

$$kA\frac{\partial\phi}{\partial n} = hA(\phi_f - \phi_b) \tag{11.34}$$

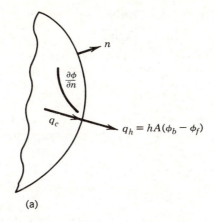

(a)

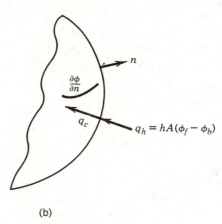

(b)

Figure 11.5.   Convection heat transfer at a surface.

and

$$k\frac{\partial \phi}{\partial n} = -h\phi_b + h\phi_f \tag{11.35}$$

We again conclude that

$$M = h \quad \text{and} \quad S = h\phi_f \tag{11.36}$$

The conclusion of the analysis is that $M = h$ and $S = h\phi_f$ regardless of whether heat is coming into or leaving a body during the process of convection heat transfer. These results hold for exterior surfaces and interior surfaces such as those in a chimney.

The above analysis assumes that a positive outward normal, n, is always directed away from the surface.

11.4.2 Heat Flux into The Body

The situation where heat is applied to a portion of the boundary is shown schematically in Figure 11.6. The heat must be conducted away from the boundary. Thus

$$q_c = kA \frac{\partial \phi}{\partial n} \qquad (11.37)$$

The heat applied to the surface is q^*A and equating heat flows gives

$$kA \frac{\partial \phi}{\partial n} = q^*A \qquad (11.38)$$

Equating (11.38) and (11.28) yields

$$M = 0 \qquad \text{and} \qquad S = q^* \qquad (11.39)$$

where q^* is the heat flux per unit area. If heat were being removed from the body, then $S = -q^*$.

The conclusion of this analysis is that a heat flux on the boundary does have a sign associated with it. Heat is positive if it is moving into the body and negative if it is being removed. The coefficient M is zero when a heat flux occurs.

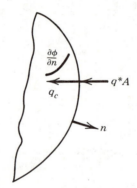

Figure 11.6. A heat flux applied to a surface.

11.4.3 Concluding Remarks

Convection heat transfer and a flux boundary condition are incorporated into a finite element analysis using the element matrices developed in Chapter 9. In the case of convection heat transfer, $M = h$ and $S = h\phi_f$ whereas $M = 0$ and $S = \pm q^*$ for the flux condition. Heat into the body is considered positive for both q^* and Q.

11.5 A COMPUTER EXAMPLE

We close this chapter with a computer example. A grid of heating cables have been embedded in a thin concrete slab for the purpose of melting the snow on the

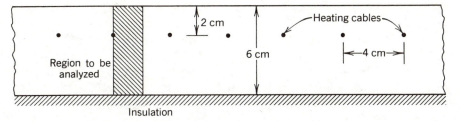

Figure 11.7. Heating cables within a conducting media.

exposed surface (Figure 11.7). The cables are 4 cm on center and 2 cm below the surface. The slab rests on a thick layer of insulation and the heat loss from the bottom can be neglected. The conductives are $k_x = k_y = 0.0180$ W/cm − °C and the surface convection coefficient is $h = 0.00340$ W/cm^2 − °C. The latter corresponds to about a 30–35 km/hr wind velocity. The objective is to determine the surface

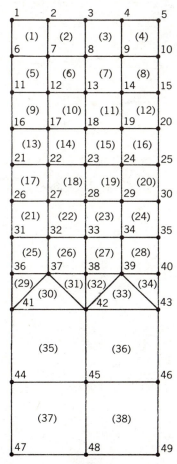

Figure 11.8. The finite element grid for the heat transfer example.

temperature when the cables produce 0.050 W/cm of heat and the air temperature is $-5°C$.

The repeated symmetry in the problem reduces the region to be analyzed to one which is 2 cm wide and 6 cm deep (Figure 11.7). The grid for this region (Figure 11.8) consists of 38 elements arranged so that node 21 is located at a heating cable. Larger elements are used in the lower portion of the grid because previous experience indicates that the temperature probably does not vary a great deal there. Elements (1), (2), (3), and (4) have the convection boundary condition. The values of M and S in each case are

$$M = h = 0.00340$$

and

$$S = h\phi_f = (0.00340)(-5) = -0.0170.$$

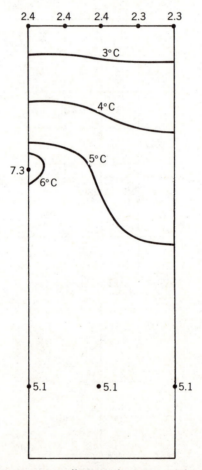

Figure 11.9. Temperature contour lines and some nodal values for the heat transfer example.

Table 11.1   Coordinate Data and Results for the Heat Transfer Example

Node	X	Y	ϕ	Node	X	Y	ϕ
1	0.0	6.0	2.4	26	0.0	3.5	5.6
2	0.5	6.0	2.4	27	0.5	3.5	5.5
3	1.0	6.0	2.4	28	1.0	3.5	5.0
4	1.5	6.0	2.3	29	1.5	3.5	4.9
5	2.0	6.0	2.3	30	2.0	3.5	4.8
6	0.0	5.5	3.1	31	0.0	3.0	5.3
7	0.5	5.5	3.1	32	0.5	3.0	5.3
8	1.0	5.5	3.0	33	1.0	3.0	5.1
9	1.5	5.5	3.0	34	1.5	3.0	5.0
10	2.0	5.5	3.0	35	2.0	3.0	5.0
11	0.0	5.0	3.9	36	0.0	2.5	5.2
12	0.5	5.0	3.9	37	0.5	2.5	5.2
13	1.0	5.0	3.7	38	1.0	2.5	5.1
14	1.5	5.0	3.6	39	1.5	2.5	5.1
15	2.0	5.0	3.6	40	2.0	2.5	5.1
16	0.0	4.5	4.9	41	0.0	2.0	5.2
17	0.5	4.5	4.8	42	1.0	2.0	5.1
18	1.0	4.5	4.3	43	2.0	2.0	5.1
19	1.5	4.5	4.2	44	0.0	1.0	5.1
20	2.0	4.5	4.1	45	1.0	1.0	5.1
21	0.0	4.0	7.3	46	2.0	1.0	5.1
22	0.5	4.0	5.3	47	0.0	0.0	5.1
23	1.0	4.0	4.8	48	1.0	0.0	5.1
24	1.5	4.0	4.6	49	2.0	0.0	5.1
25	2.0	4.0	4.5				

The nodal coordinates and calculated temperature values (to two significant digits) are given in Table 11.1. Some contour lines and nodal temperature values are shown in Figure 11.9. The top surface is close to a uniform value of 2.4°C. The lower part of the region is a uniform 5.1°C justifying the use of large elements.

PROBLEMS

11.1   Verify that $[k_G^{(e)}]$ as defined by (11.7) is equal to (11.9).

11.2   Evaluate $[k_M^{(e)}]$ and $\{f_S^{(e)}\}$ for the derivative boundary condition

$$D\frac{d\phi}{dx}=hA(\phi_b-\phi_f) \qquad \text{at} \qquad x=0$$

11.3   Determine the temperature distribution for the fin in Figure 11.1 using the grid shown in Figure P11.3.

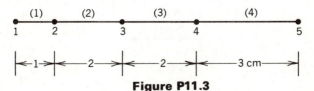

Figure P11.3

11.4 Determine the temperature distribution in the circular fin shown in Figure P11.4. Include the convection heat loss from the end of the fin.

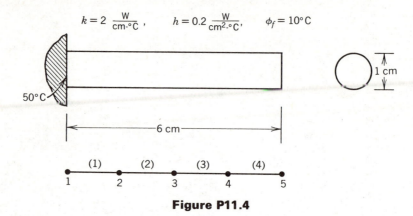

$$k = 2 \ \frac{W}{cm\cdot°C} \ , \qquad h = 0.2 \ \frac{W}{cm^2\cdot°C}, \qquad \phi_f = 10°C$$

1 cm

50°C

6 cm

(1) (2) (3) (4)

1 2 3 4 5

Figure P11.4

11.5 Determine the temperature distribution in the circular fin of Problem 11.4 using the three-element grid shown in Figure P11.5. Include the convection heat loss from the end of the fin.

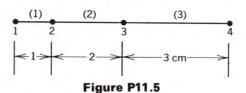

(1) (2) (3)

1 2 3 4

1 2 3 cm

Figure P11.5

11.6 Calculate the surface temperatures for the wall shown in Figure P11.6. Convection heat transfer occurs on both surfaces. Assume a unit of surface area.

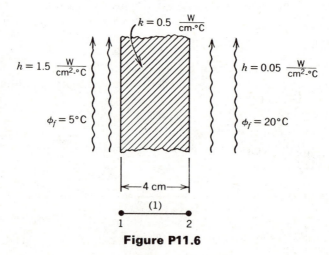

$$k = 0.5 \ \frac{W}{cm\cdot°C}$$

$$h = 1.5 \ \frac{W}{cm^2\cdot°C}$$

$$\phi_f = 5°C$$

$$h = 0.05 \ \frac{W}{cm^2\cdot°C}$$

$$\phi_f = 20°C$$

4 cm

(1)

1 2

Figure P11.6

11.7   Calculate the surface temperature at $x=4$ in Problem 11.6 when the surface temperature on the left side is specified at 5°C.

11.8   Calculate the surface temperature at $x=0$ in Problem 11.6 when the surface temperature on the right-hand side is specified at 30°C.

11.9–11.12   Evaluate $[k^{(e)}]$ and $\{f^{(e)}\}$ for the triangular elements shown in Figures P11.9 through P11.12. The conductivities are $k_x=k_y=2$ W/°C-cm

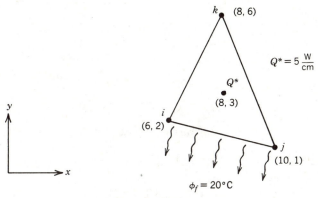

Figure P11.9

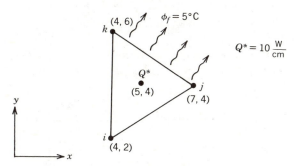

Figure P11.10

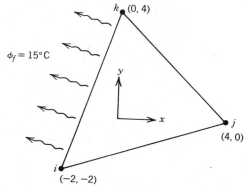

Figure P11.11

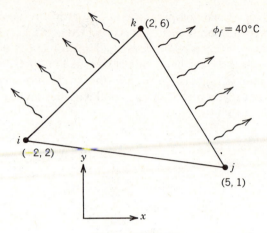

Figure P11.12

and $h = 0.2$ W/cm²-°C. The heat sources, Q^* in Problems 11.9 and 11.10, are line sources.

11.13–11.14 Use the computer program **TDFIELD**, Chapter 16, to analyze one of the two-dimensional fins shown in Figures P11.13 and P11.14. The physical parameters are $k_x = k_y = 3$ W/cm-°C, $h = 0.010$ W/cm²-°C, and $\phi_f = 20$°C. Each fin is 0.3 cm thick. The pipe temperature is 85°C.

11.15–11.19 Use the computer program **TDFIELD**, Chapter 16, to analyze one of the long two-dimensional bodies shown in Figures P11.15 to P11.19. The physical parameters and the boundary conditions are defined in the respective figures.

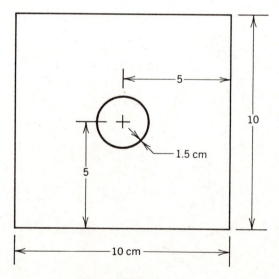

Figure P11.13

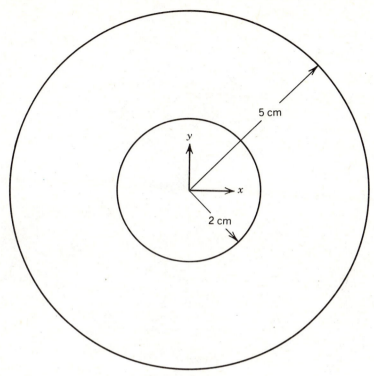

Figure P11.14

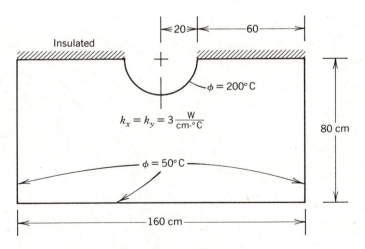

Figure P11.15

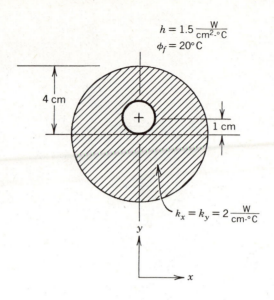

$$h = 1.5 \frac{W}{cm^2 \cdot °C}$$
$$\phi_f = 20° C$$

4 cm

1 cm

$$k_x = k_y = 2 \frac{W}{cm \cdot °C}$$

y

x

Boundary of the inner cylinder is at 140 C
Diameter of the inner cylinder, 2 cm
Diameter of the outer cylinder, 8 cm

Figure P11.16

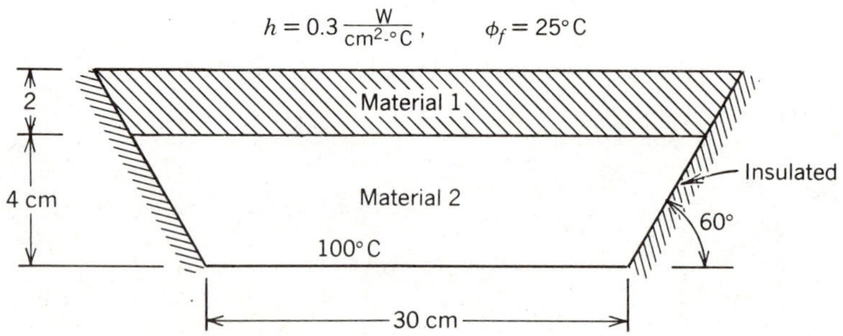

$$h = 0.3 \frac{W}{cm^2 \cdot °C} , \qquad \phi_f = 25° C$$

2

Material 1

Insulated

4 cm

Material 2

60°

100° C

30 cm

Material 1 $k_x = k_y = 0.1 \dfrac{W}{cm - °C}$

Material 2 $k_x = k_y = 4 \dfrac{W}{cm - °C}$

Figure P11.17

156

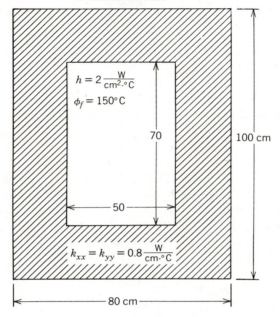

Figure P11.18

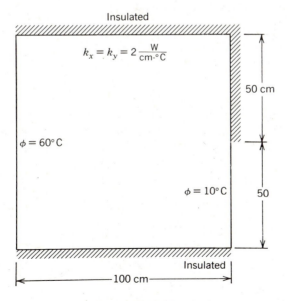

Figure P11.19

Chapter 12

ACOUSTICAL VIBRATIONS

A zero value for Q and a negative value for G in the steady-state field equation, (7.1), yields the Helmholtz equation

$$D_x \frac{\partial^2 \phi}{\partial x^2} + D_y \frac{\partial^2 \phi}{\partial y^2} + G\phi = 0 \tag{12.1}$$

Physical problems governed by (12.1) include the wave motion for shallow bodies of water and acoustical vibrations in closed rooms or compartments.

The solution of the Helmholtz equation results in the need to solve an eigenvalue problem because the boundary conditions are such that the global force vector, $\{F\}$, is zero. The global system of equations has the form $[K]\{\Phi\} = \{0\}$.

The solution of the Helmholtz equation is the topic of discussion for this chapter. Acoustical vibrations is the subject matter area used to illustrate the solution technique. We start by considering the one-dimensional problem because the calculations can be done by hand. The natural frequencies and vibration modes for a two-dimensional problem conclude the chapter.

12.1 ONE-DIMENSIONAL VIBRATIONS

The governing differential equation for the pressure field associated with acoustical vibrations in a two-dimensional room with rigid boundaries is

$$\frac{\partial^2 \phi}{\partial x^2} + \frac{\partial^2 \phi}{\partial y^2} + \frac{w^2}{c^2}\phi = 0 \tag{12.2}$$

where ϕ is the change in pressure from some ambient value, w is the wave frequency, and c is the wave velocity in the media. The condition to be satisfied on each boundary is

$$\frac{\partial \phi}{\partial n} = 0 \tag{12.3}$$

The one-dimensional analog to (12.2) is

$$\frac{d^2 \phi}{dx^2} + \frac{w^2}{c^2}\phi = 0 \tag{12.4}$$

with $d\phi/dx = 0$ at each end. Equation (12.4) has the general form

$$D\frac{d^2 \phi}{dx^2} - G\phi = 0 \tag{12.5}$$

with $D=1$ and $G=-w^2/c^2$. The element stiffness matrix for the first term, $d^2\phi/dx$ is given by (4.11). The contribution of the $-G\phi$ term to the element stiffness matrix is given by (11.9). The complete element stiffness matrix for (12.5) is

$$[k^{(e)}]=\frac{1}{L}\begin{bmatrix} 1 & -1 \\ -1 & 1 \end{bmatrix}-\frac{w^2 L}{6c^2}\begin{bmatrix} 2 & 1 \\ 1 & 2 \end{bmatrix} \qquad (12.6)$$

The element force vector, $\{f^{(e)}\}$, consists of zeros, since there is no source term in (12.5) and the $d\phi/dx=0$ boundary conditions (both ends) do not generate any nonzero terms in $\{f^{(e)}\}$.

The problem under consideration is the closed pipe in Figure 12.1. The finite element model consists of two elements. Substituting $H/2$ for L in (12.6) and multiplying through by $H/2$ gives

$$[k^{(e)}]=\begin{bmatrix} 1 & -1 \\ -1 & 1 \end{bmatrix}-Z\begin{bmatrix} 2 & 1 \\ 1 & 2 \end{bmatrix} \qquad (12.7)$$

where

$$Z=\frac{w^2 H^2}{24c^2} \qquad (12.8)$$

Combining the two elements matrices gives a system of three equations

$$\begin{bmatrix} 1 & -1 & 0 \\ -1 & 2 & -1 \\ 0 & -1 & 1 \end{bmatrix}\begin{Bmatrix} \Phi_1 \\ \Phi_2 \\ \Phi_3 \end{Bmatrix}-Z\begin{bmatrix} 2 & 1 & 0 \\ 1 & 4 & 1 \\ 0 & 1 & 2 \end{bmatrix}\begin{Bmatrix} \Phi_1 \\ \Phi_2 \\ \Phi_3 \end{Bmatrix}=\begin{Bmatrix} 0 \\ 0 \\ 0 \end{Bmatrix} \qquad (12.9)$$

or

$$([K_D]-Z[K_G])\{\Phi\}=\{0\} \qquad (12.10)$$

Both matrices, $[K_D]$ and $[K_G]$, are symmetric; $[K_G]$ is positive definite whereas $[K_D]$ is semidefinite (it has a zero determinant). Eigenvalue theory states that all of the eigenvalues, Z_i, that satisfy (12.9) are distinct, real, and positive numbers and the corresponding eigenvectors $\{\Phi\}_i$ are independent.

The eigenvalues, Z_i, are the values of Z that make the determinant of (12.9)

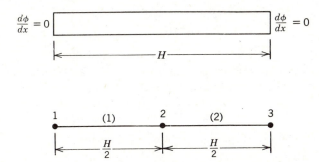

Figure 12.1. A two-element grid for a one-dimensional pipe.

zero. Combining the two matrices gives

$$
\begin{bmatrix}
(1-2Z) & -(1+Z) & 0 \\
-(1+Z) & (2-4Z) & -(1+Z) \\
0 & -(1+Z) & (1-2Z)
\end{bmatrix}
\begin{Bmatrix} \Phi_1 \\ \Phi_2 \\ \Phi_3 \end{Bmatrix}
=
\begin{Bmatrix} 0 \\ 0 \\ 0 \end{Bmatrix}
\tag{12.11}
$$

The determinant is

$$
2(1-2Z)[(1-2Z)^2 - (1+Z)^2] = 0
\tag{12.12}
$$

which has the roots

$$
Z_1 = 0, \qquad Z_2 = \tfrac{1}{2}, \qquad \text{and} \qquad Z_3 = 2
\tag{12.13}
$$

There is an eigenvector, $\{\Phi\}_i$, associated with each root in (12.13). It is impossible to uniquely determine the three components of $\{\Phi\}_i$ because the set of equations is homogeneous. The usual procedure is to assign an arbitrary value to one component and solve for the remaining components in terms of the assigned value.

The eigenvector $\{\Phi\}_1$ is determined by substituting $Z_1 = 0$ into (12.11). The resulting equations are

$$
\begin{aligned}
\Phi_1 - \Phi_2 \qquad\quad &= 0 \\
-\Phi_1 + 2\Phi_2 - \Phi_3 &= 0 \\
-\Phi_2 + \Phi_3 &= 0
\end{aligned}
\tag{12.14}
$$

The first equation says that $\Phi_1 = \Phi_2$ while the third equation states that $\Phi_2 = \Phi_3$; thus $\Phi_1 = \Phi_2 = \Phi_3$ and the eigenvector is

$$
\{\Phi\}_1^T = \begin{bmatrix} 1 & 1 & 1 \end{bmatrix}
\tag{12.15}
$$

when $\Phi_1 = 1$ is used as the arbitrary value.

Substituting $Z_2 = \tfrac{1}{2}$ into (12.11) gives

$$
\begin{aligned}
-\frac{3}{2}\Phi_2 \qquad\qquad &= 0 \\
-\frac{3}{2}\Phi_1 \qquad\qquad -\frac{3}{2}\Phi_3 &= 0 \\
-\frac{3}{2}\Phi_2 \qquad\qquad &= 0
\end{aligned}
\tag{12.16}
$$

The first and third equations state that $\Phi_2 = 0$ while the second yields $\Phi_3 = -\Phi_1$. Using $\Phi_1 = 1$ as the arbitrary value, we obtain

$$
\{\Phi\}_2^T = \begin{bmatrix} 1 & 0 & -1 \end{bmatrix}
\tag{12.17}
$$

It is left for the reader to show that the third eigenvector is

$$
\{\Phi\}_3^T = \begin{bmatrix} 1 & -1 & 1 \end{bmatrix}
\tag{12.18}
$$

The theoretical values for the natural frequencies, w_n, are given by

$$w_n = \frac{n\pi c}{H} \tag{12.19}$$

The calculated values for w_n are obtained by substituting the roots Z_1, Z_2, and Z_3 given by (12.13) into (12.8) and solving for w_n. The calculated values of w,

$$w_1 = 0, \qquad w_2 = \frac{3.464c}{H}, \qquad \text{and} \qquad w_3 = \frac{6.928c}{H}$$

compare fairly well with the theoretical values of

$$w_1 = 0, \qquad w_2 = \frac{3.142c}{H} \qquad \text{and} \qquad w_3 = \frac{6.283c}{H}$$

considering the grid consisted of only two elements.

The theoretical mode shapes have the general form $P = \cos(n\pi x/H)$. The theoretical mode shapes and the calculated eigenvectors are shown in Figure 12.2. The theoretical and calculated shapes for the first mode coincide.

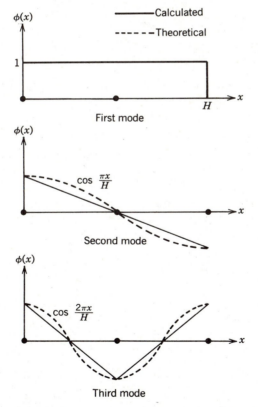

Figure 12.2. The theoretical and calculated node shapes for the one-dimensional pipe.

12.2 TWO-DIMENSIONAL VIBRATIONS

The two-dimensional acoustical vibration equation and its boundary conditions were given in (12.2) and (12.3). The element matrices for this equation are those given by (7.36) and (7.38) for the triangular element, and (7.49) and (7.55) for the rectangular element.

A rectangular room, 20 m by 10 m, is divided into four triangular elements (Figure 12.3). Defining $Z = w^2/c^2$, we find that the global system of equations is

$$([K_D] - Z[K_G])\{\Phi\} = \{0\} \tag{12.20}$$

where

$$[K_D] = \frac{1}{8} \begin{bmatrix} 10 & 3 & 0 & -3 & -10 \\ 3 & 10 & -3 & 0 & -10 \\ 0 & -3 & 10 & 3 & -10 \\ -3 & 0 & 3 & 10 & -10 \\ -10 & -10 & -10 & -10 & 40 \end{bmatrix} \tag{12.21}$$

$$[K_G] = \frac{1}{6} \begin{bmatrix} 100 & 25 & 0 & 25 & 50 \\ 25 & 100 & 25 & 0 & 50 \\ 0 & 25 & 100 & 25 & 50 \\ 25 & 0 & 25 & 100 & 50 \\ 50 & 50 & 50 & 50 & 200 \end{bmatrix} \tag{12.22}$$

and

$$\{\Phi\}^T = \Phi_1 \quad \Phi_2 \quad \Phi_3 \quad \Phi_4 \quad \Phi_5] \tag{12.23}$$

Hand computation of the values of Z that makes the determinant of $[K_D] - Z[K_G])$ zero is unreasonable; a computer program should be used. Discussions of direct and iterative methods of evaluating eigenvalues are given in many textbooks. Bathe and Wilson (1976) present a comprehensive discussion of eigenvalue calculations relative to finite element problems.

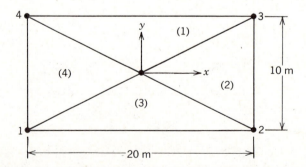

Figure 12.3. A four-element gird for a two-dimensional room.

The five eigenvalues and eigenvectors for (12.20) are

$$Z_1 = 0, \quad \{\Phi\}_1^T = [1, \quad 1, \quad 1, \quad 1, \quad 1]$$
$$Z_2 = 0.030, \quad \{\Phi\}_2^T = [1, \quad -1, \quad -1, \quad 1, \quad 0]$$
$$Z_3 = 0.120, \quad \{\Phi\}_3^T = [1, \quad 1, \quad -1, \quad -1, \quad 0] \qquad (12.24)$$
$$Z_4 = 0.150, \quad \{\Phi\}_4^T = [1, \quad -1, \quad 1, \quad -1, \quad 0]$$
$$Z_5 = 0.450, \quad \{\Phi\}_5^T = [-0.5, \quad -0.5, \quad -0.5, \quad -0.5, \quad 1]$$

The calculated eigenvalues compare reasonably well with the theoretical values of $Z = w^2/c^2$, which are

$$Z_1 = 0, \qquad Z_2 = 0.0247, \qquad Z_3 = 0.0987$$

and

$$Z_4 = 0.123, \qquad Z_5 = 0.395$$

Each of the mode shapes can be illustrated graphically as was done for the one-dimensional case. The mode shapes corresponding to $\{\Phi\}_2$ and $\{\Phi\}_4$ are shown in Figure 12.4.

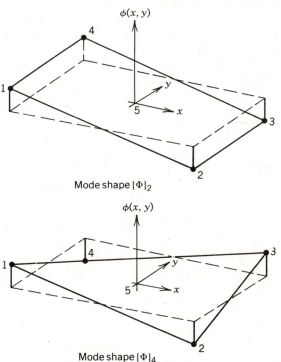

Figure 12.4. The second and fourth mode shapes for a two-dimensional room.

PROBLEMS

12.1 Calculate the eigenvalues and eigenvectors for the one-dimensional problem in Figure 12.1 using one element of length H.

12.2 Calculate the eigenvalues and eigenvectors for the one-dimensional problem in Figure 12.1 using three equal length elements.

12.3 Solve the vibration problem in Figure 12.1 using several elements and a computer program to evaluate the eigenvalues and eigenvectors.

12.4 Verify the mode shape $\{\Phi\}_3$ given in (12.18).

12.5 Verify the mode shape $\{\Phi\}_1$ given in (12.24).

12.6 Verify the mode shape $\{\Phi\}_3$ given in (12.24).

12.7 The acoustic field in an automotive compartment is an important application area for acoustical vibrations. Construct and analyze a grid similar to that given by Shuku and Ishihara (1973) using linear triangular and bilinear rectangular elements.

12.8 The buckling load of a simply supported column (Figure P12.8) is governed by the differential equation

$$\frac{d^2y}{dx^2} + \frac{P}{EI} y = 0$$

where P is the buckling load and EI is a section property. The boundary conditions are $y(0) = y(H) = 0$. Evaluate the eigenvalues and eigenvectors for the four-element grid shown in the figure. Compare the calculated critical loads with the theoretical values given by $P_{cr} = n^2\pi^2 EI/H^2$.

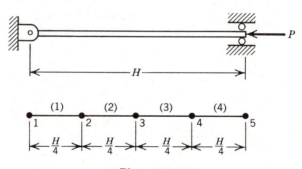

Figure P12.8

12.9 Calculate the critical buckling loads for the column in Figure P12.8 using (a) two equal-length elements and (b) three equal-length elements.

Chapter **13**

AXISYMMETRIC FIELD PROBLEMS

There is a group of three-dimensional field problems that can be solved using two-dimensional elements. These problems possess symmetry about an axis of rotation and are known as axisymmetric problems. The boundary conditions as well as as the region geometry must be independent of the circumferential direction.

The Galerkin formulation and the element equations are similar to those for two-dimensional field problems, but they do differ in some significant ways. The derivation and results differ enough from those in Chapters 7 and 9 to warrant the special consideration given in this chapter.

13.1 DIFFERENTIAL EQUATION

The field equation in a cylindrical coordinate system (r, θ, z) is

$$D_r \frac{\partial^2 \phi}{\partial r^2} + \frac{D_r}{r} \frac{\partial \phi}{\partial r} + \frac{D_\theta}{r^2} \frac{\partial^2 \phi}{\partial \theta^2} + D_z \frac{\partial^2 \phi}{\partial z^2} + Q = 0 \tag{13.1}$$

An axisymmetric problem is independent of θ; thus (13.1) reduces to

$$D_r \frac{\partial^2 \phi}{\partial r^2} + \frac{D_r}{r} \frac{\partial \phi}{\partial r} + D_z \frac{\partial^2 \phi}{\partial z^2} + Q = 0 \tag{13.2}$$

which can also be written as

$$\frac{1}{r} \left[D_r \frac{\partial}{\partial r} \left(r \frac{\partial \phi}{\partial r} \right) \right] + D_z \frac{\partial^2 \phi}{\partial z^2} + Q = 0 \tag{13.3}$$

assuming that D_r is a constant.

The boundary conditions associated with (13.3) are

$$\phi(\Gamma) = \text{specified values} \tag{13.4}$$

on a part of the boundary, call it Γ_1, and

$$D_r \frac{\partial \phi}{\partial r} \cos \theta + D_z \frac{\partial \phi}{\partial z} \sin \theta = -M\phi_b + S \tag{13.5}$$

on the rest of the boundary, Γ_2. Both of these conditions must be independent of the circumferential direction. Equation 13.5 is similar to (9.1) and the numerical values for M and S are determined similar to the examples in Chapters 10 and 11.

165

13.2 AXISYMMETRIC ELEMENTS

The axisymmetric element is obtained by rotating a two-dimensional element about the z-axis to obtain a torus. The idea is illustrated with the triangular element in Figure 13.1.

A single triangular element in the $r-z$ plane is shown in Figure 13.2. This element is identical to the triangular element discussed in Chapter 5 except that the coordinate variables are r and z instead of x and y. The element shape functions are identical to those given in (5.7) through (5.9) with x and y replaced by r and z.

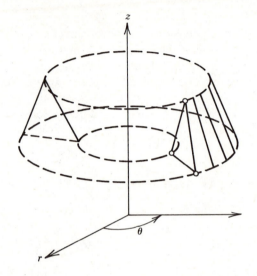

Figure 13.1. The axisymmetric triangular element.

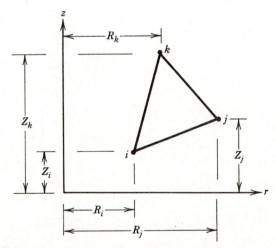

Figure 13.2. The axisymmetric triangular element in the r-z plane.

The variable ϕ and the triangular shape functions in the new coordinate system are

$$\phi = N_i \Phi_i + N_j \Phi_j + N_k \Phi_k \qquad (13.6)$$

where

$$N_i = \frac{1}{2A}(a_i + b_i r + c_i z)$$

$$N_j = \frac{1}{2A}(a_j + b_j r + c_j z)$$

$$N_k = \frac{1}{2A}(a_k + b_k r + c_k z) \qquad (13.7)$$

where

$$
\begin{aligned}
a_i &= R_j Z_k - R_k Z_j, & b_i &= Z_j - Z_k, & c_i &= R_k - R_j \\
a_j &= R_k Z_i - R_i Z_k, & b_j &= Z_k - Z_i, & c_j &= R_i - R_k \\
a_k &= R_i Z_j - R_j Z_i, & b_k &= Z_i - Z_j, & c_k &= R_j - R_i
\end{aligned}
$$

A single rectangular element in the r-z plane is shown in Figure 13.3. This element is similar to the one in Figure 5.10 and was discussed in Section 5.3. The shape functions, however, are different from those given in Section 5.3 because they must be referenced relative to the origin of the r-z coordinate system. All of the rectangular shape functions in Chapter 5 were written relative to local coordinate systems.

Noting that

$$r = R_i + s \qquad \text{and} \qquad z = Z_i + t \qquad (13.8)$$

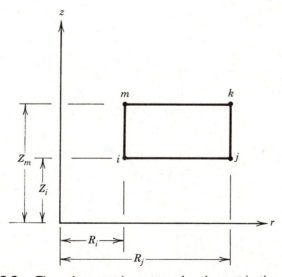

Figure 13.3. The axisymmetric rectangular element in the r-z plane.

rearranging into

$$s = r - R_i \qquad \text{and} \qquad t = z - Z_i \tag{13.9}$$

and substituting for s and t in (5.17) gives the rectangular shape functions in the r-z coordinate system

$$N_i = \frac{1}{4ab} (R_j - r)(Z_m - z)$$

$$N_j = \frac{1}{4ab} (r - R_i)(Z_m - z)$$

$$N_k = \frac{1}{4ab} (r - R_i)(z - Z_i)$$

$$N_m = \frac{1}{4ab} (R_j - r)(z - Z_i) \tag{13.10}$$

13.3 GALERKIN'S METHOD

The weighted residual integral for an axisymmetric field problem is the volume integral

$$\{R^{(e)}\} = - \int_V [N]^T \left(\frac{D_r}{r} \frac{\partial}{\partial r} \left(r \frac{\partial \phi}{\partial r} \right) + D_z \frac{\partial^2 \phi}{\partial z^2} + Q \right) dV \tag{13.11}$$

The derivative terms must be transformed into lower-order forms using the product rule for differentiation and Gauss's theorem. The term involving D_z reduces in a manner identical to (7.12).

The product rule for differentiation gives

$$\frac{\partial}{\partial z} \left([N]^T \frac{\partial \phi}{\partial z} \right) = [N]^T \frac{\partial^2 \phi}{\partial z^2} + \frac{\partial [N]^T}{\partial z} \frac{\partial \phi}{\partial z} \tag{13.12}$$

Rearranging gives

$$[N]^T \frac{\partial^2 \phi}{\partial z^2} = \frac{\partial}{\partial z} \left([N]^T \frac{\partial \phi}{\partial z} \right) - \frac{\partial [N]^T}{\partial z} \frac{\partial \phi}{\partial z} \tag{13.13}$$

The first term in (13.11) is replaced once it is determined that

$$\frac{1}{r} \frac{\partial}{\partial r} \left([N]^T r \frac{\partial \phi}{\partial r} \right) = \frac{1}{r} \left(\frac{\partial [N]^T}{\partial r} r \frac{\partial \phi}{\partial r} \right) + \frac{1}{r} [N]^T \frac{\partial}{\partial r} \left(r \frac{\partial \phi}{\partial r} \right) \tag{13.14}$$

Rearranging gives

$$[N]^T \left(\frac{1}{r} \frac{\partial}{\partial r} \left(r \frac{\partial \phi}{\partial r} \right) \right) = \frac{1}{r} \frac{\partial}{\partial r} \left([N]^T r \frac{\partial \phi}{\partial r} \right) - \frac{\partial [N]^T}{\partial r} \frac{\partial \phi}{\partial r} \tag{13.15}$$

Substitution of (13.13) and (13.15) into (13.11) produces

$$\{R^{(e)}\} = \int_V \left(D_r \frac{\partial [N]^T}{\partial r} \frac{\partial \phi}{\partial r} + D_z \frac{\partial [N]^T}{\partial z} \frac{\partial \phi}{\partial z} - [N]^T Q \right) dV$$
$$- \int_V \left(\frac{D_r}{r} \frac{\partial}{\partial r} \left([N]^T r \frac{\partial \phi}{\partial r} \right) + D_z \frac{\partial}{\partial z} \left([N]^T \frac{\partial \phi}{\partial z} \right) \right) dV \qquad (13.16)$$

The second volume integral can be transformed into a surface integral using Gauss's theorem (Olmstead, 1961). The result is

$$\int_\Gamma \left(\frac{D_r}{r} \left([N]^T r \frac{\partial \phi}{\partial r} \right) \cos \theta + D_z [N]^T \frac{\partial \phi}{\partial z} \sin \theta \right) d\Gamma \qquad (13.17)$$

which simplifies to

$$\int_\Gamma [N]^T \left(D_r \frac{\partial \phi}{\partial r} \cos \theta + D_z \frac{\partial \phi}{\partial z} \sin \theta \right) d\Gamma \qquad (13.18)$$

The complete residual integral is

$$\{R^{(e)}\} = \int_V \left(D_r \frac{\partial [N]^T}{\partial r} \frac{\partial \phi}{\partial r} + D_z \frac{\partial [N]^T}{\partial z} \frac{\partial \phi}{\partial z} - [N]^T Q \right) dV$$
$$- \int_\Gamma [N]^T \left(D_r \frac{\partial \phi}{\partial r} \cos \theta + D_z \frac{\partial \phi}{\partial z} \sin \theta \right) d\Gamma \qquad (13.19)$$

Since $\phi^{(e)} = [N]\{\Phi^{(e)}\}$, $\partial \phi / \partial r$ and $\partial \phi / \partial z$ in the first integral of (13.19) can be replaced by

$$\frac{\partial \phi}{\partial r} = \frac{\partial [N]}{\partial r} \{\Phi^{(e)}\} \qquad \text{and} \qquad \frac{\partial \phi}{\partial z} = \frac{\partial [N]}{\partial z} \{\Phi^{(e)}\} \qquad (13.20)$$

giving

$$\{R^{(e)}\} = \left(\int_V \left(D_r \frac{\partial [N]^T}{\partial r} \frac{\partial [N]}{\partial r} + D_z \frac{\partial [N]^T}{\partial z} \frac{\partial [N]}{\partial z} \right) dV \right) \{\Phi^{(e)}\}$$
$$- \int_V Q [N]^T dV$$
$$- \int_\Gamma [N]^T \left(D_r \frac{\partial \phi}{\partial r} \cos \theta + D_z \frac{\partial \phi}{\partial z} \sin \theta \right) d\Gamma \qquad (13.21)$$

The first integral in (13.21) multiplies $\{\Phi^{(e)}\}$; thus it is the element stiffness matrix. The integral containing Q becomes $\{f^{(e)}\}$, while the surface integral is the interelement requirement for interior element boundaries and the derivative boundary conditions for element boundaries on Γ_2. The general form of $\{R^{(e)}\}$ is

$$\{R^{(e)}\} = \{I^{(e)}\} + [k^{(e)}]\{\Phi^{(e)}\} - \{f_Q^{(e)}\} \qquad (13.22)$$

where

$$\{I^{(e)}\} = -\int_{\Gamma} [N]^T \left(D_r \frac{\partial \phi}{\partial r} \cos \theta + D_z \frac{\partial \phi}{\partial z} \sin \theta \right) d\Gamma \tag{13.23}$$

$$[k^{(e)}] = \int_V \left(D_r \frac{\partial [N]^T}{\partial r} \frac{\partial [N]}{\partial r} + D_z \frac{\partial [N]^T}{\partial z} \frac{\partial [N]}{\partial z} \right) dV \tag{13.24}$$

and

$$\{f_Q^{(e)}\} = \int_V Q[N]^T \, dV \tag{13.25}$$

The terms in the integral for $[k^{(e)}]$ are similar to those in (7.20), and (13.24) can be written in the compact form

$$[k^{(e)}] = \int_V [B]^T [D][B] \, dV \tag{13.26}$$

where

$$\{gv\} = \begin{Bmatrix} \dfrac{\partial \phi}{\partial r} \\[2mm] \dfrac{\partial \phi}{\partial z} \end{Bmatrix} = [B]\{\Phi^{(e)}\} \tag{13.27}$$

and

$$[D] = \begin{bmatrix} D_r & 0 \\ 0 & D_z \end{bmatrix} \tag{13.28}$$

13.4 ELEMENT MATRICES

The immediate objective is to evaluate the volume integrals that give $[k^{(e)}]$ and $\{f^{(e)}\}$. The contribution of the derivative boundary condition to these integrals is discussed in the next section. The discussion is limited to the triangular element because the integrals are relatively easy to evaluate. The integrals for the rectangular element are usually evaluated using numerical techniques.

The coefficients in $[B]$ are obtained by differentiating the shape functions relative to r and z. This produces a matrix identical to the one in (7.34)

$$[B] = \frac{1}{2A} \begin{bmatrix} b_i & b_j & b_k \\ c_i & c_j & c_k \end{bmatrix} \tag{13.29}$$

Each coefficient in $[B]$ is a constant. Since $[D]$, as given by (13.28), also consists of constant coefficients

$$[k^{(e)}] = \int_V [B]^T [D][B] \, dV = [B]^T [D][B] \int_V dV$$
$$= [B]^T [D][B] V \tag{13.30}$$

The volume of an area revolved about the z-axis is $V = 2\pi \bar{r} A$, where $\bar{r}$ is the radial distance to the centroid of the area. The element stiffness matrix is

$$[k^{(e)}] = 2\pi \bar{r} A [B]^T [D][B] \tag{13.31}$$

This matrix product is easily evaluated to give

$$[k^{(e)}] = \frac{2\pi \bar{r} D_r}{4A} \begin{bmatrix} b_i^2 & b_i b_j & b_i b_k \\ b_i b_j & b_j^2 & b_j b_k \\ b_i b_k & b_j b_k & b_k^2 \end{bmatrix} + \frac{2\pi \bar{r} D_z}{4A} \begin{bmatrix} c_i^2 & c_i c_j & c_i c_k \\ c_i c_j & c_j^2 & c_j c_k \\ c_i c_k & c_j c_k & c_k^2 \end{bmatrix} \tag{13.32}$$

The radial distance to the centroid of a triangular element is

$$\bar{r} = \frac{R_i + R_j + R_k}{3} \tag{13.33}$$

The element force vector $\{f_Q^{(e)}\}$ is

$$\{f_Q^{(e)}\} = \int_V Q[N]^T \, dV = 2\pi Q \int_A \begin{Bmatrix} N_i r \\ N_j r \\ N_k r \end{Bmatrix} dA \tag{13.34}$$

since $dV = 2\pi r \, dA$. The shape functions in (13.34) can be replaced by area coordinates, and the radial distance r can be written as

$$r = N_i R_i + N_j R_j + N_k R_k = L_1 R_i + L_2 R_j + L_3 R_k \tag{13.35}$$

and the integral for $\{f_Q^{(e)}\}$ is

$$\{f_Q^{(e)}\} = 2\pi Q \int_A \begin{Bmatrix} L_1(L_1 R_i + L_2 R_j + L_3 R_k) \\ L_2(L_1 R_i + L_2 R_j + L_3 R_k) \\ L_3(L_1 R_i + L_2 R_j + L_3 R_k) \end{Bmatrix} dA$$

Evaluating the integrals of the area coordinate products using (6.29) produces

$$\{f_Q^{(e)}\} = \frac{2\pi Q A}{12} \begin{bmatrix} 2 & 1 & 1 \\ 1 & 2 & 1 \\ 1 & 1 & 2 \end{bmatrix} \begin{Bmatrix} R_i \\ R_j \\ R_k \end{Bmatrix} \tag{13.36}$$

A uniform Q within the element is not distributed equally among the nodes as occurred with the two-dimensional element. Each node receives an amount related to its radial distance from the origin.

ILLUSTRATIVE EXAMPLE

An axisymmetric triangular element with the nodal coordinates given in Figure 13.4 has a uniform heat generation of $Q = 3$ W/cm^3. Calculate $[k^{(e)}]$ and $\{f_Q^{(e)}\}$ when $D_r = D_z = 1.5$ W/cm-°C.

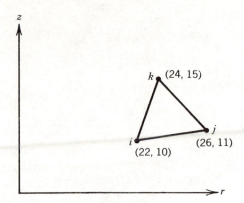

Figure 13.4.   The triangular element for the example problem.

By using (13.7), the determinant equation (5.6) and (13.33) gives

$$b_i = -4, \qquad b_j = 5, \qquad b_k = -1$$

$$c_i = -2, \qquad c_j = -2, \qquad c_k = 4$$

$$\bar{r} = \frac{22 + 26 + 24}{3} = 24 \text{ cm}$$

and

$$2A = 18 \text{ cm}^2$$

The multiplier coefficients are

$$\frac{2\pi \bar{r} D_r}{4A} = \frac{2\pi \bar{r} D_z}{4A} = \frac{2\pi(24)(1.5)}{2(18)} = 2\pi$$

$$\frac{2\pi Q A}{12} = \frac{2\pi(3)(9)}{12} = 4.5\,\pi$$

The element stiffness matrix is

$$[k^{(e)}] = 2\pi \begin{bmatrix} 16 & -20 & 4 \\ -20 & 25 & -5 \\ 4 & -5 & 1 \end{bmatrix} + 2\pi \begin{bmatrix} 4 & 4 & -8 \\ 4 & 4 & -8 \\ -8 & -8 & 16 \end{bmatrix}$$

or

$$[k^{(e)}] = \pi \begin{bmatrix} 40 & -32 & -8 \\ -32 & 58 & -26 \\ -8 & -26 & 34 \end{bmatrix}$$

The element force vector is

$$\{f_Q^{(e)}\} = 4.5\pi \begin{bmatrix} 2 & 1 & 1 \\ 1 & 2 & 1 \\ 1 & 1 & 2 \end{bmatrix} \begin{Bmatrix} R_i \\ R_j \\ R_k \end{Bmatrix}$$

$$\{f_Q^{(e)}\} = 4.5\pi \begin{bmatrix} 2 & 1 & 1 \\ 1 & 2 & 1 \\ 1 & 1 & 2 \end{bmatrix} \begin{Bmatrix} 22 \\ 26 \\ 24 \end{Bmatrix}$$

$$\{f_Q^{(e)}\} = \pi \begin{Bmatrix} 423 \\ 441 \\ 432 \end{Bmatrix}$$

Since Q is a constant within the element, the total heat generated is

$$\int_V Q\,dV = Q\int_V dV = 2\pi\bar{r}AQ = 1296\pi \text{ W}$$

The components of $\{f_Q^{(e)}\}$ sum to this value, but the quantity is not distributed equally among the nodes.

13.5 THE DERIVATIVE BOUNDARY CONDITION

The element matrices (13.32) and (13.36) for $[k^{(e)}]$ and $\{f_Q^{(e)}\}$, respectively, are valid for interior elements and boundary elements when $\phi(r, z)$ is specified on the boundary. When the derivative boundary condition (13.5) is specified, there are additional contributions to $[k^{(e)}]$ and $\{f^{(e)}\}$. These contributions are considered in this section.

The element contributions due to the derivative boundary condition come from the interelement vector (13.23) after substituting the relationship in (13.5). Assuming that Γ_{bc} is the surface of the element with the boundary condition

$$\{I_{bc}^{(e)}\} = -\int_{\Gamma_{bc}} [N]^T(-M\phi_b + S)\,d\Gamma \tag{13.37}$$

or

$$\{I_{bc}^{(e)}\} = \int_{\Gamma_{bc}} M[N]^T\phi_b\,d\Gamma - \int_{\Gamma_{bc}} S[N]^T\,d\Gamma \tag{13.38}$$

The value of ϕ on the boundary, ϕ_b, is given by $\phi^{(e)} = [N]\{\Phi^{(e)}\}$; therefore,

$$\{I_{bc}^{(e)}\} = \left(\int_{\Gamma_{bc}} M[N]^T[N]\,d\Gamma\right)\{\Phi^{(e)}\} - \int_{\Gamma_{bc}} S[N]^T\,d\Gamma \tag{13.39}$$

The first integral contributes to $[k^{(e)}]$, since it multiplies $\{\Phi^{(e)}\}$; the second is part of $\{f^{(e)}\}$. Both integrals in (13.39) are surface integrals and are evaluated similarly to those discussed in Chapter 10. The integrals in (13.39) are

$$[k_M^{(e)}] = \int_{\Gamma_{bc}} M[N]^T[N]\,d\Gamma \tag{13.40}$$

and

$$\{f_S^{(e)}\} = \int_{\Gamma_{bc}} S[N]^T\,d\Gamma \tag{13.41}$$

The integral in (13.41) is the easiest to evaluate giving

$$\{f_S^{(e)}\} = S \int_{\Gamma_{bc}} \begin{Bmatrix} N_i \\ N_j \\ N_k \end{Bmatrix} d\Gamma = L_{jk} \int_0^1 S \begin{Bmatrix} L_1 \\ L_2 \\ L_3 \end{Bmatrix} 2\pi r \, d\ell_2 \qquad (13.42)$$

assuming that the integration is along side jk. The area coordinates reduce to $L_1 = 0$, $L_2 = \ell_1$, and $L_3 = \ell_2$ along this side [see (6.34)] and the integral becomes

$$\{f_S^{(e)}\} = 2\pi S L_{jk} \int_0^1 \begin{Bmatrix} 0 \\ \ell_1 r \\ \ell_2 r \end{Bmatrix} d\ell_2 \qquad (13.43)$$

The radial distance to a point on the boundary is

$$r = N_i R_i + N_j R_j + N_k R_k = \ell_1 R_j + \ell_2 R_k \qquad (13.44)$$

since $N_i = 0$. Substitution for r in (13.43) produces

$$\{f_S^{(e)}\} = 2\pi S L_{jk} \int_0^1 \begin{Bmatrix} 0 \\ \ell_1(\ell_1 R_j + \ell_2 R_k) \\ \ell_2(\ell_1 R_j + \ell_2 R_k) \end{Bmatrix} d\ell_2 \qquad (13.45)$$

The integrals of ℓ_1^2, $\ell_1\ell_2$, and ℓ_2^2 are evaluated using (6.17). The final result is

$$\{f_S^{(e)}\} = \frac{2\pi S L_{jk}}{6} \begin{Bmatrix} 0 \\ 2R_j + R_k \\ R_j + 2R_k \end{Bmatrix} \qquad (13.46)$$

The other two results for $\{f_S^{(e)}\}$ are

$$\{f_S^{(e)}\} = \frac{2\pi S L_{ij}}{6} \begin{Bmatrix} 2R_i + R_j \\ R_i + 2R_j \\ 0 \end{Bmatrix}, \qquad \frac{2\pi S L_{ik}}{6} \begin{Bmatrix} 2R_i + R_k \\ 0 \\ R_i + 2R_k \end{Bmatrix} \qquad (13.47)$$

for sides ij and ik, respectively.

The surface integral (13.40) is evaluated in a similar manner. Considering side jk, (13.40) becomes

$$\int_{\Gamma_{bc}} M[N]^T[N] \, d\Gamma = 2\pi M L_{jk} \int_0^1 \begin{Bmatrix} 0 \\ \ell_1 \\ \ell_2 \end{Bmatrix} \begin{bmatrix} 0 & \ell_1 & \ell_2 \end{bmatrix} r \, d\ell_2$$

$$= 2\pi M L_{jk} \int_0^1 \begin{bmatrix} 0 & 0 & 0 \\ 0 & r\ell_1^2 & r\ell_1\ell_2 \\ 0 & r\ell_1\ell_2 & r\ell_2^2 \end{bmatrix} d\ell_2 \qquad (13.48)$$

Substituting (13.44) for r and using the factorial formula (6.17) yields

$$[k_M^{(e)}] = \frac{2\pi M L_{jk}}{12} \begin{bmatrix} 0 & 0 & 0 \\ 0 & (3R_j + R_k) & (R_j + R_k) \\ 0 & (R_j + R_k) & (R_j + 3R_k) \end{bmatrix} \tag{13.49}$$

The other two results for (13.40) are

$$[k_M^{(e)}] = \frac{2\pi M L_{ij}}{12} \begin{bmatrix} (3R_i + R_j) & (R_i + R_j) & 0 \\ (R_i + R_j) & (R_i + 3R_j) & 0 \\ 0 & 0 & 0 \end{bmatrix} \tag{13.50}$$

$$[k_M^{(e)}] = \frac{2\pi M L_{ik}}{12} \begin{bmatrix} (3R_i + R_k) & 0 & (R_i + R_k) \\ 0 & 0 & 0 \\ (R_i + R_k) & 0 & (R_i + 3R_k) \end{bmatrix} \tag{13.51}$$

for sides ij and ik, respectively.

ILLUSTRATIVE EXAMPLE

Evaluate $[k_M^{(e)}]$ and $\{f_S^{(e)}\}$ for side ij of the element shown in Figure 13.4 when $M = 4$ and $S = 3$.

The pertinent quantities are

$$R_i = 22, \qquad R_j = 26$$

and

$$L_{ij} = \sqrt{(26 - 22)^2 + (11 - 10)^2} = 4.12 \text{ cm}$$

The multiplying coefficients are

$$\frac{2\pi M L_{ij}}{12} = \frac{2(3.14)(4)(4.12)}{12} = 8.62$$

$$\frac{2\pi S L_{ij}}{6} = \frac{2(3.14)(3)(4.12)}{6} = 12.9$$

while

$$3R_i + R_j = 3(22) + 26 = 92$$
$$R_i + R_j = 22 + 26 = 48$$
$$R_i + 3R_j = 22 + 3(26) = 100$$
$$2R_i + R_j = 2(22) + 26 = 70$$
$$R_i + 2R_j = 22 + 2(26) = 74$$

The element quantities are

$$[k_M^{(e)}] = 8.62 \begin{bmatrix} 92 & 48 & 0 \\ 48 & 100 & 0 \\ 0 & 0 & 0 \end{bmatrix} = \begin{bmatrix} 793 & 414 & 0 \\ 414 & 862 & 0 \\ 0 & 0 & 0 \end{bmatrix}$$

and

$$\{f_S^{(e)}\} = 12.9 \begin{Bmatrix} 70 \\ 74 \\ 0 \end{Bmatrix} = \begin{Bmatrix} 903 \\ 955 \\ 0 \end{Bmatrix}$$

Note that the diagonal values in $[k_M^{(e)}]$ as well as the values in $\{f_S^{(e)}\}$ are not the same.

PROBLEMS

13.1   Evaluate (13.40) for side jk of a triangular element.

13.2   Evaluate (13.40) for side km of a rectangular element.

13.3   Evaluate (13.41) for side ki of a triangular element.

13.4   Evaluate (13.41) for side km of a rectangular element.

13.5   Evaluate (13.25) for a rectangular element using the shape functions defined by (13.10).

13.6   Verify that (13.46) and (13.49) behave similarly to (9.19) and (9.33) when nodes j and k are on a vertical side, that is, $R_j = R_k$.

13.7–13.11   The nodal coordinate and coefficient values for five axisymmetric elements follow. Evaluate $[k_D^{(e)}]$ and $\{f_Q^{(e)}\}$ for one of the elements. Assume that $D_x = D_y$.

Element Quantity	Problem Number				
	13.7	13.8	13.9	13.10	13.11
D_x	2	3	2	4	5
Q	20	30	40	50	60
R_i	2	10	8	20	6
R_j	4	14	12	20	10
R_k	3	14	10	16	10
Z_i	0	10	6	20	12
Z_j	0	12	10	24	8
Z_k	4	16	14	24	14

13.12   Use the element data for Problem 13.7 and the values $M = 6$ and $S = 2$. Evaluate $[k_M^{(e)}]$ and $\{f_S^{(e)}\}$ along side ij.

13.13   Use the element data for Problem 13.8 and the values $M = 10$ and $S = 5$. Evaluate $[k_M^{(e)}]$ and $\{f_S^{(e)}\}$ along side ik.

13.14   Use the element data for Problem 13.10 and the values $M = 6$ and $S = -2$. Evaluate $[k_M^{(e)}]$ and $\{f_S^{(e)}\}$ along side jk.

13.15   The governing differential equation for a radial symmetric problem is

$$D_r \frac{\partial^2 \phi}{\partial r^2} + \frac{D_r}{r} \frac{\partial \phi}{\partial r} + Q = 0$$

Obtain the general equations for $[k_D^{(e)}]$ and $\{f_Q^{(e)}\}$ by evaluating the residual integral

$$\{R^{(e)}\} = -\int_A [N]^T \left(D_r \frac{\partial^2 \phi}{\partial r^2} + \frac{D_r}{r} \frac{\partial \phi}{\partial r} + Q \right) dA$$

Chapter 14

TIME-DEPENDENT FIELD PROBLEMS: THEORETICAL CONSIDERATIONS

The field problems considered in the previous chapters were steady-state problems. An equally important set of physical problems are those that are time-dependent. The finite element solution of time-dependent problems is introduced in this chapter with the primary emphasis placed on the theoretical aspects. Some practical aspects of solving time-dependent problems are discussed in Chapter 15. The discussion in each chapter is build around the one-dimensional equation

$$D\frac{\partial^2 \phi}{\partial x^2} + Q = \lambda \frac{\partial \phi}{\partial t} \tag{14.1}$$

because the numerical calculations are relatively easy and three-dimensional plots of ϕ-x-t can be constructed.

The two-dimensional time-dependent equation is

$$D_x\frac{\partial^2 \phi}{\partial x^2} + D_y\frac{\partial^2 \phi}{\partial y^2} - G\phi + Q = \lambda \frac{\partial \phi}{\partial t} \tag{14.2}$$

The only new term in each of (14.1) and (14.2) is $\lambda\partial\phi/\partial t$. We shall see that it behaves the same in each dimension so that a discussion of (14.1) is adequate.

14.1 GALERKIN METHOD

The general procedure for analyzing (14.1) or (14.2) is to evaluate the Galerkin residual integral with respect to the space coordinate(s) for a fixed instant of time. This yields a system of differential equations that are solved to obtain the variation of ϕ with time.

By rearranging (14.1) into

$$D\frac{\partial^2 \phi}{\partial x^2} + Q - \lambda \frac{\partial \phi}{\partial t} = 0 \tag{14.3}$$

the residual integral is

$$\{R^{(e)}\} = -\int_{X_i}^{X_j} [W]^T \left(D\frac{\partial^2 \phi}{\partial x^2} + Q - \lambda \frac{\partial \phi}{\partial t} \right) dx \tag{14.4}$$

where $[W]^T$ contains the Galerkin weighting functions. Equation 14.4 separates into

$$\{R^{(e)}\} = -\int_{X_i}^{X_j} [W]^T \left(D\frac{\partial^2 \phi}{\partial x^2} + Q \right) dx + \int_{X_i}^{X_j} [W]^T \left(\lambda \frac{\partial \phi}{\partial t} \right) dx \qquad (14.5)$$

or

$$\{R^{(e)}\} = \{R_D^{(e)}\} + \{R_\lambda^{(e)}\} \qquad (14.6)$$

Defining $[W]^T = [N]^T$ in the first integral of (14.5) yields the same integrals that were analyzed in Chapter 3 and

$$\{R_D^{(e)}\} = \{I^{(e)}\} + [k^{(e)}]\{\Phi^{(e)}\} - \{f^{(e)}\} \qquad (14.7)$$

The only new term is

$$\{R_\lambda^{(e)}\} = \int_{X_i}^{X_j} [W]^T \left(\lambda \frac{\partial \phi}{\partial t} \right) dx \qquad (14.8)$$

Equation 14.8 has two different solutions. One solution is called the consistent formulation and the other is referred to as the lumped formulation. We shall consider each of these solutions starting with the consistent formulation.

14.2 THE CONSISTENT FORMULATION

A set of weighting functions and the variation of $\partial\phi/\partial t$ with respect to x must be defined before the integral in (14.8) can be evaluated. The parameter $\partial\phi/\partial t$ is shown schematically in Figure 14.1. One equation that defines the variation states that the time derivative varies linearly between the nodal values. If we denote the nodal values of $\partial\phi/\partial t$ by $\dot{\Phi}_1, \dot{\Phi}_2, \ldots, \dot{\Phi}_p$, the linear variation of $\partial\phi/\partial t$ within an

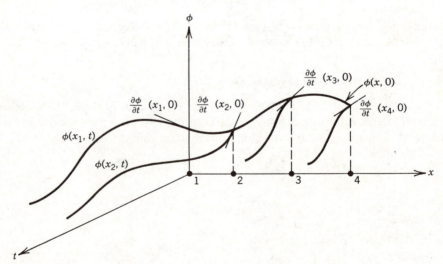

Figure 14.1. The $\partial\phi/\partial t$ as a function of x.

element can be written as

$$\frac{\partial \phi}{\partial t} = [N_i \ N_j] \begin{Bmatrix} \dot{\Phi}_i \\ \dot{\Phi}_j \end{Bmatrix} \tag{14.9}$$

or

$$\frac{\partial \phi}{\partial t} = [N]\{\dot{\Phi}^{(e)}\} \tag{14.10}$$

If we also use the linear shape functions for the weighting coefficients, that is, $[W]^T = [N]^T$, the residual integral is

$$\{R_\lambda^{(e)}\} = \int_{X_i}^{X_j} \lambda[N]^T[N]\{\dot{\Phi}^{(e)}\} \, dx$$

$$= \left(\int_{X_i}^{X_j} \lambda[N]^T[N] \, dx \right) \{\dot{\Phi}^{(e)}\} \tag{14.11}$$

$$= [c^{(e)}]\{\dot{\Phi}^{(e)}\} \tag{14.12}$$

A similar analysis for (14.2) yields

$$[c^{(e)}] = \int_A \lambda[N]^T[N] \, dA \tag{14.13}$$

where $[N]$ contains the triangular or rectangular shape functions.

When $[c^{(e)}]$ is combined with the other element matrices and summed over all the elements using the direct stiffness procedure, the final result is a system of first-order differential equations given by

$$[C]\{\dot{\Phi}\} + [K]\{\Phi\} - \{F\} = \{0\} \tag{14.14}$$

where the boundary conditions in $\{I\}$ have been incorporated into $[K]$ and $\{F\}$ and the interelement requirements discarded. The vector $\{\dot{\Phi}\}$ is

$$\{\dot{\Phi}\}^T = \left[\frac{\partial \Phi_1}{\partial t} \frac{\partial \Phi_2}{\partial t} \cdots \frac{\partial \Phi_p}{\partial t} \right] \tag{14.15}$$

The new matrix $[C]$, is usually called the capacitance matrix.

The integrals that define $[c^{(e)}]$ are the same as those associated with the G term in the steady-state form of the differential equation. We immediately conclude that

$$[c^{(e)}] = \frac{\lambda L}{6} \begin{bmatrix} 2 & 1 \\ 1 & 2 \end{bmatrix} \tag{14.16}$$

for the one-dimensional linear element,

$$[c^{(e)}] = \frac{\lambda A}{12} \begin{bmatrix} 2 & 1 & 1 \\ 1 & 2 & 1 \\ 1 & 1 & 2 \end{bmatrix} \tag{14.17}$$

for the two-dimensional triangular element, and

$$[c^{(e)}] = \frac{\lambda A}{36} \begin{bmatrix} 4 & 2 & 1 & 2 \\ 2 & 4 & 2 & 1 \\ 1 & 2 & 4 & 2 \\ 2 & 1 & 2 & 4 \end{bmatrix} \tag{14.18}$$

for the two-dimensional rectangular element [see (11.9), (7.38), and (7.49)].

The formulation discussed in this section is called the consistent formulation because the linear variation of $\partial \phi / \partial t$ with respect to x within an element is identical or consistent with the linear variation assumed for $\phi(x)$. The same set of weighting functions, namely $[N]^T$, is used in both of the integrals in (14.5).

14.3 THE LUMPED FORMULATION

An alternate approach to defining the variation in $\partial \phi / \partial t$ with respect to x is to assume that it is constant between the midpoints of adjacent elements. This concept is shown schematically in Figure 14.2a for a single element and in Figure 14.2b for a grid of five elements.

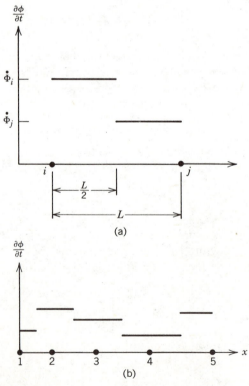

(a)

(b)

Figure 14.2. The step variation of $\partial \phi / \partial t$ as a function of x (a) within an element (b) for a grid.

The variation of $\partial\phi/\partial t$ within an element can be written explicitly using step functions

$$\frac{\partial\phi}{\partial t}(x) = \left[1 - h\left(s - \frac{L}{2}\right)\right]\frac{\partial\Phi_i}{\partial t} + h\left(s - \frac{L}{2}\right)\frac{\partial\Phi_j}{\partial t} \tag{14.19}$$

where

$$h\left(s - \frac{L}{2}\right) = \begin{cases} 0 & s < \dfrac{L}{2} \\[2mm] 1 & s > \dfrac{L}{2} \end{cases} \tag{14.20}$$

and s is measured from node i. The quantities multiplying Φ_i and Φ_j in (14.19) can be thought of as new shape functions, N_i^* and N_j^*, where

$$N_i^* = 1 - h\left(s - \frac{L}{2}\right)$$

$$N_j^* = h\left(s - \frac{L}{2}\right) \tag{14.21}$$

Upon using this concept, (14.19) becomes

$$\frac{\partial\phi}{\partial t} = N_i^*\frac{\partial\Phi_i}{\partial t} + N_j^*\frac{\partial\Phi_j}{\partial t} = [N^*]\{\dot{\Phi}^{(e)}\} \tag{14.22}$$

Returning to the residual integral, $\{R_\lambda^{(e)}\}$, and using (14.22) gives

$$\{R_\lambda^{(e)}\} = \int_{X_i}^{X_j} \lambda[W]^T\frac{\partial\phi}{\partial t}\,dx$$

$$= \int_0^L \lambda[W]^T[N^*]\{\dot{\Phi}^{(e)}\}\,ds \tag{14.23}$$

Using the same functions in $[W]^T$ that are used in the relationship for $\partial\phi/\partial t$ gives

$$\{R_\lambda^{(e)}\} = \left(\int_0^L \lambda[N^*]^T[N^*]\,ds\right)\{\dot{\Phi}^{(e)}\} = [c^{(e)}]\{\dot{\Phi}^{(e)}\} \tag{14.24}$$

The element capacitance matrix is defined by

$$[c^{(e)}] = \int_0^L \lambda[N^*]^T[N^*]\,ds \tag{14.25}$$

which is easily evaluated because

$$N_i^*N_i^* = N_j^*N_j^* = 1 \quad\text{and}\quad N_i^*N_j^* = 0$$

The element capacitance matrix is

$$[c^{(e)}] = \frac{\lambda L}{2}\begin{bmatrix} 1 & 0 \\ 0 & 1 \end{bmatrix} \tag{14.26}$$

for the one-dimensional linear element. It is a diagonal matrix, which means that $[C]$ is also diagonal. Since the formulation produces a diagonal matrix, it is usually referred to as a lumped formulation.

The generalization of the step function variation for two-dimensional problems is accomplished by defining functions similar to those shown in Figure 14.3. The shaded region for N_i^* has corners at node i, at the midpoints of the sides that touch node i and at the center of the element. Defining the functions in this manner produces $N_\alpha^* N_\beta^*$ products whose value is zero. The resulting diagonal matrices are

$$[c^{(e)}] = \frac{\lambda A}{3} \begin{bmatrix} 1 & 0 & 0 \\ 0 & 1 & 0 \\ 0 & 0 & 1 \end{bmatrix} \tag{14.27}$$

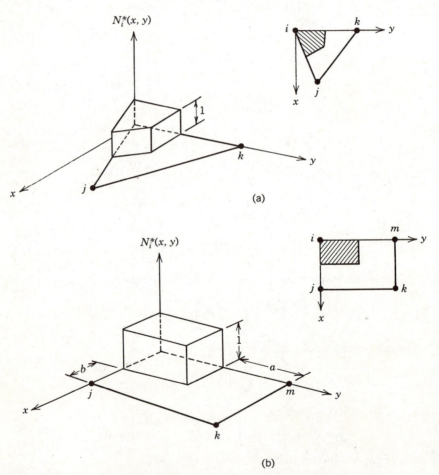

(a)

(b)

Figure 14.3. The step-function variation for two-dimensional elements. (a) A triangular element. (b) A rectangular element.

for the triangular element and

$$[c^{(e)}] = \frac{\lambda A}{4} \begin{bmatrix} 1 & 0 & 0 & 0 \\ 0 & 1 & 0 & 0 \\ 0 & 0 & 1 & 0 \\ 0 & 0 & 0 & 1 \end{bmatrix} \tag{14.28}$$

for the rectangular element.

14.4 FINITE DIFFERENCE SOLUTION IN TIME

The finite element solution of time-dependent field problems produces a system of linear first-order differential equations in the time domain. These equations must be solved before the variation of ϕ in space and time is known. There are several procedures for numerically solving the equations given by (14.14). We shall concentrate on the methods associated with the solution of heat transfer problems that use a finite difference approximation in the time domain to generate a numerical solution.

Given a function $\phi(t)$ and the interval $[a, b]$, we can use the mean value theorem for differentiation to develop an equation for $\phi(t)$. The mean value theorem states that there is a value of t, call it ξ, such that

$$\phi(b) - \phi(a) = (b - a)\frac{d\phi}{dt}(\xi) \tag{14.29}$$

Rearranging gives

$$\frac{d\phi}{dt}(\xi) = \frac{\phi(b) - \phi(a)}{\Delta t} \tag{14.30}$$

where $\Delta t = (b - a)$ is the length of the interval. The location of ξ in (14.29) and (14.30) is not known. This aspect will be dealt with in a few moments.

The value of ϕ at $t = a$, $\phi(a)$, can be approximated as shown in Figure 14.4.

$$\phi(a) = \phi(\xi) - (\xi - a)\frac{d\phi}{dt}(\xi) \tag{14.31}$$

Rearranging gives

$$\phi(\xi) = \phi(a) + (\xi - a)\frac{d\phi}{dt}(\xi) \tag{14.32}$$

and after substituting (14.30)

$$\phi(\xi) = \phi(a) + \frac{\phi(b) - \phi(a)}{\Delta t}(\xi - a) \tag{14.33}$$

If we define the ratio θ as

$$\theta = \frac{(\xi - a)}{\Delta t} \tag{14.34}$$

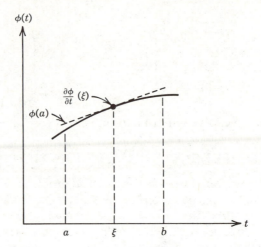

Figure 14.4.   An approximation for ϕ (a) given $\partial\phi/\partial t$.

then (14.33) can be written

$$\phi(\xi) = (1-\theta)\phi(a) + \theta\phi(b) \tag{14.35}$$

Equations (14.30) and (14.35) can be generalized to a set of nodal values by replacing $\phi(a)$ and $\phi(b)$ by column vectors containing the nodal values. Define $\{\Phi\}_a$ and $\{\Phi\}_b$ as the vectors containing the nodal values at times a and b, then

$$\frac{d\{\Phi\}}{dt} = \frac{\{\Phi\}_b - \{\Phi\}_a}{\Delta t} \tag{14.36}$$

and

$$\{\Phi\} = (1-\theta)\{\Phi\}_a + \theta\{\Phi\}_b \tag{14.37}$$

at the point $t = \xi$.

An equation for $\{F\}$ at $t = \xi$ can be derived using the same procedure that was used to obtain (14.35). The result is

$$\{F\} = (1-\theta)\{F\}_a + \theta\{F\}_b \tag{14.38}$$

Equation (14.14) can now be written in terms of $\{\Phi\}_a$, $\{\Phi\}_b$, $\{F\}_a$ and $\{F\}_b$ by substituting (14.36), (14.37), and (14.38). The result is

$$([C] + \theta\Delta t\,[K])\{\Phi\}_b = ([C] - (1-\theta)\,\Delta t\,[K])\{\Phi\}_a + \Delta t((1-\theta)\{F\}_a + \theta\{F\}_b) \tag{14.39}$$

Equation (14.39) gives the nodal values, $\{\Phi\}_b$, in terms of a set of known values, $\{\Phi\}_a$, the force vector at times a and b and the ratio θ. The force vectors, $\{F\}_a$ and $\{F\}_b$, must be known. The value of θ must also be specified. This is equivalent to specifying the location of ξ at which the mean value theorem is applied.

There are four popular choices for θ. These choices and their associated names are

1. $\theta = 0$, $\xi = a$ Forward difference method

2. $\theta = \dfrac{1}{2}$, $\xi = \dfrac{\Delta t}{2}$ Central difference method

3. $\theta = \dfrac{2}{3}$, $\xi = \dfrac{2\Delta t}{3}$ Galerkin's method

4. $\theta = 1$, $\xi = b$ Backward difference method

The finite difference equation for each value of θ follows.

1. $\theta = 0$,   Forward difference method

$$[C]\{\Phi\}_b = ([C] - \Delta t\,[K])\{\Phi\}_a + \Delta t\,\{F\}_a \tag{14.40}$$

3. $\theta = \dfrac{1}{2}$,   Central difference method

$$\left([C] + \frac{\Delta t}{2}[K]\right)\{\Phi\}_b = \left([C] - \frac{\Delta t}{2}[K]\right)\{\Phi\}_a + \frac{\Delta t}{2}(\{F\}_a + \{F\}_b) \tag{14.41}$$

3. $\theta = \dfrac{2}{3}$,   Galerkin's method

$$\left([C] + \frac{2\Delta t}{3}[K]\right)\{\Phi\}_b = \left([C] - \frac{\Delta t}{3}[K]\right)\{\Phi\}_a + \frac{\Delta t}{3}(\{F\}_b + 2\{F\}_a) \tag{14.42}$$

4. $\theta = 1$,   Backward difference method

$$([C] + \Delta t\,[K])\{\Phi\}_b = [C]\{\Phi\}_a + \Delta t\{F\}_b$$

Regardless of which value is specified for θ, the final system of equations has the general form

$$[A]\{\Phi\}_b = [P]\{\Phi\}_a + \{F^*\} \tag{14.44}$$

where $[A]$ and $[P]$ are combinations of $[C]$ and $[K]$ and are dependent on the time step Δt. If Δt and the material properties are independent of time or $\{\Phi\}$, then $[A]$ is the same for all time points. If either Δt or the material parameters change during the solution process, $[A]$ and $[P]$ are re-evaluated for the new time step. Each solution in time is equivalent to solving a single steady-state problem when $[A]$ and $[P]$ are evaluated for each time step. The vector, $\{F^*\}$, in (14.44) is a combination of $\{F\}_a$ and $\{F\}_b$.

14.5 HEAT FLOW IN A ROD

The solution procedure along with some of the computational problems that arise when solving time-dependent field problems are illustrated in this section using the heat flow in an insulated rod as the example.

The solution of (14.39) for any value of θ is accompanied by problems. Euler's method (the forward difference method) is known to be unstable when Δt exceeds a certain value. Solutions obtained using $\theta = \frac{1}{2}$ or $\theta = \frac{2}{3}$ contain numerical oscillations when Δt is too large. These methods, however, are unconditionally stable.

The purely implicit method (backward difference method) is unconditionally stable, and the calculated values do not oscillate about the correct values. The method, however, becomes less accurate for large values of time. A good discussion of the problems that accompany the solution of (14.14) is given by Myers (1971). We shall discuss how to avoid these difficulties in Chapter 15.

Consider the insulated rod (Figure 14.5) that is modeled by two elements. We shall solve this problem using the consistent formulation, the central difference method, and a time step of 1 sec. This time step is below the value that introduces numerical oscillations into the solution. The rod is initially at a temperature of zero with heat applied at the left end.

The element matrices are given by (3.9) and (14.16) and are

$$[k^{(e)}] = \begin{bmatrix} 2 & -2 \\ -2 & 2 \end{bmatrix} \quad \text{and} \quad [c^{(e)}] = \begin{bmatrix} 8 & 4 \\ 4 & 8 \end{bmatrix}$$

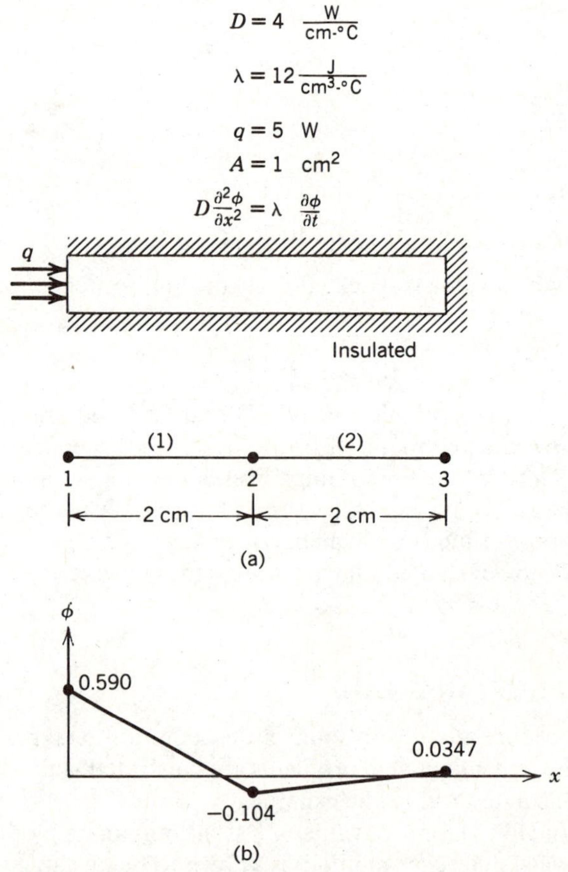

$$D = 4 \ \frac{W}{cm \cdot {}^\circ C}$$

$$\lambda = 12 \frac{J}{cm^3 \cdot {}^\circ C}$$

$$q = 5 \ W$$

$$A = 1 \ cm^2$$

$$D \frac{\partial^2 \phi}{\partial x^2} = \lambda \ \frac{\partial \phi}{\partial t}$$

Figure 14.5. Heat transfer in a one-dimensional insulated bar. (a) the bar and finite element grid. (b) The calculated temperatures for the first time step.

The global matrices are

$$[C] = \begin{bmatrix} 8 & 4 & 0 \\ 4 & 16 & 4 \\ 0 & 4 & 8 \end{bmatrix}, \quad [K] = \begin{bmatrix} 2 & -2 & 0 \\ -2 & 4 & -2 \\ 0 & -2 & 2 \end{bmatrix}$$

and

$$\{F\} = \begin{Bmatrix} 5 \\ 0 \\ 0 \end{Bmatrix}$$

where $\{F\}$ is obtained by noting that the heat input is a point source.

Constructing $[A]$ and $[P]$ using (14.41) yields

$$\begin{bmatrix} 9 & 3 & 0 \\ 3 & 18 & 3 \\ 0 & 3 & 9 \end{bmatrix} \{\Phi\}_b = \begin{bmatrix} 7 & 5 & 0 \\ 5 & 14 & 5 \\ 0 & 5 & 7 \end{bmatrix} \{\Phi\}_a + \begin{Bmatrix} 5 \\ 0 \\ 0 \end{Bmatrix} \qquad (14.45)$$

which has the general form

$$[A]\{\Phi\}_b = [P]\{\Phi\}_a + \{F^*\} \qquad (14.46)$$

Multiplying by $[A]^{-1}$ gives

$$\{\Phi\}_b = [A]^{-1}[P]\{\Phi\}_a + [A]^{-1}\{F^*\} \qquad (14.47)$$

Since $\{\Phi\}_a = \{0\}$ (initial values are zero)

$$\{\Phi\}_b = [A]^{-1}\{F^*\} \qquad (14.48)$$

The matrix, $[A]^{-1}$, is

$$[A]^{-1} = \begin{bmatrix} 0.1180 & -0.0208 & 0.0069 \\ -0.0208 & 0.0625 & -0.0208 \\ 0.0069 & -0.0208 & 0.1180 \end{bmatrix} \qquad (14.49)$$

and multiplying gives

$$\{\Phi\}_b = \begin{Bmatrix} 0.5900 \\ -0.1040 \\ 0.0347 \end{Bmatrix} \qquad (14.50)$$

These values are shown in Figure 14.5b. The calculated values violate physical reality. Heat is moving into the rod, but the temperature of node two is decreasing. This type of result occurs quite often when using the consistent formulation. The abnormality disappears after several time steps, but this is not helpful if one is interested in the temperature history near $t = 0$.

An explanation for the abnormal result goes as follows. The matrix $[A]$ in (14.45) consists entirely of positive coefficients; thus $[A]^{-1}$ must contain some negative coefficients in each column in order for $[A][A]^{-1} = [I]$. When $\{\Phi\}_a = \{0\}$,

$\{\Phi\}_b$ is just the first column of $[A]^{-1}$ multiplied by five and the negative value for node two is one of the results.

It would appear that one way to avoid the undesirable negative value is to increase the time step to a point where the negative coefficients in $\Delta t/2[K]$ exceeds the positive coefficients in $[C]$. The time step that accomplishes this, however, exceeds the time step necessary to prevent numerical oscillations. A procedure for avoiding this conflict is outlined in Chapter 15.

PROBLEMS

14.1 Solve the problem in Section 14.5 using a lumped capacitance matrix. Observe that the temperatures at nodes two and three always increase with time.

14.2 Solve the problem in Section 14.5 using a lumped capacitance matrix and $\theta=0$. Obtain the nodal temperatures for the first three time steps.

14.3 An insulated rod (Figure P14.3) is initially at 50°C. At time zero, the temperature at each end of the rod is fixed at 10°C. Obtain $[A]$ and $[P]$ for the four-element grid shown and modify them to account for the fact that the temperatures at nodes one and five are known, that is, the equations for nodes one and five must be eliminated. Use the consistent formulation and $\theta=\frac{1}{2}$. Do not solve for any of the temperature values. Use $\Delta t=1$ sec.

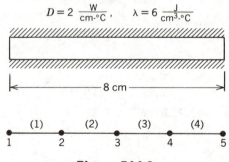

$$D = 2\ \frac{W}{cm\text{-}°C}, \qquad \lambda = 6\ \frac{J}{cm^3\text{-}°C}$$

Figure P14.3

14.4 Solve Problem 14.3 using the consistent formulation and $\theta=\frac{2}{3}$.

14.5 Solve Problem 14.3 using the lumped formulation and $\theta=\frac{1}{2}$.

14.6 Solve Problem 14.3 using the lumped formulation and $\theta=0$. Use a time step of 1 sec and calculate the nodal temperature for the first two time steps.

14.7 An insulated rod (Figure P14.7) is initially at 30°C when the left end is reduced to 5°C and the right end is insulated. Obtain $[A]$ and $[P]$ for the four-element grid shown and modify them to account for the fact that the temperature at node one is known; that is, the equation for node one must be eliminated. Use the consistent formulation and $\theta=\frac{1}{2}$. Do not solve for any of the temperature values. Use $\Delta t=1$ sec.

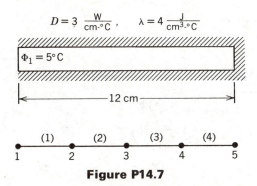

Figure P14.7

14.8 Solve Problem 14.7 using the consistent formulation and $\theta = 0$.

14.9 Solve Problem 14.7 using a lumped formulation and $\theta = \frac{2}{3}$.

14.10 Solve Problem 14.7 using a lumped formulation and $\theta = 0$. Use a time step of 1 sec and calculate the nodal temperatures for the first two time steps.

14.11 Solve the problem in Section 14.5 assuming that Φ_1 is fixed at 20°C while $\Phi_2 = \Phi_3 = 0$ initially. Modify $[A]$ and $[P]$ and solve for the nodal temperatures for three time steps. Use the consistent formulation and the central difference method with Δt equal to 1 sec.

14.12 Solve the problem in Section 14.5 assuming that $\Phi_1 = 10$°C while $\Phi_2 = \Phi_3 = 0$ initially. Use a lumped formulation and $\theta = \frac{2}{3}$. Modify $[A]$ and $[P]$ and solve for the nodal temperatures for three time steps. Use Δt equal to 1 sec.

Chapter 15

TIME-DEPENDENT FIELD PROBLEMS: PRACTICAL CONSIDERATIONS

The element matrices for the time-dependent field problems and the numerical solution of the system of differential equations were discussed in the previous chapter. One might assume that the solution process is a straightforward application of this information. Several numerical difficulties, however, arise during the solution process. The two most important are the failure of the calculated values to satisfy physical requirements and the problem of numerical oscillations. The first of these was observed in the calculations for heat input into the insulated rod (Section 14.5). Adding heat to the rod caused a nodal temperature to decrease. Numerical oscillations is a name given to the phenomenon where the calculated results oscillate around the correct solution. For one time step the calculated value is below the correct value. For the next time step, the calculated value is above the correct value. The stability of the calculations is also a concern, but there are many references which prove that the solutions in time using (14.39) are unconditionally stable when $\theta \geqslant \frac{1}{2}$.

The objective in this chapter is to establish the requirements necessary to satisfy the physical reality of the problem and to avoid numerical oscillations. We then use these requirements to make decisions relative to the element sizes and shapes which lend themselves to efficient and meaningful calculations.

15.1 PHYSICAL REALITY

The general form of the finite difference solution for the system of differential equations is

$$[A]\{\Phi\}_b = [P]\{\Phi\}_a + \{F^*\} \tag{15.1}$$

Multiplying by $[A]^{-1}$, we obtain

$$\{\Phi\}_b = [A]^{-1}[P]\{\Phi\}_a + [A]^{-1}\{F^*\} \tag{15.2}$$

The requirements imposed on $[A]$ and $[P]$ such that the physical reality of the problem is satisfied are determined by the $[A]^{-1}\{F^*\}$ term because $\{\Phi\}_a$ can always be assumed zero for the first time step.

Positive nonzero coefficients in $\{F^*\}$ come from heat input at a node or the

190

$$[A] = \begin{bmatrix} + & - & - & - & \cdot & \cdot & - \\ - & + & - & - & \cdot & \cdot & - \\ - & - & + & - & \cdot & \cdot & - \\ - & - & - & + & - & \cdot & - \\ \cdot & \cdot & \cdot & \cdot & \cdot & \cdot & \cdot \\ \cdot & \cdot & \cdot & \cdot & \cdot & \cdot & \cdot \\ - & - & \cdot & - & - & - & + \end{bmatrix}$$

Figure 15.1. The sign requirements for the entries of [A].

specifying of a nodal temperature that is greater than any of the other initial nodal values. In either case, it is expected that the surrounding nodes will increase in value. Let us analyze a single equation from the $[A]^{-1}\{F^*\}$ product, call it equation β. The equation is

$$\phi_\beta = A_{\beta 1}^{-1} F_1^* + A_{\beta 2}^{-1} F_2^* + \cdots + A_{\beta p}^{-1} F_p^* \tag{15.3}$$

Given that one or more of the F^* terms are positive, the only way to insure that ϕ_β increases is to require that each coefficient in $[A]^{-1}$ be positive. This requirement also insures that ϕ_β decreases if one or more of the F^* are negative.

It is not immediately clear what the positive coefficient requirement on $[A]^{-1}$ means relative to $[A]$. Maadooliat (1983) has shown that one way of satisfying this requirement is for the diagonal coefficients of $[A]$ to be positive and all of the off-diagonal entries to be negative. The signs of the coefficients in $[A]$ should be as shown in Figure 15.1.

Since the finite element analyst never sees $[A]^{-1}$, the requirement for satisfying physical reality is written in terms of $[A]$. Physical reality is satisfied when $[A]$ is a matrix whose diagonal entries are positive and whose off-diagonal entries are negative. The significance of this requirement relative to the lumped or consistent formulations and element shapes is discussed later in this chapter.

15.2 NUMERICAL OSCILLATIONS

Numerical oscillations in the values of $\{\Phi\}_b$ from one time step to the next are related to the eigenvalues of the matrix product $[A]^{-1}[P]$ (Myers, 1971). The possible situations involving the eigenvalues are

1. All eigenvalues positive; no oscillations, stable calculations.
2. Some eigenvalues negative but greater than -1; stable calculations, numerical oscillations.
3. One or more eigenvalues less than -1; unstable calculations.

The criterion for avoiding numerical oscillations is quite clear. All of the eigenvalues of $[A]^{-1}[P]$ must be positive. The question at hand is "How do we satisfy these criteria?." The discussion in this section is related to answering this question.

Let the eigenvalues of $[A]^{-1}[P]$ be denoted by β_i. An eigenvalue is the value of β that satisfies

$$\det([A]^{-1}[P]-\beta[I])=0 \tag{15.4}$$

where det ( ) indicates the determinant of the matrix within the parentheses. Since

$$\det([H][J])=\det([H])\det([J])$$

(15.4) is multiplied by $\det([A])$ to obtain

$$\det([A])\det([A]^{-1}[P]-\beta[I])=0 \tag{15.5}$$

or

$$\det([A][A]^{-1}[P]-\beta[A][I])=0 \tag{15.6}$$

and

$$\det([P]-\beta[A])=0 \tag{15.7}$$

which is in a more workable form because we have eliminated $[A]^{-1}$.

Hildebrand (1965) states that (15.7) has positive eigenvalues when both $[A]$ and $[P]$ are positive definite. The study of the eigenvalues of $[A]^{-1}[P]$ is now reduced to a study of $[A]$ and $[P]$, which are related to $[C]$ and $[K]$ by

$$[A]=[C]+\theta\,\Delta t[K] \tag{15.8}$$
$$[P]=[C]-(1-\theta)\,\Delta t[K] \tag{15.9}$$

Before analyzing $[A]$ and $[P]$, we state a fact developed by Fried (1979). The minimum eigenvalue for a global matrix, $[R]$, is greater than the minimum eigenvalue for all of its element matrices, that is,

$$\min_{e}(\beta_{\min}^{(e)})<\beta_{\min}^{[R]} \tag{15.10}$$

The positive definite characteristic of $[A]$ is easily established because $[C]$ is known to be positive definite and $[K]$ is at worst singular. The addition of a part of $[K]$ to $[C]$ does not change the characteristic that $[C]$ and $[A]$ is positive definite.

The requirement for avoiding numerical oscillations reduces to the analysis of $[P]$. The question to be answered is "How much of $[K]$ can be subtracted from $[C]$ before the resulting matrix is singular?"

Define a new element matrix $[p^{(e)}]$ given by

$$[p^{(e)}]=[c^{(e)}]-\alpha[k^{(e)}] \tag{15.11}$$

where $\alpha=(1-\theta)\,\Delta t$. We are looking for the value of α that makes $[p^{(e)}]$ singular, that is, makes the minimum eigenvalue zero. Using Fried's rule, if the minimum eigenvalue of $[p^{(e)}]$ is zero, then we know that $\beta_{\min}^{[P]}>0$ and $[P]$ is positive definite. The value of $\beta_{\min}^{(e)}$ is the value of β satisfying

$$\det([p^{(e)}]-\beta[I])=0 \tag{15.12}$$

but since we are looking for a singular matrix, we set $\beta=0$ and (15.12) becomes

$$\det([c^{(e)}]-\alpha[k^{(e)}])=0 \tag{15.13}$$

Numerical oscillations are avoided when

$$\Delta t < \frac{\alpha}{1-\theta} \tag{15.14}$$

and α is the smallest value satisfying

$$\det\left([c^{(e)}] - \alpha[k^{(e)}]\right) = 0 \tag{15.15}$$

The smallest value of α occurs in the smallest element of the grid. The estimate in (15.14) is (by experience) conservative for nonuniform grids, but it is still a good guide for selecting a numerical value for the time step. Equation 15.14 is an excellent estimate for uniform grids.

ILLUSTRATIVE EXAMPLE

Determine the time step necessary to avoid numerical oscillations when using a one-dimensional linear element and the lumped formulation.

The element matrices are

$$[c^{(e)}] = \frac{\lambda L}{2}\begin{bmatrix} 1 & 0 \\ 0 & 1 \end{bmatrix} \quad \text{and} \quad [k^{(e)}] = \frac{D}{L}\begin{bmatrix} 1 & -1 \\ -1 & 1 \end{bmatrix}$$

Applying (15.15), we obtain

$$\det\left(\frac{\lambda L}{2}\begin{bmatrix} 1 & 0 \\ 0 & 1 \end{bmatrix} - \frac{\alpha D}{L}\begin{bmatrix} 1 & -1 \\ -1 & 1 \end{bmatrix}\right) = 0$$

or

$$\det\begin{bmatrix} \left(\dfrac{\lambda L}{2} - \dfrac{\alpha D}{L}\right) & \dfrac{\alpha D}{L} \\ \dfrac{\alpha D}{L} & \left(\dfrac{\lambda L}{2} - \dfrac{\alpha D}{L}\right) \end{bmatrix} = 0$$

Evaluating the determinant gives

$$\left(\frac{\lambda L}{2} - \frac{\alpha D}{L}\right)^2 - \left(\frac{\alpha D}{L}\right)^2 = \frac{\lambda L^2}{4} - \alpha D = 0$$

and

$$\alpha = \frac{\lambda L^2}{4D}$$

The time step is given by (15.14)

$$\Delta t \leqslant \frac{\alpha}{1-\theta} \leqslant \frac{\lambda L^2}{4D(1-\theta)}$$

The fact that (15.14) is not valid when $\theta = 1$ has the following interpretation. When $\theta = 1$,

$$[P] = [C] - (1 - 1)\Delta t[K] = [C]$$

Since $[C]$ is positive definite, $[P]$ is positive definite and there are no numerical oscillations. One of the advantages of the backward difference method, $\theta = 1$, is that it is free of numerical oscillations.

We have established two criteria that should be satisfied when solving time-dependent field problems. The first specifies the signs of the entries in $[A]$ while the second requires that $[P]$ be positive definite. We now use these criteria to make some decisions about element size and formulations.

15.3   LUMPED OR CONSISTENT FORMULATION

There are two popular methods of formulating the element capacitance matrix: the consistent formulation and the lumped formulation. We now analyze these formulations using the criteria developed in the previous sections. We shall use a uniform grid of one-dimensional elements for the analysis. The analysis of two-dimensional grids proceeds in the same way with very similar results.

15.3.1   Consistent Formulation

The element matrices are

$$[c^{(e)}] = \frac{\lambda L}{6}\begin{bmatrix} 2 & 1 \\ 1 & 2 \end{bmatrix} \quad \text{and} \quad [k^{(e)}] = \frac{D}{L}\begin{bmatrix} 1 & -1 \\ -1 & 1 \end{bmatrix}$$

and the equation for row s in $[A]$ is

$$\left(\frac{\lambda L}{6} - \theta \Delta t \frac{D}{L}\right)\Phi_{s-1} + \left(\frac{4\lambda L}{6} + \theta \Delta t \frac{2D}{L}\right)\Phi_s + \left(\frac{\lambda L}{6} - \theta \Delta t \frac{D}{L}\right)\Phi_{s+1} \quad (15.16)$$

The diagonal coefficient in $[A]$ is positive for all values of Δt. The off-diagonal values are negative only if

$$\frac{\lambda L}{6} - \theta \Delta t \frac{D}{L} < 0$$

or

$$\Delta t \gtrless \frac{\lambda L^2}{6D\theta} \quad (15.17)$$

The positive definite characteristic of $[P]$ exists only if

$$\det \begin{bmatrix} \left(\frac{\lambda L}{3} - \alpha \frac{D}{L}\right) & \left(\frac{\lambda L}{6} + \alpha \frac{D}{L}\right) \\ \left(\frac{\lambda L}{6} + \alpha \frac{D}{L}\right) & \left(\frac{\lambda L}{3} - \alpha \frac{D}{L}\right) \end{bmatrix} = 0 \quad (15.18)$$

which occurs if

$$\left(\frac{\lambda L}{3} - \alpha \frac{D}{L}\right)^2 - \left(\frac{\lambda L}{6} + \alpha \frac{D}{L}\right)^2 = 0$$

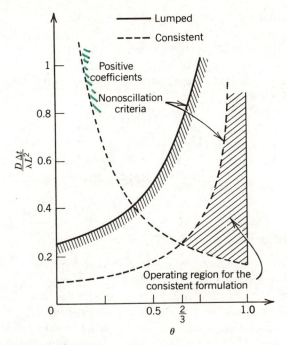

Figure 15.2. The operating regions for lumped and consistent finite element formulations in one dimension.

and

$$\alpha = \frac{\lambda L^2}{12D} \tag{15.19}$$

which gives

$$\Delta t < \frac{\lambda L^2}{12D(1-\theta)} \tag{15.20}$$

Equations (15.17) and (15.20) are two criteria for the time step. The first insures that physical reality is satisfied; the second insures that the calculated values do not oscillate. The two criteria are shown in Figure 15.2. The region in which both are satisfied is relatively small and exists for $\theta \geqslant \frac{2}{3}$.

15.3.2 Lumped Formulation

The element matrices for the one-dimensional lumped formulation were given and analyzed in the example problem in the previous section. The time step must satisfy

$$\Delta t < \frac{\lambda L^2}{4D(1-\theta)} \tag{15.21}$$

if $[P]$ is to be positive definite.

The general equation for row s in $[A]$ is

$$-\left(\theta\,\Delta t\,\frac{D}{L}\right)\Phi_{s-1}+\left(\frac{\lambda L}{2}+\theta\,\Delta t\,\frac{D}{L}\right)\Phi_s-\left(\theta\,\Delta t\,\frac{D}{L}\right)\Phi_{s+1} \qquad (15.22)$$

The off-diagonal coefficients are negative and the diagonal coefficient is positive for all values of Δt; thus

$$\Delta t > 0 \qquad (15.23)$$

satisfies the requirement for $[A]$.

The criteria (15.21) and (15.23) are also shown in Figure 15.2. It is readily apparent that the operating region for the lumped formulation is much larger than that for the consistent formulation.

15.3.3 Summary

The consistent finite element formulation is accompanied by restraints on the time step that significantly reduce its operating range. The method cannot be used with Euler's forward difference method or with the central difference method without violating physical reality and being accompanied by numerical oscillations. The lumped formulation has a full operating range for θ and allows one to use larger time steps. If a solution is to be obtained using θ in the range of $\frac{1}{2}$ to $\frac{2}{3}$, then the lumped formulation is the only choice available.

15.4 TWO-DIMENSIONAL ELEMENTS

The criteria for $[A]$ and $[P]$ is now used to develop some useful information for the two-dimensional triangular and rectangular elements. The criteria on $[A]$ requires that all of the off-diagonal coefficients in $[k^{(e)}]$ be negative. The criteria for $[P]$ is used to calculate some time step estimates.

15.4.1 The Rectangular Element

The element stiffness matrix for the bilinear rectangular element is given by (7.55) as

$$[k^{(e)}]=\frac{D_x a}{6b}\begin{bmatrix} 2 & -2 & -1 & 1 \\ -2 & 2 & 1 & -1 \\ -1 & 1 & 2 & -2 \\ 1 & -1 & -2 & 2 \end{bmatrix}+\frac{D_y b}{6a}\begin{bmatrix} 2 & 1 & -1 & -2 \\ 1 & 2 & -2 & -1 \\ -1 & -2 & 2 & 1 \\ -2 & -1 & 1 & 2 \end{bmatrix}$$

If each off-diagonal coefficient is to be negative, then

$$-\frac{2D_x a}{6b}+\frac{D_y b}{6a}<0 \qquad (15.25)$$

and

$$\frac{D_x a}{6b} - \frac{2D_y b}{6a} < 0 \qquad (15.26)$$

These inequalities translate into the requirement that

$$\sqrt{\frac{D_y}{2D_x}} < \frac{a}{b} < \sqrt{\frac{2D_y}{D_x}} \qquad (15.27)$$

The aspect ratio a/b has to be within certain limits.

The allowable time step for a square with $D_x = D_y = D$ and a lumped formulation is related to

$$\det\left(\frac{\lambda A}{4} \begin{bmatrix} 1 & 0 & 0 & 0 \\ 0 & 1 & 0 & 0 \\ 0 & 0 & 1 & 0 \\ 0 & 0 & 0 & 1 \end{bmatrix} - \frac{\alpha D}{6} \begin{bmatrix} 4 & -1 & -2 & -1 \\ -1 & 4 & -1 & -2 \\ -2 & -1 & 4 & -1 \\ -1 & -2 & -1 & 4 \end{bmatrix} \right) = 0$$

which after a considerable amount of algebra yields

$$\alpha = \frac{\lambda A}{4D} \qquad (15.28)$$

and

$$\Delta t < \frac{A\lambda}{4D(1-\theta)} \qquad (15.29)$$

15.4.2 The Triangular Element

The calculations for the triangle are not within reason unless we assume that $D_x = D_y = D$. If we use the triangle in Figure 15.3, $[k^{(e)}]$ is

$$[k^{(e)}] = \frac{D}{4A} \begin{bmatrix} d^2 + (c-b)^2 & -d^2 - c(c-b) & b(c-b) \\ -d^2 - c(c-b) & d^2 + c^2 & -cb \\ b(c-b) & -cb & b^2 \end{bmatrix} \qquad (15.30)$$

The diagonal coefficients are always positive so that we need to investigate only the off-diagonal entries. The inequalities are

$$-d^2 - c(c-b) < 0 \qquad (15.31)$$

$$b(c-b) < 0 \qquad (15.32)$$

$$-cb < 0 \qquad (15.33)$$

The last inequality states that both c and b must be positive, which means that the angle α (Figure 15.3) cannot be greater than 90°. The middle inequality, (15.29), is satisfied only if $c \leqslant b$, which means that the angle θ (Figure 15.3) cannot be

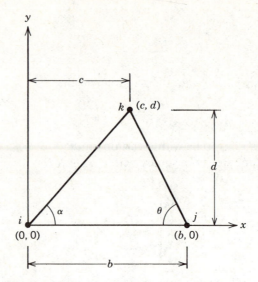

Figure 15.3. A general triangular element.

greater than 90°. The first inequality is equivalent to

$$d^2 > c(b-c)$$

which gives the aspect ratio for the triangle.

The solution of the inequalities for the isotropic triangular element can be summarized by the statement: *No interior angle of the triangle should exceed 90°*.

The allowable time step is calculated for a right triangle to allow a comparison with the results for the square. The element matrices for a triangle with a side of length b $(A = b^2/2)$ are

$$[c^{(e)}] = \frac{\lambda A}{3} \begin{bmatrix} 1 & 0 & 0 \\ 0 & 1 & 0 \\ 0 & 0 & 1 \end{bmatrix}, \qquad [k^{(e)}] = \frac{D}{2} \begin{bmatrix} 2 & -1 & -1 \\ -1 & 1 & 0 \\ -1 & 0 & 1 \end{bmatrix}$$

For $[P]$ to be positive definite

$$\det \begin{bmatrix} \left(\dfrac{\lambda A}{3} - \alpha D\right) & -\dfrac{D\alpha}{2} & -\dfrac{D\alpha}{2} \\[2ex] -\dfrac{D\alpha}{2} & \left(\dfrac{\lambda A}{3} - \alpha D\right) & 0 \\[2ex] -\dfrac{D\alpha}{2} & 0 & \left(\dfrac{\lambda A}{3} - \alpha D\right) \end{bmatrix} = 0 \qquad (15.34)$$

and

$$\alpha = \frac{2\lambda A}{9D} \qquad (15.35)$$

which yields a time step of

$$\Delta t < \frac{2\lambda A}{9D(1-\theta)} \tag{15.36}$$

Remember that the result in (15.36) is for a right triangle with both sides having a length b. The multiplying coefficient changes with the ratio of the sides and the magnitude of the interior angles.

15.4.3 Summary

Several conclusions can be drawn from the calculations done in this section. First, the element shape is important. Triangular elements should not have any interior angles greater than $90°$, and the aspect ratio, a/b, for a bilinear rectangular element has to satisfy (15.27). Second, the maximum allowable time step for a right triangle whose sides have a length b is less than one-half the value for a square of side b. This fact comes from comparing (15.29) and (15.36). Since the area of the triangle is one-half that of the square, (15.36) becomes $\Delta t \leqslant A^*\lambda/9D(1-\theta)$ compared with $\Delta t \leqslant A^*\lambda/4D(1-\theta)$ for the rectangle, where A^* represents the area of the rectangle. This comparison shows that rectangular or square elements should not be divided into triangles. The triangular element should only be used to model irregular boundaries.

The third important fact is that the length dimension is always in the numerator of the time step calculation. One way of increasing the time step is to increase the size of the element. A good rule of thumb is "Make the elements as large as is reasonably possible." Decreasing the element size while keeping the time step the same does not improve the calculated results. The time step must be reduced as the size of the element is reduced.

15.5 DERIVATIVE BOUNDARY CONDITIONS

The last topic of this chapter concerns the derivative boundary conditions. These boundary conditions as well as the G term in (7.1) produce element matrices that add to $[k_D^{(e)}]$. For example,

$$[k_G^{(e)}] = \frac{GA}{12}\begin{bmatrix} 2 & 1 & 1 \\ 1 & 2 & 1 \\ 1 & 1 & 2 \end{bmatrix} \tag{15.37}$$

for the fin problem of Chapter 11 and

$$[k_M^{(e)}] = \frac{ML_{ij}}{6}\begin{bmatrix} 2 & 1 & 0 \\ 1 & 2 & 0 \\ 0 & 0 & 0 \end{bmatrix} \tag{15.38}$$

for the convection boundary condition from side ij of a triangular element (Chapter 9).

Each of the matrices in (15.37) and (15.38) have positive off-diagonal coefficients. If G or M become large, positive off-diagonal entries occur in $[A]$. The matrices associated with $[k_G^{(e)}]$ and $[k_M^{(e)}]$ should be lumped to avoid this difficulty. Another way of stating this fact is that the step function type of shape function should be used with every Galerkin residual integral except the one associated with the second-derivative terms.

The lumping of the matrices in (15.37) and (15.38) produce results of the type

$$[k_G^{(e)}] = \frac{GA}{3} \begin{bmatrix} 1 & 0 & 0 \\ 0 & 1 & 0 \\ 0 & 0 & 1 \end{bmatrix} \tag{15.39}$$

and

$$[k_M^{(e)}] = \frac{ML_{ij}}{2} \begin{bmatrix} 1 & 0 & 0 \\ 0 & 1 & 0 \\ 0 & 0 & 0 \end{bmatrix} \tag{15.40}$$

The lumping of matrices associated with convection heat loss is consistent with finite difference methods, which have used this approach for many years.

PROBLEMS

15.1 Calculate the time step necessary to avoid numerical oscillations when using the one-dimensional linear element and a lumped formulation to analyze convection heat loss from a fin. The element matrices are $[c^{(e)}]$ as given by (14.26) and

$$[k^{(e)}] = \frac{D}{L} \begin{bmatrix} 1 & -1 \\ -1 & 1 \end{bmatrix} + \frac{hPL}{2} \begin{bmatrix} 1 & 0 \\ 0 & 1 \end{bmatrix}$$

The parameters are

(a) $D = 2 \dfrac{\text{W}}{\text{cm-°C}}$, $hP = 0.10 \dfrac{\text{W}}{\text{cm-°C}}$,

$L = 0.5$ cm, $\lambda = 6 \dfrac{\text{J}}{\text{cm}^3\text{-°C}}$

(b) $D = 3 \dfrac{\text{W}}{\text{cm-°C}}$, $hP = 0.05 \dfrac{\text{W}}{\text{cm-°C}}$,

$L = 1.5$ cm, $\lambda = 4 \dfrac{\text{J}}{\text{cm}^3\text{-°C}}$

(c) $D = 2 \dfrac{\text{W}}{\text{cm-°C}}$, $hP = 0.15 \dfrac{\text{W}}{\text{cm-°C}}$,

$L = 2$ cm, $\lambda = 5 \dfrac{\text{J}}{\text{cm}^3\text{-°C}}$

15.2 The time step necessary to avoid numerical oscillations in a one-dimensional lumped formulation using $\theta = \frac{1}{2}$ is $\Delta t < \lambda L^2/2D$. Assume that $\lambda = 1$ and show that $[p^{(e)}]$ has a negative eigenvalue when $\Delta t > \lambda L^2/2D$.

15.3 Calculate the time step criteria for a equilateral triangle whose base is b. Assume a lumped formulation. Note that the value is greater than the allowable value for a right triangle given by (15.36).

15.4 Develop an equation similar to (15.36) for the consistent formulation; that is, $[c^{(e)}]$ is given by (14.17). Assume a right triangle with both sides having a length b.

15.5 Obtain inequalities similar to those in (15.31) to (15.33) for the situation where $D_x \neq D_y$. Use the triangle in Figure 15.3. Show that the 90° rule is still valid (no interior angle should be greater than 90°).

Chapter 16

COMPUTER PROGRAM FOR TWO-DIMENSIONAL FIELD PROBLEMS

The finite element method involves large systems of linear equations and has a limited usefulness if a digital computer is not available. A computer program for solving the two-dimensional steady-state field equation is discussed in this chapter. It is written in standard FORTRAN IV and it is written for educational use rather than commercial problem solving. The program contains several diagnostic checks to locate errors that are common to the beginning user. The primary advantage of the program from a teaching point of view is that the program options are at a minimum and the input data is limited to one type of problem. These characteristics allow the input data to be discussed in one class period.

The program is not accompanied by a grid-generating program. This author has found that students do not understand or appreciate grid-generating programs until they have generated the element data by hand.

16.1 THE EQUATIONS

The computer program TDFIELD and its subroutines solve problems governed by the two-dimensional differential equation

$$D_x \frac{\partial^2 \phi}{\partial x^2} + D_y \frac{\partial^2 \phi}{\partial y^2} - G\phi + Q = 0 \tag{16.1}$$

with the boundary conditions

$$\phi = \phi(\Gamma_1) \tag{16.2}$$

on Γ_1 and

$$D_x \frac{\partial \phi}{\partial x} \cos \theta + D_y \frac{\partial \phi}{\partial y} \sin \theta = -M\phi_b + S \tag{16.3}$$

on Γ_2, where Γ_1 and Γ_2 comprise the boundary of the region.

The computer program allows the user to analyze grids consisting of linear triangular elements and bilinear rectangular elements. The coefficients D_x, D_y, G, and Q may differ between elements. The coefficients M and S may vary along the boundary. Equations (7.36) and (7.38) are used for $[k_D^{(e)}]$ and $[k_G^{(e)}]$ for the tri-

202

angular element while (7.49) and (7.55) are used for the rectangular element. The element force vectors are given by (7.40) and (7.46). The boundary condition contributions to $[k_M^{(e)}]$ and $\{f_S^{(e)}\}$ are those given in (9.28) through (9.34) and (9.18) and (9.19). The gradient values are calculated using (7.34) and (7.51).

16.2 PROGRAM LISTING

A listing of the main program **TDFIELD** and the four subroutines **ELSTMF, MODIFY, DCMPBD,** and **SLVBD** are given on the following pages. The program logic is similar to the flow chart in Figure 4.6.

The input data for **TDFIELD** and **MODIFY** are described by comment cards in the respective programs. The input data is free-format (independent of the column location on a card). The numbers must be separated by a space or a comma. The data organization is illustrated for a sample problem later in this chapter.

The input and output units are defined by integer variables within the **READ** and **WRITE** statements. The numerical values of **IN** and **IO** are defined at the beginning of each program. The values of **IN** and **IO** must be changed to match the values for the computer system being used.

This program stores the coefficients of $\{\Phi\}$, $\{F\}$, and $[K]$ in a single vector $\{A\}$ denoted by $A(\ )$ in the program. Vector storage eliminates the need to change the dimension of $[K]$ each time a new problem is solved. The vector storage concept can be illustrated using the system of equations

$$\begin{bmatrix} 2 & 1 & 0 & 1 & 0 & 0 \\ 1 & 6 & 1 & 2 & 2 & 0 \\ 0 & 1 & 2 & 0 & 1 & 0 \\ 1 & 2 & 0 & 6 & 2 & 1 \\ 0 & 2 & 1 & 2 & 6 & 1 \\ 0 & 0 & 0 & 1 & 1 & 2 \end{bmatrix} \begin{Bmatrix} \Phi_1 \\ \Phi_2 \\ \Phi_3 \\ \Phi_4 \\ \Phi_5 \\ \Phi_6 \end{Bmatrix} = \begin{Bmatrix} 900 \\ 5500 \\ 2600 \\ 4900 \\ 6500 \\ 2000 \end{Bmatrix} \tag{16.4}$$

Conventional storage of this system would require 48 memory spaces, 36 for $[K]$ and 6 each for $\{\Phi\}$ and $\{F\}$. The system has a bandwidth of four and is symmetric; therefore, only part of the coefficients in $[K]$ are needed to obtain a solution. These coefficients can be stored in the rectangular array

$$\begin{bmatrix} 2 & 1 & 0 & 1 \\ 6 & 1 & 2 & 2 \\ 2 & 0 & 1 & 0 \\ 6 & 2 & 1 & x \\ 6 & 1 & x & x \\ 2 & x & x & x \end{bmatrix} \tag{16.5}$$

where the x's indicate nonexistent numbers. The first column of (16.5) is the main diagonal of $[K]$, the second column is the first off-diagonal (upper), and so forth.

$$
\left\{
\begin{array}{c}
\Phi_1 \\
\Phi_2 \\
\Phi_3 \\
\Phi_4 \\
\Phi_5 \\
\Phi_6 \\
\hline
900 \\
5500 \\
2600 \\
4900 \\
6500 \\
2000 \\
\hline
2 \\
6 \\
2 \\
6 \\
6 \\
2 \\
\hline
1 \\
1 \\
0 \\
2 \\
1 \\
\hline
0 \\
2 \\
1 \\
1 \\
\hline
1 \\
2 \\
0
\end{array}
\right\}
$$

Figure 16.1.   Storage of the system of equations in the *A* vector.

When using a vector storage, the nodal values $\{\Phi\}$ are located at the top followed by $\{F\}$ and then the columns of $[K]$ as they appear in (16.5). The storage of (6.4) using a one-dimensional array is illustrated in Figure 16.1.

The location of the first coefficient of $\{\Phi\}$, $\{F\}$, and $[K]$ is calculated within the program as well as the length of $A(\ )$ needed for a particular problem. All of the calculations associated with the solution of $[K]\{\Phi\}=\{F\}$ are done within $\{A\}$.

The derivative boundary conditions occur on one or more sides of an element. The sides have been numbered to facilitate the input. These numbers are shown in Figure 16.2.

The subroutines perform the following functions.

Subroutine ELSTMF. The subroutine **ELSTMF** evaluates the element stiffness matrix and force vector using the equations referenced in the previous section. The element matrices can be printed to allow students to check hand calculations.

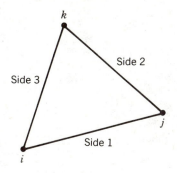

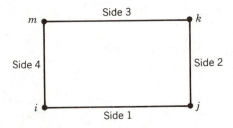

Figure 16.2. Procedure for denoting the sides of the element.

Subroutine MODIFY. The subroutine **MODIFY** incorporates the specified nodal values into the system of equations using the method of deletion of rows and columns (see Appendix III). The subroutine is also used to add point source or sink values directly to $\{F\}$ as discussed in Section 9.3. It is important that the user realize that this subroutine has **READ** statements.

Subroutine DCMPBD. The subroutine **DCMPBD** decomposes the global stiffness matrix $[K]$ into an upper triangular form using the method of Gaussian elimination (Conte and deBoor, 1980). This subroutine assumes that $[K]$ is symmetric and only those elements within the bandwidth and on or above the main diagonal are stored. The programming logic is not easy to follow because the coefficients of $[K]$ are stored in a vector rather than a two-dimensional array.

Subroutine SLVBD. The subroutine **SLVBD** is a companion program to **DCMPBD**. This subroutine decomposes the global force vector, $\{F\}$, and solves the system of equations using back substitution. The solution of the system of equations is separated into two subroutines so that they can be used to solve time-dependent problems where each new time step requires the decomposition of $\{F\}$ but may not require that $[K]$ be converted into an upper triangular form.

```
              PROGRAM TDFIELD (INPUT,OUTPUT,TAPE60=INPUT,TAPE61=OUTPUT)
              COMMON/ELMATX/ESM(4,4),EF(4),X(4),Y(4),KL
              COMMON/MATL/DXE,DYE,GE,QE
              COMMON/HCV/IDBC(50,2),DBC(50,2),NDBC
    5         COMMON/AV/A(5000),JGF,JGSM,NP,NBW
              COMMON/TLE/TITLE(20)
              DIMENSION NEL(300,4),NMTL(300),XC(250),YC(250)
              DIMENSION NS(4),PHI(4),GDX(4),GDY(4),GDN(4)
              DIMENSION DX(5),DY(5),G(5),Q(5),GRDC(5,2)
   10         DIMENSION B(3),C(3),ICK(250)
              DATA GRDC /-1.,-1.,1.,-1.,-1.,1.,1.,1.,-1.,-1./
              DATA IN/60/,IO/61/,IFE/0/,VOL/0./
      C
      C*********
   15 C*********
      C
      C  DEFINITION OF THE INPUT PARAMETERS
      C
      C*********
   20 C*********
      C
      C  TITLE AND PARAMETERS
      C
      C         TITLE - A DESCRIPTIVE STATEMENT OF THE PROBLEM
   25 C                 BEING SOLVED
      C         NP - NUMBER OF EQUATIONS (ALSO NUMBER OF NODES)
      C         NE - NUMBER OF ELEMENTS
      C         NCOEF - NUMBER OF SETS OF EQUATION COEFFICIENTS
      C                 MAXIMUM OF FIVE
   30 C         NDBC - NUMBER OF ELEMENT SIDES WITH A DERIVATIVE
      C                 BOUNDARY CONDITION
      C         ITYP - TYPE OF FIELD PROBLEM BEING RUN
      C                 1 - TORSION PROBLEM
      C                 2 - IDEAL FLUID, STREAMLINE FLOW
   35 C                 3 - IDEAL FLUID, POTENTIAL FLOW
      C                 4 - GROUNDWATER FLOW
      C                 5 - HEAT TRANSFER
      C         IPLVL - PRINT LEVEL
      C                 0 - DON'T WRITE THE ELEMENT MATRICES
   40 C                 1 - WRITE THE ELEMENT MATRICES
      C                     IPLVL IS SET TO ZERO WHEN NE EXCEEDS 10
      C
      C  EQUATION COEFFICIENTS
      C     THE NUMBER OF SETS MUST EQUAL NCOEF
   45 C
      C         DX(I) - MATERIAL PROPERTY IN THE X DIRECTION
      C         DY(I) - MATERIAL PROPERTY IN THE Y DIRECTION
      C         G(I)  - COEFFICIENT MULTIPLYING PHI IN THE DIFF. EQUATION
      C         Q(I)  - CONSTANT COEFFICIENT IN THE DIFF. EQUATION
   50 C
      C  NODAL COORDINATE VALUES
      C
      C         XC(I) - X COORDINATES OF THE NODES
      C         YC(I) - Y COORDINATES OF THE NODES
   55 C                 THE COORDINATES MUST BE IN NUMERICAL SEQUENCE
      C                 RELATIVE TO THE NODE NUMBERS
      C
```

```
       C   ELEMENT DATA
       C
       C         N - ELEMENT NUMBER
       C         NMTL - INTEGER SPECIFYING THE EQUATION COEFFICIENT SET
       C         NEL(N,1) - NUMERICAL VALUE OF NODE I
       C         NEL(N,2) - NUMERICAL VALUE OF NODE J
       C         NEL(N,3) - NUMERICAL VALUE OF NODE K
       C         NEL(N,4) - NUMERICAL VALUE OF NODE M
       C                    NEL(N,4) IS SET EQUAL TO ZERO FOR
       C                    THE TRIANGULAR ELEMENT
       C
       C   DERIVATIVE BOUNDARY CONDITIONS
       C     THE NUMBER OF VALUES MUST EQUAL NDBC
       C
       C         IDBC(I,1) - ELEMENT NUMBER WITH A DERIVATIVE
       C                         BOUNDARY CONDITION
       C         IDBC(I,2) - SIDE OF THE ELEMENT WITH THE
       C                         DERIVATIVE BOUNDARY CONDITION
       C         DBC(I,1) - (M COEFFICIENT)*(LENGTH OF THE SIDE)
       C         DBC(I,2) - (S COEFFICIENT)*(LENGTH OF THE SIDE)
       C
       C
       C   DATA IS READ BY THE SUBROUTINE MODIFY
       C
       C*********
       C*********
       C
       C   DATA INPUT SECTION OF THE PROGRAM
       C
       C*********
       C*********
       C
       C   INPUT OF THE TITLE CARD AND CONTROL PARAMETERS
       C
             READ(IN,3) TITLE
           3 FORMAT(20A4)
             READ(IN,*) NP,NE,NCOEF,NDBC,ITYP,IPLVL
             IF(NE.GT.10) IPLVL=0
             IF(IPLVL.GT.1) IPLVL=0
       C
       C   COMPARISON OF NP, NE, NDBC, AND ITYP WITH
       C     THE VALUES IN THE DIMENSION STATEMENTS
       C
             ISTOP=0
       C   CHECK OF NP
             IF(NP.LE.250) GOTO600
             WRITE(IO,10)
          10 FORMAT(10X,27HNUMBER OF NODES EXCEEDS 250/
            +/10X,16HINPUT TERMINATED)
             ISTOP=1
       C   CHECK OF NE
         600 IF(NE.LE.300) GOTO601
             WRITE(IO,2)
           2 FORMAT(10X,30HNUMBER OF ELEMENTS EXCEEDS 300/
            +10X,16HINPUT TERMINATED)
             ISTOP=1
       C   CHECK OF NDBC
```

```
115        601   IF(NDBC.LE.50) GOTO602
                 WRITE(10,47)
           47    FORMAT(10X,34HDERIVATIVE BOUNDARY CONDITION DATA/
                 +23HEXCEEDS DIMENSION OF 50/10X,16HINPUT TERMINATED)
                 ISTOP=1
120        C   CHECK OF ITYP
           602   IF(ITYP.LE.5) GOTO603
                 WRITE(10,101)
           101   FORMAT(10X,14HITYP EXCEEDS 5/10X,
                 +16HINPUT TERMINATED)
125              ISTOP=1
           C   CHECK OF NCOEF
           603   IF(NCOEF.LE.5) GOTO109
                 WRITE(10,604)
           604   FORMAT(10X,15HNCOEF EXCEEDS 5/10X,
130              +16HINPUT TERMINATED)
                 ISTOP=1
           109   IF(ISTOP.EQ.1) STOP
           C
           C   INPUT OF EQUATION COEFFICIENTS AND THE NODAL
135        C       COORDINATES
           C
                 READ(IN,*)  (DX(I),DY(I),G(I),Q(I),I=1,NCOEF)
                 READ(IN,*)  (XC(I),I=1,NP)
                 READ(IN,*)  (YC(I),I=1,NP)
140        C
           C   OUTPUT OF THE TITLE AND PARAMETERS
           C
                 WRITE(10,4)  TITLE,NP,NE,ITYP,IPLVL
           4     FORMAT(1H1///10X,20A4//10X,5HNP = ,I5/10X,5HNE = ,I5/
145              +10X,8HITYP  = ,I2/10X,8HIPLVL = ,I2)
           C
           C   OUTPUT OF THE EQUATION COEFFICIENTS
           C
                 WRITE(10,48)
150        48    FORMAT(//10X,21HEQUATION COEFFICIENTS,/10X,
                 +8HMATERIAL/13X,3HSET,8X,2HDX,13X,2HDY,13X,1HG,14X,1HQ)
                 WRITE(10,16)  (I,DX(I),DY(I),G(I),Q(I),I=1,NCOEF)
           16    FORMAT(14X,I2,4E15.5)
           C
155        C   OUTPUT OF THE NODAL COORDINATES
           C
                 WRITE(10,11)
           11    FORMAT(//10X,17HNODAL COORDINATES/10X,
                 +4HNODE,5X,1HX,14X,1HY)
160              WRITE(10,12)  (I,XC(I),YC(I),I=1,NP)
           12    FORMAT(10X,I4,2E15.5)
           C
           C   INPUT AND ECHO PRINT OF THE ELEMENT NODAL DATA
           C
165              WRITE(10,8)  TITLE
           8     FORMAT(1H1///10X,20A4//10X,12HELEMENT DATA/
                 +15X,3HNEL,4X,4HNMTL,4X,12HNODE NUMBERS)
                 NID=0
                 DO9KK=1,NE
170              READ(IN,*) N,NMTL(KK),(NEL(N,I),I=1,4)
                 IF((N-1).NE.NID) WRITE(10,17) N
```

```
17      FORMAT(10X,7HELEMENT,I4,16H NOT IN SEQUENCE)
        NID=N
        IF(NEL(N,4).EQ.0)WRITE(IO,7)N,NMTL(KK),(NEL(N,I),I=1,3)
      9 IF(NEL(N,4).NE.0) WRITE(IO,7) N,NMTL(KK),(NEL(N,I),I=1,4)
7       FORMAT(15X,I3,5X,I3,2X,4I4)
C
C  INPUT AND ECHO PRINT OF THE DERIVATIVE
C      BOUNDARY CONDITION DATA
C
        IF(NDBC.EQ.0) GOTO72
        WRITE(IO,49)
49      FORMAT(//10X,34HDERIVATIVE BOUNDARY CONDITION DATA/
       +15X,7HELEMENT,4X,4HSIDE,7X,2HML,13X,2HSL)
        DO451=1,NDBC
        READ(IN,*) IDBC(I,1),IDBC(I,2),DBC(I,1),DBC(I,2)
45      WRITE(IO,71) IDBC(I,1),IDBC(I,2),DBC(I,1),DBC(I,2)
71      FORMAT(15X,I4,9X,I1,2E15.5)
C*********
C*********
C
C  ANALYSIS OF THE NODE NUMBERS
C
C*********
C*********
C
C  INITIALIZATION OF A CHECK VECTOR
C
72      DO5001=1,NP
500     ICK(I)=0
C
C  CHECK TO SEE IF ANY NODE NUMBER EXCEEDS NP
C
        DO5011=1,NE
        KL=4
        IF(NEL(I,4).EQ.0) KL=3
        DO502J=1,KL
        K=NEL(I,J)
        ICK(K)=1
502     IF(K.GT.NP) WRITE(IO,503) J,I,NP
503     FORMAT(/10X,4HNODE,I4,11H OF ELEMENT,I4,
       +13H EXCEEDS NP =,I4)
501     CONTINUE
C
C  CHECK TO SEE IF ALL NODE NUMBERS THROUGH
C      NP ARE INCLUDED
C
        DO5051=1,NE
505     IF(ICK(I).EQ.0) WRITE(IO,506) I
506     FORMAT(/10X,4HNODE,I4,15H DOES NOT EXIST)
C
C*********
C*********
C
C  CREATION AND INITIALIZATION OF THE A VECTOR
C
C*********
C*********
```

```
                C
230             C   CALCULATION OF THE BANDWIDTH
                C
                      INBW=0
                      NBW=0
                      DO20KK=1,NE
235                   KL=4
                      IF(NEL(KK,4).EQ.0) KL=3
                      DO25I=1,KL
                   25 NS(I)=NEL(KK,I)
                      LK=KL-1
240                   DO21I=1,LK
                      IJ=I+1
                      DO21J=IJ,KL
                      NB=IABS(NS(I)-NS(J))
                      IF(NB.EQ.0) WRITE(10,26) KK
245                26 FORMAT(/10X,7HELEMENT,I3,18HHAS TWO NODES WITH/
                     +10X,25HWITH THE SAME NODE NUMBER)
                      IF(NB.LE.NBW) GOTO21
                      INBW=KK
                      NBW=NB
250                21 CONTINUE
                   20 CONTINUE
                      NBW=NBW+1
                      WRITE(10,27) NBW,INBW
                   27 FORMAT(//10X,12HBANDWIDTH IS,I4,11H IN ELEMENT,I4)
255             C
                C   CALCULATION OF POINTERS AND INITIALIZATION OF
                C       THE COLUMN VECTOR A( )
                C
                      JGF=NP
260                   JGSM=JGF+NP
                      JEND=JGSM+NP*NBW
                      IF(JEND.GT.5000) GOTO22
                      DO24I=1,JEND
                   24 A(I)=0.0
265                   GOTO30
                   22 WRITE(10,23)
                      STOP
                   23 FORMAT(10X,30HDIMENSION OF A VECTOR EXCEEDED)
                C********
270             C********
                C
                C   GENERATION OF THE SYSTEM OF EQUATIONS
                C
                C********
275             C********
                   30 DO32KK=1,NE
                      KL=4
                      IF(NEL(KK,4).EQ.0) KL=3
                C
280             C   RETRIEVAL OF NODAL COORDINATES AND NODE NUMBERS
                C
                      DO31I=1,KL
                      NS(I)=NEL(KK,I)
                      J=NS(I)
285                   X(I)=XC(J)
                   31 Y(I)=YC(J)
```

210

```
      C
      C   ELEMENT COEFFICIENTS
      C
290         II=NMTL(KK)
            DXE=DX(II)
            DYE=DY(II)
            GE=G(II)
            QE=Q(II)
295   C
      C   CALCULATION OF THE ELEMENT STIFFNESS MATRIX
      C     AND ELEMENT FORCE VECTOR
      C
            CALL ELSTMF(KK,IPLVL)
300   C
      C   DIRECT STIFFNESS PROCEDURE
      C
            DO33I=1,KL
            II=NS(I)
305         A(JGF+II)=A(JGF+I.I)+EF(I)
            DO34J=1,KL
            JJ=NS(J)+1-II
            IF(JJ.LE.0) GOTO34
            J1=JGSM+(JJ-1)*NP+II-(JJ-1)*(JJ-2)/2
310         A(J1)=A(J1)+ESM(I,J)
         34 CONTINUE
         33 CONTINUE
         32 CONTINUE
      C*********
315   C*********
      C
      C   MODIFICATION AND SOLUTION OF THE SYSTEM OF EQUATIONS
      C       OUTPUT OF THE CALCULATED NODAL VALUES
      C
320   C*********
      C*********
            WRITE(10,62) TITLE
         62 FORMAT(1H1//10X,20A4)
            CALL MODIFY(IFE)
325         CALL DCMPBD
            CALL SLVBD
      C
      C   OUTPUT OF THE CALCULATED VALUES
      C
330         WRITE(10,65)
         65 FORMAT(//10X,21HCALCULATED QUANTITIES/
           +12X,20HNODAL VALUES FOR PHI )
            WRITE(10,66)  (I,A(I),I=1,NP)
         66 FORMAT(12X,13,E14.5,3X,13,E14.5,3X,13,E14.5)
335   C*********
      C*********
      C
      C   EVALUATION OF THE VOLUME UNDER THE PHI SURFACE AND THE
      C    ELEMENT GRADIENTS
340   C
      C*********
      C*********
      C   START OF THE LOOP ON THE ELEMENTS
      C
```

```
345            ILINE=0
               DO83KK=1,NE
               IF(ILINE.GT.0) GOTO110
       C
       C   OUTPUT OF THE CORRECT GRADIENT HEADING
350    C
               WRITE(10,43) TITLE
       43      FORMAT(1H1///10X,20A4)
               IF(ITYP.EQ.1) WRITE(10,44)
       44      FORMAT(//10X,7HELEMENT,4X,8HLOCATION,7X,7HTAU(ZX),
355            +8X,7HTAU(ZY))
               IF(ITYP.NE.1.AND.ITYP.NE.5) WRITE(10,147)
       147     FORMAT(///10X,7HELEMENT,4X,8HLOCATION,8X,6HVEL(X),
               +10X,6HVEL(Y))
               IF(ITYP.EQ.5) WRITE(10,146)
360    146     FORMAT(///10X,7HELEMENT,4X,8HLOCATION,10X,4HQ(X),
               +11X,4HQ(Y))
       C
       C   INCREASE THE LINE COUNT
       C
365    110     KL=4
               IF(NEL(KK,4).EQ.0) KL=3
               IF(KL.EQ.4) ILINE=ILINE+4
               IF(KL.EQ.3) ILINE=ILINE+2
               IF(ILINE.GT.50) ILINE=0
370    C
       C   RETRIEVAL OF THE NODAL COORDINATES, THE NODE NUMBERS
       C     AND THE NODAL VALUES OF PHI
       C
               SP=0.0
375            DO40I=1,KL
               NS(I)=NEL(KK,I)
               J=NS(I)
               X(I)=XC(J)
               Y(I)=YC(J)
380            PHI(I)=A(J)
       40      SP=SP+PHI(I)
       C
       C   ELEMENT COEFFICIENTS FOR THE GRADIENT VALUES
       C
385            II=NMTL(KK)
               DXE=DX(II)
               DYE=DY(II)
       C
       C   EVALUATION OF THE ELEMENT GRADIENTS
390    C
               IF(KL.EQ.4) GOTO51
       C
       C   TRIANGULAR ELEMENT
       C
395            B(1)=Y(2)-Y(3)
               B(2)=Y(3)-Y(1)
               B(3)=Y(1)-Y(2)
               C(1)=X(3)-X(2)
               C(2)=X(1)-X(3)
400            C(3)=X(2)-X(1)
               AR2=X(2)*Y(3)+X(3)*Y(1)+X(1)*Y(2)
               +-X(2)*Y(1)-X(3)*Y(2)-X(1)*Y(3)
```

```
                    GRADX=(B(1)*PHI(1)+B(2)*PHI(2)+B(3)*PHI(3))/AR2
                    GRADY=(C(1)*PHI(1)+C(2)*PHI(2)+C(3)*PHI(3))/AR2
405                 GRADX=DXE*GRADX*GRDC(ITYP,1)
                    GRADY=DYE*GRADY*GRDC(ITYP,2)
      C
      C   OUTPUT FOR TORSION AND STREAMLINE FLOW
      C
410                 IF(ITYP.LE.2)  WRITE(IO,52) KK,GRADY,GRADX
      C
      C   OUTPUT FOR POTENTIAL FLOW, GROUNDWATER FLOW,
      C      AND HEAT TRANSFER
      C
415                 IF(ITYP.GE.3)  WRITE(IO,52) KK,GRADX,GRADY
                 52 FORMAT(/13X,I3,5X,6HCENTER,2X,2E15.5)
      C
      C   CALCULATION OF THE VOLUME UNDER THE ELEMENT
      C
420                 VOL=VOL+SP*AR2/6.0
                    GOTO83
      C
      C   RECTANGULAR ELEMENT
      C
425        51       AA=Y(4)-Y(1)
                    BB=X(2)-X(1)
                    AR=AA*BB
                    GDX(1)=(PHI(2)-PHI(1))/BB
                    GDX(2)=(-PHI(1)+PHI(2)+PHI(3)-PHI(4))/(2.*BB)
430                 GDX(3)=(PHI(3)-PHI(4))/BB
                    GDY(1)=(PHI(4)-PHI(1))/AA
                    GDY(2)=(-PHI(1)-PHI(2)+PHI(3)+PHI(4))/(2.*AA)
                    GDY(3)=(PHI(3)-PHI(2))/AA
                    DO 82 I=1,3
435                 GDX(I)=DXE*GDX(I)*GRDC(ITYP,1)
           82       GDY(I)=DYE*GDY(I)*GRDC(ITYP,2)
                    IF(ITYP.GE.3) GOTO85
      C
      C   OUTPUT FOR TORSION AND STREAMLINE FLOW
440   C
                    WRITE(IO,53) KK,NS(1),GDY(1),GDX(1)
                    WRITE(IO,54) GDY(2),GDX(2)
                    WRITE(IO,55) NS(3),GDY(3),GDX(3)
                    GOTO86
445   C
      C   OUTPUT FOR POTENTIAL FLOW, GROUNDWATER FLOW,
      C      AND HEAT TRANSFER
      C
           85       WRITE(IO,53) KK,NS(1),GDX(1),GDY(1)
450                 WRITE(IO,54) GDX(2),GDY(2)
                    WRITE(IO,55) NS(3),GDX(3),GDY(3)
                 53 FORMAT(/13X,I3,5X,5HNODE ,I3,2E15.5)
                 54 FORMAT(21X,6HCENTER,2X,2E15.5)
                 55 FORMAT(21X,5HNODE ,I3,2E15.5)
455   C
      C   CALCULATION OF THE VOLUME UNDER THE ELEMENT
      C
           86       VOL=VOL+SP*AR/4.0
           83       CONTINUE
```

213

```
460          C*********
             C*********
             C
             C   OUTPUT OF THE INTEGRAL VALUE
             C
465          C*********
             C*********
                   VOL=VOL*2
                   IF(ITYP.EQ.1) WRITE(10,56) VOL
             56    FORMAT(//10X,29HTHE TORQUE FOR THE SECTION IS,E15.5)
470                STOP
                   END


                   SUBROUTINE ELSTMF(KK,IPLVL)
                   COMMON/ELMATX/ESM(4,4),EF(4),X(4),Y(4),KL
                   COMMON/MATL/DXE,DYE,GE,QE
                   COMMON/HCV/IDBC(50,2),DBC(50,2),NDBC
5                  DIMENSION ES(4,4),ET(4,4),EG(4,4)
                   DIMENSION B(3),C(3)
                   REAL LG
                   DATA ES/2.,-2.,-1.,1.,-2.,2.,1.,-1.,
                  +-1.,1.,2.,-2.,1.,-1.,-2.,2./
10                 DATA ET/2.,1.,-1.,-2.,1.,2.,-2.,-1.,
                  + -1.,-2.,2.,1.,-2.,-1.,1.,2./
                   DATA EG/4.,2.,1.,2.,2.,4.,2.,1.,
                  +1.,2.,4.,2.,2.,1.,2.,4./
                   IO=61
15                 IF(KL.EQ.4) GOTO2
             C*********
             C*********J
             C
             C   LINEAR TRIANGULAR ELEMENT WITHOUT THE DERIVATIVE
20           C       BOUNDARY CONDITION
             C
             C*********
             C*********
                   B(1)=Y(2)-Y(3)
25                 B(2)=Y(3)-Y(1)
                   B(3)=Y(1)-Y(2)
                   C(1)=X(3)-X(2)
                   C(2)=X(1)-X(3)
                   C(3)=X(2)-X(1)
30                 AR2=X(2)*Y(3)+X(3)*Y(1)+X(1)*Y(2)-X(2)*Y(1)
                  +-X(3)*Y(2)-X(1)*Y(3)
                   IF(ABS(AR2).LT.0.0001) GOTO5
                   DO11I=1,3
                   EF(I)=QE*AR2/6.
35                 DO1J=1,3
                   A=1.0
                   IF(I.EQ.J) A=2.0
             1     ESM(I,J)=((DXE*B(I)*B(J)+DYE*C(I)*C(J))/(AR2*2.))
                  ++A*GE*AR2/24.
40                 IF(NDBC.EQ.0) GOTO7
                   GOTO4
             C*********
             C*********
             C
```

214

```
     C  BILINEAR RECTANGULAR ELEMENT WITHOUT THE DERIVATIVE
     C      BOUNDARY CONDITION
     C
     C*********
     C*********
        2 AA=Y(4)-Y(1)
          BB=X(2)-X(1)
          AR=AA*BB
          IF(ABS(AR).LT.O.0001) GOTO5
          DO3I=1,4
          EF(I)=QE*AR/4.
          DO3J=1,4
        3 ESM(I,J)=DXE*AA*ES(I,J)/(6.*BB)+DYE*BB*ET(I,J)/(6.*AA)
         ++GE*AR*EG(I,J)/36.
        4 IF(NDBC.EQ.0) GO TO 7
     C*********
     C*********
     C
     C  DERIVATIVE BOUNDARY CONDITION
     C
     C*********
     C*********
          DO11I=1,NDBC
          IF(IDBC(I,1).NE.KK) GO TO 11
          J=IDBC(I,2)
          K=J+1
          IF(J.EQ.KL)K=1
          EF(J)=EF(J)+DBC(I,2)/2.
          EF(K)=EF(K)+DBC(I,2)/2.
          ESM(J,J)=ESM(J,J)+DBC(I,1)/3.
          ESM(J,K)=ESM(J,K)+DBC(I,1)/6.
          ESM(K,J)=ESM(K,J)
          ESM(K,K)=ESM(K,K)+DBC(I,1)/3.
       11 CONTINUE
     C*********
     C*********
     C
     C  OUTPUT OF THE ELEMENT MATRICES
     C
     C*********
     C*********
        7 IF(IPLVL.EQ.0) RETURN
          WRITE(IO,8) KK
        8 FORMAT(/10X,7HELEMENT,I4/10X,12HFORCE VECTOR,10X,
         +16HSTIFFNESS MATRIX)
          DO9I=1,KL
        9 WRITE(IO,10) EF(I),(ESM(I,J),J=1,KL)
       10 FORMAT(10X,E12.5,10X,4E13.5)
          RETURN
     C*********
     C*********
     C
     C  DIAGNOSTIC OUTPUT
     C
     C*********
     C*********
```

```
      5 WRITE(10,6) KK
      6    FORMAT(//10X,19HTHE AREA OF ELEMENT,I4,
           +20H IS LESS THAN 0.0001/
           +10X,39HTHE NODE NUMBERS ARE IN THE WRONG ORDER/
           +10X,33HOR THE NODES FORM A STRAIGHT LINE/
           +10X,20HEXECUTION TERMINATED)
           STOP
           END



           SUBROUTINE MODIFY(IFE)
           COMMON/AV/A(5000),JGF,JGSM,NP,NBW
           DATA IN/60/,IO/61/
C*********
C*********
C
C    INPUT OF THE NODAL FORCE VALUES
C         FOR FIELD PROBLEMS
C             IB - NODE NUMBER
C             BV - SOURCE OR SINK VALUE
C         FOR SOLID MECHANICS PROBLEMS
C             IB - DEGREE OF FREEDOM OF THE FORCE
C             BV - VALUE OF THE FORCE
C
C    INPUT OF IB AND BV IS TERMINATED BY
C         INPUTTING A ZERO VALUE FOR IB
C
C*********
C*********
           NIW =0
      202  READ(IN,*) IB
           IF(IB.LE.0) GOTO216
           IF(NIW.EQ.0.AND.IFE.EQ.0) WRITE(10,200)
           IF(NIW.EQ.0.AND.IFE.EQ.1) WRITE(10,201)
      200  FORMAT(//10X,22HSOURCE AND SINK VALUES)
      201  FORMAT(//10X,31HCONCENTRATED FORCES AND MOMENTS)
           NIW=1
           READ(IN,*) BV
           A(JGF+IB)=A(JGF+IB)+BV
           WRITE(10,203) IB,BV
      203  FORMAT(10X,I3,E15.5)
           GOTO202
C*********
C*********
C
C    INPUT OF THE PRESCRIBED NODAL VALUES
C         FOR FIELD PROBLEMS
C             IB - NODE NUMBER
C             BV - KNOWN VALUE OF PHI
C         FOR SOLID MECHANICS PROBLEMS
C             IB - DEGREE OF FREEDOM OF THE KNOWN DISPLACEMENT
C             BV - THE VALUE OF THE DISPLACEMENT
C
C    INPUT OF IB AND BV IS TERMINATED BY INPUTTING
C         A ZERO VAUE FOR IB
C
C*********
```

216

```
C********
216   NIW=0
209   READ(IN,*) IB
      IF(IB.LE.0) RETURN
      IF(NIW.EQ.0.AND.IFE.EQ.0) WRITE(IO,212)
      IF(NIW.EQ.0.AND.IFE.EQ.1) WRITE(IO,208)
212   FORMAT(//IOX,25HKNOWN NODAL VALUES OF PHI)
208   FORMAT(//IOX,25HKNOWN DISPLACEMENT VALUES)
      NIW=1
      READ(IN,*) BV
C
C   MODIFICATION OF THE GLOBAL STIFFNESS MATRIX AND
C        THE GLOBAL FORCE VECTOR USING THE METHOD
C          OF DELETION OF ROWS AND COLUMNS
C
      K=IB-1
      DO211J=2,NBW
      M=IB+J-1
      IF(M.GT.NP) GOTO210
      IJ=JGSM+(J-1)*NP+IB-(J-1)*(J-2)/2
      A(JGF+M)=A(JGF+M)-A(IJ)*BV
      A(IJ)=0.0
210   IF(K.LE.0) GOTO 211
      KJ=JGSM+(J-1)*NP+K-(J-1)*(J-2)/2
      A(JGF+K)=A(JGF+K)-A(KJ)*BV
      A(KJ)=0.0
      K=K-1
211   CONTINUE
      A(JGF+IB)=A(JGSM+IB)*BV
221   CONTINUE
      WRITE(IO,203) IB,BV
      GOTO209
      END


      SUBROUTINE DCMPBD
      COMMON/AV/A(5000),JGF,JGSM,NP,NBW
      IO=61
C*********
C*********
C
C   DECOMPOSITION OF A BANDED MATRIX INTO AN UPPER
C        TRIANGULAR FORM USING GAUSSIAN ELIMINATION
C
C*********
C*********
      NP1=NP-1
      DO226I=1,NP1
      MJ=I+NBW-1
      IF(MJ.GT.NP) MJ=NP
      NJ=I+1
      MK=NBW
      IF((NP-I+1).LT.NBW) MK=NP-I+1
      ND=0
      DO225J=NJ,MJ
      MK=MK-1
      ND=ND+1
      NL=ND+1
```

217

```
                    DO225K=1,MK
25                  NK=ND+K
                    JK=JGSM+(K-1)*NP+J-(K-1)*(K-2)/2
                    INL=JGSM+(NL-1)*NP+I-(NL-1)*(NL-2)/2
                    INK=JGSM+(NK-1)*NP+I-(NK-1)*(NK-2)/2
                    II=JGSM+I
30          225     A(JK)=A(JK)-A(INL)*A(INK)/A(II)
            226     CONTINUE
                    RETURN
                    END




                    SUBROUTINE SLVBD
                    COMMON/AV/A(5000),JGF,JGSM,NP,NBW
                    NP1=NP-1
            C*********
5           C*********
            C
            C   DECOMPOSITION OF THE GLOBAL FORCE VECTOR
            C
            C*********
10          C*********
                    DO250I=1,NP1
                    MJ=I+NBW-1
                    IF(MJ.GT.NP)  MJ=NP
                    NJ=I+1
15                  L=1
                    DO250J=NJ,MJ
                    L=L+1
                    IL=JGSM+(L-1)*NP+I-(L-1)*(L-2)/2
            250     A(JGF+J)=A(JGF+J)-A(IL)*A(JGF+I)/A(JGSM+I)
20          C*********
            C*********
            C
            C   BACKWARD SUBSTITUTION FOR DETERMINATION OF
            C     THE NODAL VALUES
25          C
            C*********
            C*********
                    A(NP)=A(JGF+NP)/A(JGSM+NP)
                    DO252K=1,NP1
30                  I=NP-K
                    MJ=NBW
                    IF((I+NBW-1).GT.NP)  MJ=NP-I+1
                    SUM=0.0
                    DO251J=2,MJ
35                  N=I+J-1
                    IJ=JGSM+(J-1)*NP+I-(J-1)*(J-2)/2
            251     SUM=SUM+A(IJ)*A(N)
            252     A(I)=(A(JGF+I)-SUM)/A(JGSM+I)
                    RETURN
40                  END
```

16.3 AN EXAMPLE PROBLEM

The data and computer solution of a heat transfer problem is given in this section. The problem was designed to illustrate all of the possible data inputs and has little or no practical use.

The problem configuration is shown in Figure 16.3. It consists of a five-sided region with the upper and lower surfaces insulated, $\partial\phi/\partial y = 0$. The temperatures

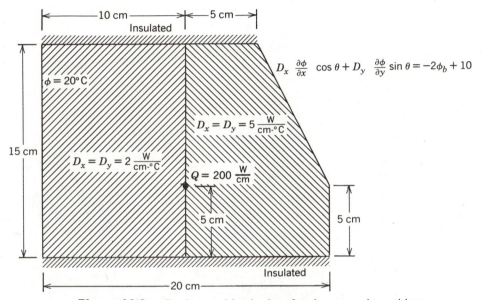

Figure 16.3. Region and basic data for the example problem.

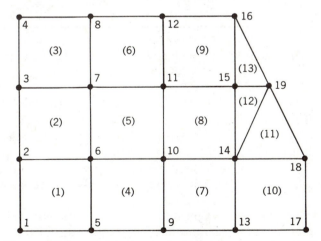

Figure 16.4. Finite element grid for the example problem.

Table 16.1 Computer Data

Title	TEXTBOOK EXAMPLE PROBLEM
Parameters	19 13 2 3 5 0
Material coefficients	2. 2. 0. 0. 5. 5. 0. 0.
X-Coordinates	0. 0. 0. 0. 5. 5. 5. 5. 10. 10. 10. 10. 15. 15. 15. 15. 20. 20. 17.5
Y-Coordinates	0. 5. 10. 15. 0. 5. 10. 15. 0. 5. 10. 15. 0. 5. 10. 15. 0. 5. 10.
Element data	1 1 1 5 6 2 2 1 2 6 7 3 3 1 3 7 8 4 4 1 5 9 10 6 5 1 6 10 11 7 6 1 7 11 12 8 7 2 9 13 14 10 8 2 10 14 15 11 9 2 11 15 16 12 10 2 13 17 18 14 11 2 14 18 19 0 12 2 14 19 15 0 13 2 19 16 15 0
Derivative boundary conditions	10 2 10. 50. 11 2 11.18 55.90 12 1 11.18 55.90
Source	10 200. 0
Known ϕ values	1 20. 2 20. 3 20. 4 20. 0

along the left vertical edge are maintained at 20°C while the boundary condition

$$D_x \frac{\partial \phi}{\partial x} \cos \theta + D_y \frac{\partial \phi}{\partial y} \sin = -2\phi_b + 10 \qquad (16.6)$$

is imposed on the right vertical surface and the sloping surface. There is a point source of 200 W/cm located within the region. The region is long in the z-direction; thus the analysis is performed on 1 cm of thickness.

The grid used to solve the problem is given in Figure 16.4. The grid consists of 13 elements and 19 nodes. The grid has not been refined to increase the accuracy of the solution.

The computer program data is given to the right of the solid line in Table 16.1. The titles on the left of the line indicate the nature of the data. Each line represents a card of input. The computer output is presented as a sequence entitled Figure 16.5.

TEXTBOOK EXAMPLE PROBLEM

NP =   19
NE =   13
ITYP = 5
IPLVL = 0

EQUATION COEFFICIENTS
MATERIAL

SET	DX	DY	G	Q
1	.20000E+01	.20000E+01	0.	0.
2	.50000E+01	.50000E+01	0.	0.

NODAL COORDINATES

NODE	X	Y
1	0.	0.
2	0.	.50000E+01
3	0.	.10000E+02
4	0.	.15000E+02
5	.50000E+01	0.
6	.50000E+01	.50000E+01
7	.50000E+01	.10000E+02
8	.50000E+01	.15000E+02
9	.10000E+02	0.
10	.10000E+02	.50000E+01
11	.10000E+02	.10000E+02
12	.10000E+02	.15000E+02
13	.15000E+02	0.
14	.15000E+02	.50000E+01
15	.15000E+02	.10000E+02
16	.15000E+02	.15000E+02
17	.20000E+02	0.
18	.20000E+02	.50000E+01
19	.17500E+02	.10000E+02

TEXTBOOK EXAMPLE PROBLEM

ELEMENT DATA

NEL	NMTL	NODE NUMBERS			
1	1	1	5	6	2
2	1	2	6	7	3
3	1	3	7	8	4
4	1	5	9	10	6
5	1	6	10	11	7
6	1	7	11	12	8
7	2	9	13	14	10
8	2	10	14	15	11
9	2	11	15	16	12
10	2	13	17	18	14
11	2	14	18	19	
12	2	14	19	15	
13	2	19	16	15	

DERIVATIVE BOUNDARY CONDITION DATA

ELEMENT	SIDE	ML	SL
10	2	.10000E+02	.50000E+02
11	2	.11180E+02	.55900E+02
12	1	.11180E+02	.55900E+02

BANDWIDTH IS   6 IN ELEMENT   1.

TEXTBOOK EXAMPLE PROBLEM

SOURCE AND SINK VALUES
   10   .20000E+03

KNOWN NODAL VALUES OF PHI
   1   .20000E+02
   2   .20000E+02
   3   .20000E+02
   4   .20000E+02

CALCULATED QUANTITIES
   NODAL VALUES FOR PHI

1	.20000E+02	2	.20000E+02	3	.20000E+02
4	.20000E+02	5	.27428E+02	6	.25018E+02
7	.23703E+02	8	.21293E+02	9	.25224E+02
10	.42083E+02	11	.21709E+02	12	.19521E+02
13	.20764E+02	14	.15293E+02	15	.16297E+02
16	.18851E+02	17	.91043E+01	18	.85150E+01
19	.10237E+02				

ELEMENT	LOCATION		Q(X)	Q(Y)
1	NODE	1	-.29713E+01	0.
	CENTER		-.24894E+01	.48199E+00
	NODE	6	-.20074E+01	.96398E+00
2	NODE	2	-.20074E+01	0.
	CENTER		-.17443E+01	.26306E+00
	NODE	7	-.14812E+01	.52612E+00
3	NODE	3	-.14812E+01	0.
	CENTER		-.99925E+00	.48199E+00
	NODE	8	-.51726E+00	.96398E+00
4	NODE	5	.88192E+00	.96398E+00
	CENTER		-.29720E+01	-.28900E+01
	NODE	10	-.68260E+01	-.67439E+01
5	NODE	6	-.68260E+01	.52612E+00
	CENTER		-.30141E+01	.43379E+01
	NODE	11	.79766E+00	.81497E+01

6	NODE 7	.79766E+00	.96398E+00
	CENTER	.75325E+00	.91957E+00
	NODE 12	.70884E+00	.87516E+00
7	NODE 9	.44598E+01	-.16860E+02
	CENTER	.15625E+02	-.56944E+01
	NODE 14	.26791E+02	.54710E+01
8	NODE 10	.26791E+02	.20374E+02
	CENTER	.16101E+02	.96850E+01
	NODE 15	.54118E+01	-.10044E+01
9	NODE 11	.54118E+01	.21879E+01
	CENTER	.30408E+01	-.18309E+00
	NODE 16	.66980E+00	-.25541E+01
10	NODE 13	.11659E+02	.54710E+01
	CENTER	.92186E+01	.30301E+01
	NODE 18	.67778E+01	.58931E+00
11	CENTER	.67778E+01	.16665E+01
12	CENTER	.12120E+02	-.10044E+01
13	CENTER	.12120E+02	-.25541E+01

PART THREE
STRUCTURAL AND SOLID MECHANICS

The application of the finite element method to the solution of structural and solid mechanics problems is discussed in the next nine chapters. These chapters provide an introduction to the analysis of plane structures and elasticity problems. Chapters 1 and 2 contain the prerequisite material for Chapters 17 through 21. Chapters 5 and 6 should be covered before reading Chapters 22 through 25. Chapters 17 and 18 duplicate some of the material covered in Chapters 3 and 4 and can be covered rather quickly if Chapters 3 and 4 have already been read.

Chapter 17

AXIAL FORCE MEMBER

The finite element method is applicable to the analysis of both discrete and continuous structures. Discrete structures are those with individual members such as trusses, beams, and rigid frames. Continuous structures are plate- and shell-type structures as well as machine and structural components that must be analyzed using the theory of elasticity. The analysis of both discrete and continuous structures can be approached from several points of view. Only one of these, the principle of minimum potential energy is used in this book. The objective of this chapter is to introduce the principle and show how it can be applied to the solution of some simple statically indeterminate problems involving axial force members. These problems are similar to those encountered in a first course in the mechanics of deformable bodies.

17.1 THE ONE-DIMENSIONAL MODEL

The finite element grid for a system of axial force members is identical to that discussed relative to a differential equation in Chapter 2. It consists of a straight-line segment with nodes wherever there is a change in the material properties or the cross-sectional area. The only new rule involves externally applied forces. A node is placed wherever there is an external axial force. This is done to simplify the calculation of the work term in the potential energy equation. By placing a node where the force is applied, the work done by that force can be written as a force times a displacement. This idea is illustrated in Figure 17.1. The system is divided into three elements even though the member does not possess a change in area or material properties. A node is placed at each point where a force is applied.

A major difference between the grids for axial force members and those for an approximate solution to (2.1) is the concept of grid refinement. A finite element solution for the displacements in a discrete structure yields the correct values. No improvement is obtained by subdividing each member into several smaller elements. Each member is represented by a single element except when there are applied loads between its end points.

The quantities calculated in a finite element analysis of a discrete or continuous structure are displacements. The nodal displacements and externally applied forces are often indicated using arrows (Figure 17.1). A positive displacement is always in a positive coordinate direction. Both translational and rotational displacements are denoted by U in this book.

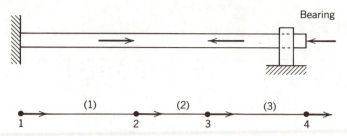

Figure 17.1.   A node is located at every external force.

17.2   PRINCIPLE OF MINIMUM POTENTIAL ENERGY

The equations that yield the joint displacements of a structural system can be derived using the principle of minimum potential energy. A statement of the principle and a discussion of its important points is presented here, and illustrations of its implementation are given in a later section. The derivation is a topic in continuum mechanics courses and is not discussed here. The reader who is interested in the derivation should see Fung (1965). The statement of the theorem as given here is similar to that given by Cook (1982).

The principle of minimum potential energy states: *Among all the displacement equations that satisfy internal compatibility and the boundary conditions, those that also satisfy the equations of equilibrium make the potential energy a minimum in a stable system.*

The above principle implies the following:

1.   The writing of a displacement equation for each member. These equations must be compatible. The equations must ensure that all members connecting at a rigid joint rotate the same amount or that a beam has a continuous first derivative.
2.   Incorporation of the boundary (support) conditions so that the displacement equations satisfy all of the physical support conditions.
3.   The writing of an equation for the potential energy within the structural system in terms of some unknown displacements.
4.   Minimization of the potential energy with respect to the undetermined displacements within the displacement equations.

The completion of these four steps leads to a system of equilibrium equations that are solved for the joint displacements. Once the joint displacements are known, the internal force and/or moment in each member is calculated.

The minimization process clearly implies the need for a potential energy equation written in terms of the displacements. The potential energy in an elastic structure is the energy contained in the elastic distortions and the capacity of the loads to do work. The potential energy contained in the elastic distortions is the strain energy. The capacity of a concentrated load to do work is $P \cdot U$, where P is the magnitude of the concentrated load and U is the displacement. A positive force and the corresponding positive displacement always have the same directions.

The total potential energy in the axial force member is

$$\Pi = \Lambda - W = \Lambda - PU \tag{17.1}$$

where Λ represents the strain energy and W is the work done by an external force. When there are several members and external loads

$$\Pi = \sum_{e=1}^{n} \Lambda^{(e)} - \sum_{i=1}^{p} P_i U_i \tag{17.2}$$

where the strain energy is summed over the number of elements, n, and the work is summed over the number of nodes, p. The negative sign appears with the work term because each force looses some of its capacity for doing work when it displaces in the direction it acts.

Equation (17.2) is used for the systems considered in this chapter. The work term in (17.2) involves a displacement, but the strain energy must be written in terms of the nodal displacements.

17.3 THE STRAIN ENERGY EQUATION

The equation that gives the strain energy in an axial force member is derived in most introductory textbooks covering the mechanics of deformable bodies. The reader unfamiliar with these equations may wish to consult Popov (1976), Higdon, et al. (1976), or other similar books. The equation is

$$\Lambda^{(e)} = \int_{V} \frac{\sigma_{xx} \varepsilon_{xx}}{2} dV \tag{17.3}$$

where σ_{xx} and ε_{xx} represent the normal stress and strain components, respectively. This equation can be written in terms of either the stress component or the strain component using Hooke's law

$$\sigma_{xx} = E\varepsilon_{xx} \tag{17.4}$$

The equivalent equations are

$$\Lambda^{(e)} = \int_{V} \frac{\sigma_{xx}^2}{2E} dV = \int_{V} \frac{E\varepsilon_{xx}^2}{2} dV \tag{17.5a, b}$$

The objective of this section is to evaluate one of the integrals of (17.5) such that a strain energy equation written in terms of the nodal displacements is obtained.

The axial force member is modeled by a straight-line segment (Figure 17.2), with a displacement at each end, U_i and U_j. The member has a cross-sectional area A; an elastic modulus E; a coefficient of thermal expansion α; a length L; and an internal force $F^{(e)}$, which is developed by the externally applied loads.

Information available from a first course in the mechanics of deformable bodies includes

$$e_{xx} = \frac{du}{dx} = \text{constant} \tag{17.6}$$

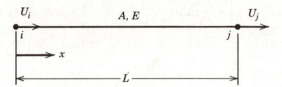

Figure 17.2.   The axial force member.

where e_{xx} is the total strain and u is the displacement equation. The total strain, e_{xx}, is not the same as ε_{xx} in (17.5b). The two are related by

$$e_{xx} = \varepsilon_{xx} + \varepsilon_T \qquad (17.7)$$

The total strain is the sum of the elastic strain, ε_{xx}, resulting from the applied loads and the strain produced by a thermal change, ε_T. A simple rearrangement gives

$$\varepsilon_{xx} = e_{xx} - \varepsilon_T \qquad (17.8)$$

and the substitution of (17.6) yields

$$\varepsilon_{xx} = \frac{du}{dx} - \alpha \delta T \qquad (17.9)$$

since $\varepsilon_T = \alpha \delta T$, where δT is the temperature change.

Substitution of (17.9) into (17.5b) gives the strain energy equation

$$\Lambda^{(e)} = \int_V \frac{E}{2} \left(\frac{du}{dx} - \alpha \delta T \right)^2 dV$$

$$= \int_V \frac{E}{2} \left(\frac{du}{dx} \right)^2 dV - \int_V E \alpha \delta T \frac{du}{dx} dV + \int_V \frac{E(\alpha \delta T)^2}{2} dV \qquad (17.10)$$

The incremental volume dV is $dV = dA\,dx$ and the volume integrals can be replaced by

$$\int_V f(x)\,dV = \int_0^L \int_A f(x)\,dA\,dx = A \int_0^L f(x)\,dx \qquad (17.11)$$

assuming that the cross-sectional area is constant. If we use this replacement, the integrals in (17.10) are

$$\Lambda^{(e)} = \frac{AE}{2} \int_0^L \left(\frac{du}{dx} \right)^2 dx - AE\alpha\delta T \int_0^L \left(\frac{du}{dx} \right) dx + \frac{AE(\alpha\delta T)^2}{2} \int_0^L dx \qquad (17.12)$$

The final step is to select an element displacement equation. The constant value for e_{xx}, (17.6), implies a linear equation for the axial displacement. The general form for a linear equation was developed in Chapter 2. Writing (2.6) in terms of the nodal displacements gives

$$u = \left(\frac{X_j - x}{L} \right) U_i + \left(\frac{x - X_i}{L} \right) U_j \qquad (17.13)$$

In this situation, $X_i=0$ and $X_j=L$.

The derivative of the displacement equation is

$$\frac{du}{dx}=\frac{-U_i+U_j}{L} \tag{17.14}$$

Since du/dx is a constant, this term can be taken outside the integrals in (17.12). Completion of the integration produces

$$\Lambda^{(e)}=\frac{AEL}{2}\left(\frac{du}{dx}\right)^2-AEL\,\alpha\delta T\left(\frac{du}{dx}\right)+\frac{AEL(\alpha\delta T)^2}{2} \tag{17.15}$$

Substitution of (17.14) yields the strain energy equation for an axial force member written in terms of the nodal displacements.

$$\Lambda^{(e)}=\frac{AE}{2L}(U_i^2-2U_iU_j+U_j^2)-AE\,\alpha\delta T(-U_i+U_j)+\frac{AEL(\alpha\delta T)^2}{2} \tag{17.16}$$

The nodal displacements are the unknowns in the potential energy formulation. Their magnitudes are determined by finding the set of values that makes the potential energy a minimum. The last term of (17.16) is not related to the nodal values and disappears during the minimization process. Since the constant does not influence the final results, it is usually discarded and (17.16) is written as

$$\Lambda^{(e)}=\frac{AE}{2L}(U_i^2-2U_iU_j+U_j^2)-AE\,\alpha\delta T(-U_i+U_j) \tag{17.17}$$

17.4 A SYSTEM OF AXIAL FORCE MEMBERS

The application of (17.17) in conjunction with (17.2) is illustrated by working through a problem consisting of three axial force members subjected to a pair of concentrated forces and a temperature change (Figure 17.3). The system is modeled by three elements and four nodes. The labeling of the model provides for a displacement and an externally applied force at each node. A positive force acts in the same direction as a positive displacement.

Analyzing the physical problem reveals that $U_1=U_4=0$ because of the rigid walls and that $P_1=P_4=0$, $P_2=10000$ N, and $P_3=-20000$ N. The negative sign for P_3 results because the load is opposite to the direction of a positive displacement.

The potential energy of the system is given by (17.2) with $n=3$ and $p=4$. Expanding (17.2), we obtain

$$\Pi=\Lambda^{(1)}+\Lambda^{(2)}+\Lambda^{(3)}-P_1U_1-P_2U_2-P_3U_3-P_4U_4 \tag{17.18}$$

or

$$\Pi=\Lambda^{(1)}+\Lambda^{(2)}+\Lambda^{(3)}-(10000)U_2-(-20000)U_3 \tag{17.19}$$

since the values of U_1, U_4, P_2, and P_3 are known. The strain energy in each member

$E = 20(16^6)$ N/cm^2 $\delta T = 10°C$
$\alpha = 11(10^{-6})/°C$

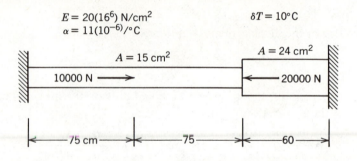

$A = 24$ cm^2

$A = 15$ cm^2

10000 N $\longrightarrow$ $\longleftarrow$ 20000 N

$\longleftarrow$ 75 cm $\longrightarrow$ $\longleftarrow$ 75 $\longrightarrow$ $\longleftarrow$ 60 $\longrightarrow$

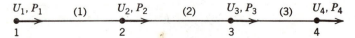

U_1, P_1 (1) U_2, P_2 (2) U_3, P_3 (3) U_4, P_4

1   2   3   4

Figure 17.3.   A system of three axial force members.

is given by (17.17). The element information needed for (17.17) is summarized in the following table.

e	i	j	A	L	$\dfrac{AE}{2L}$	$AE\,\alpha\delta T$
1	1	2	15	75	$2(10^6)$	33000
2	2	3	15	75	$2(10^6)$	33000
3	3	4	24	60	$4(10^6)$	52800

Utilizing the element information gives

$$\Lambda^{(1)} = 2(10^6)(U_1^2 - 2U_1 U_2 + U_2^2) - 33000(-U_1 + U_2) \tag{17.20}$$

$$\Lambda^{(2)} = 2(10^6)(U_2^2 - 2U_2 U_3 + U_3^2) - 33000(-U_2 + U_3) \tag{17.21}$$

and

$$\Lambda^{(3)} = 4(10^6)(U_3^2 - 2U_3 U_4 + U_4^2) - 52800(-U_3 + U_4) \tag{17.22}$$

The equations for $\Lambda^{(1)}$ and $\Lambda^{(3)}$ simplify to

$$\Lambda^{(1)} = 2(10^6)U_2^2 - 33000U_2 \tag{17.23}$$

and

$$\Lambda^{(3)} = 4(10^6)U_3^2 + 52800U_3 \tag{17.24}$$

because $U_1 = U_4 = 0$.

Substituting (17.21), (17.23), and (17.24) into (17.19) yields

$$\Pi = 2(10^6)U_2^2 - 33000U_2 + 2(10^6)(U_2^2 - 2U_2 U_3 + U_3^2)$$
$$- 33000(-U_2 + U_3) + 4(10^6)U_3^2 + 52800U_3 - 10000U_2 + 20000U_3 \tag{17.25}$$

or

$$\Pi = 4(10^6)U_2^2 - 4(10^6)U_2U_3 + 6(10^6)U_3^2 - 10000U_2 + 39800U_3 \qquad (17.26)$$

The values of U_2 and U_3 that make Π a minimum satisfy the equations

$$\frac{\partial \Pi}{\partial U_2} = 8(10^6)U_2 - 4(10^6)U_3 - 10000 = 0$$

and $\qquad\qquad\qquad\qquad\qquad\qquad\qquad\qquad\qquad (17.27)$

$$\frac{\partial \Pi}{\partial U_3} = -4(10^6)U_2 + 12(10^6)U_3 + 39800 = 0$$

Solving this pair of equations gives

$$U_2 = -0.0004900 \text{ cm} \qquad \text{and} \qquad U_3 = -0.003480 \text{ cm} \qquad (17.28)$$

The negative signs indicate that both nodes move to the left.

One objective of any structural analysis is to calculate the stresses within the individual members. The axial force acting in each member must be known to do this. The axial forces can be calculated once the nodal displacements are known. Since this calculation occurs in every analysis, it is convenient to have a general equation that gives the axial force in terms of the nodal displacements.

The axial force in an element is

$$S^{(e)} = \sigma_{xx} A \qquad (17.29)$$

where σ_{xx} is the normal stress. The normal stress is related to the normal strain by Hooke's law

$$\sigma_{xx} = E\varepsilon_{xx} \qquad (17.30)$$

and

$$S^{(e)} = AE\varepsilon_{xx} \qquad (17.31)$$

The normal strain is related to the displacements and thermal change by (17.9); thus

$$S^{(e)} = AE\left(\frac{du}{dx} - \alpha\delta T\right) \qquad (17.32)$$

Replacing the derivative term by (17.14) gives

$$S^{(e)} = \frac{AE}{L}(U_j - U_i) - AE\,\alpha\delta T \qquad (17.33)$$

Equation (17.33) gives the internal axial force $S^{(e)}$ in terms of the nodal displacement and the temperature change.

Application of (17.33) to the system in Figure 17.2 gives

$$S^{(1)} = \frac{AE}{L}(U_2 - U_1) - AE\,\alpha\delta T$$

$$= 4(10^6)(-0.000490 - 0) - 33000$$

$$= -34960 \text{ N}$$

$$S^{(2)} = \frac{AE}{L}(U_3 - U_2) - AE\,\alpha\delta T$$

$$= 4(10^6)[-0.003840 - (-0.0004900)] - 33000$$

$$= -44960 \text{ N}$$

and

$$S^{(3)} = \frac{AE}{L}(U_4 - U_3) - AE\,\alpha\delta T$$

$$= 8(10^6)[0 - (-0.003480)] - 52800$$

$$= -24960 \text{ N}$$

The negative values for $S^{(1)}$, $S^{(2)}$, and $S^{(3)}$ indicate that each member is in compression.

The solution of structural problems that involve the discrete elements can

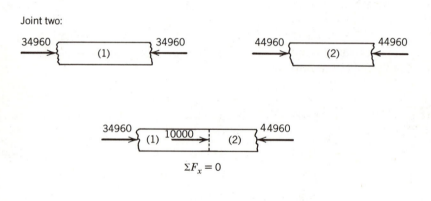

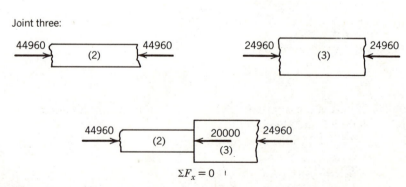

Figure 17.4. Free body diagrams of the individual members of the example problem.

always be checked by analyzing the equilibrium of the internal joints and/or the equilibrium of the complete structure. If a joint is not in equilibrium, then some of the member forces (or moments) are wrong. Free body diagrams of the joints in this example are shown in Figure 17.4. The forces do sum to zero.

The displacements should be calculated to four significant digits to insure an accurate equilibrium analysis.

17.5 MATRIX NOTATION

The system of equations (17.27) can be written using matrix notation. The matrix equation for structural problems is

$$[K]\{U\} - \{F\} - \{P\} = \{0\} \tag{17.34}$$

where $[K]$ is the stiffness matrix and $\{F\}$ and $\{P\}$ are force vectors. The force vector $\{F\}$ comes from element contributions and is usually a result of thermal changes. The force vector $\{P\}$ contains the external forces applied at the joints. The vector $\{U\}$ contains the nodal displacements.

The equations in (17.27) can be written in the form of (17.34) by separating the thermal forces from the external forces. Returning to (17.25), we find that Π can be written as

$$\Pi = 4(10^6)U_2^2 - 4(10^6)U_2U_3 + 6(10^6)U_3^2$$
$$- (0U_2 - 19800U_3) - (10000U_2 - 20000U_3) \tag{17.35}$$

The first set of forces, the 0 and 19800 values, come from the thermal term $AE\,\alpha\delta T$. The last pair, 10000 and -20000, are the external forces applied at the joints. Differentiation of (17.35) yields

$$8(10^6)U_2 - 4(10^6)U_3 - 0 - 10000 = 0$$

and $\tag{17.36}$

$$-4(10^6)U_2 + 12(10^6)U_3 - (-19800) - (-20000) = 0$$

This pair of equations can be written

$$10^6 \begin{bmatrix} 8 & -4 \\ -4 & 12 \end{bmatrix} \begin{Bmatrix} U_2 \\ U_3 \end{Bmatrix} - \begin{Bmatrix} 0 \\ -19800 \end{Bmatrix} - \begin{Bmatrix} 10000 \\ -20000 \end{Bmatrix} = \begin{Bmatrix} 0 \\ 0 \end{Bmatrix} \tag{17.37}$$

which is in the same form as (17.34).

PROBLEMS

17.1–17.5 Calculate the axial force in each member of the structural system shown in the corresponding figure. In each problem, $E = 20(10^6)\,\text{N/cm}^2$ and $\alpha = 11(10^{-6})/°\text{C}$.

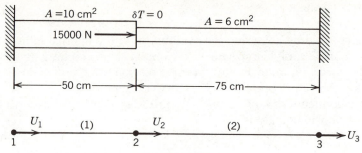

$A = 10 \text{ cm}^2$ $\delta T = 0$ $A = 6 \text{ cm}^2$

15000 N →

|←———50 cm———→|←———75 cm———→|

U_1 (1) U_2 (2)   →U_3
1   2   3

Figure P17.1

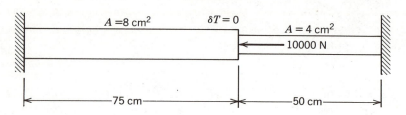

$A = 8 \text{ cm}^2$ $\delta T = 0$ $A = 4 \text{ cm}^2$

← 10000 N

|←———75 cm———→|←———50 cm———→|

U_1 (1) U_2 (2)   →U_3
1   2   3

Figure P17.2

$A = 5 \text{ cm}^2$, $\delta T = +10°C$

10000 N →   10000 N →

|←——50 cm——→|←——50——→|←——50——→|

U_1 (1) U_2 (2) U_3 (3) U_4
1   2   3   4

Figure P17.3

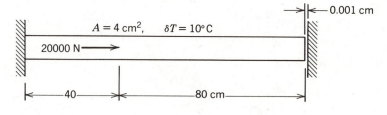

→|←— 0.001 cm

$A = 4 \text{ cm}^2$, $\delta T = 10°C$

20000 N →

|←——40——→|←————80 cm————→|

U_1 (1) U_2 (2)   →U_3
1   2   3

Figure P17.4

236

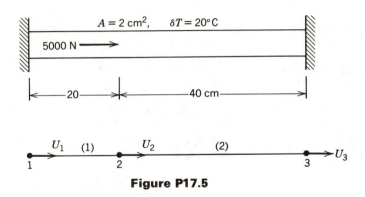

Figure P17.5

Chapter 18

ELEMENT MATRICES: POTENTIAL ENERGY FORMULATIONS

The concepts of an element stiffness matrix and an element force vector were introduced in Chapter 4. These quantities were used with the direct stiffness method to generate the coefficients in $[K]$ and $\{F\}$ without writing the residual equation for each node. The same concept can be implemented for structural and solid mechanics problems. The determination of the element matrices eliminates the need to write the potential energy equation.

The element stiffness matrix and the force vector for the axial force member are developed in this chapter and are used to solve the axial force system considered in Chapter 17. A general procedure for determining the element matrices is derived after the example problem. The results of this derivation are used to establish the element matrices in the structural and solid mechanics chapters that follow.

18.1 THE AXIAL FORCE ELEMENT

The system of equations associated with the potential energy formulation is obtained by minimizing the potential energy. If we assume that every displacement is unknown, the minimum value results when

$$\frac{\partial \Pi}{\partial U_1}=0, \qquad \frac{\partial \Pi}{\partial U_2}=0, \ldots, \qquad \frac{\partial \Pi}{\partial U_{p-1}}=0, \qquad \frac{\partial \Pi}{\partial U_p}=0 \qquad (18.1)$$

These derivatives can be written as a column vector. The derivative of Π with respect to the vector of displacements, $\{U\}$, is

$$\frac{\partial \Pi}{\partial \{U\}}=\begin{Bmatrix} \dfrac{\partial \Pi}{\partial U_1} \\[2mm] \dfrac{\partial \Pi}{\partial U_2} \\[1mm] \vdots \\[1mm] \dfrac{\partial \Pi}{\partial U_p} \end{Bmatrix} \qquad (18.2)$$

This vector is similar to the residual vector (4.1). Each component represents a single equation.

The potential energy in a system of axial force members is given by (17.2) and rewritten here for convenience:

$$\Pi = \sum_{e=1}^{n} \Lambda^{(e)} - \sum_{i=1}^{p} P_i U_i \tag{18.3}$$

The derivative of Π with respect to an arbitrary displacement U_β gives

$$\frac{\partial \Pi}{\partial U_\beta} = \sum_{e=1}^{n} \frac{\partial \Lambda^{(e)}}{\partial U_\beta} - P_\beta \tag{18.4}$$

This is equation β in the final system of equations. The element contribution to this equation is contained within the summation. Expanding the summation gives

$$\sum_{e=1}^{n} \frac{\partial \Lambda^{(e)}}{\partial U_\beta} = \frac{\partial \Lambda^{(1)}}{\partial U_\beta} + \frac{\partial \Lambda^{(2)}}{\partial U_\beta} + \cdots + \frac{\partial \Lambda^{(n)}}{\partial U_\beta} \tag{18.5}$$

The strain energy in an axial force element is

$$\Lambda^{(e)} = \frac{AE}{2L} (U_i^2 - 2U_i U_j + U_j^2) - AE\,\alpha\delta T(-U_i + U_j) \tag{18.6}$$

and is a function of only two displacements, U_i and U_j. The derivative $\partial \Lambda^{(e)}/\partial U_\beta$, therefore, is zero unless $\beta = i$ or $\beta = j$. If the strain energy in the element is not a function of U_β, the element contributes nothing to equation β.

The element contribution to the system of equations is obtained by evaluating the derivatives of $\Lambda^{(e)}$ with respect to U_i and U_j. This operation yields

$$\frac{\partial \Lambda^{(e)}}{\partial U_i} = \frac{AE}{L} (U_i - U_j) + AE\,\alpha\delta T$$

and $\tag{18.7}$

$$\frac{\partial \Lambda^{(e)}}{\partial U_j} = \frac{AE}{L} (-U_i + U_j) - AE\,\alpha\delta T$$

The equations in (18.7) can be written as

$$\begin{Bmatrix} \dfrac{\partial \Lambda^{(e)}}{\partial U_i} \\[2mm] \dfrac{\partial \Lambda^{(e)}}{\partial U_j} \end{Bmatrix} = \frac{AE}{L} \begin{bmatrix} 1 & -1 \\ -1 & 1 \end{bmatrix} \begin{Bmatrix} U_i \\ U_j \end{Bmatrix} - \begin{Bmatrix} -AE\,\alpha\delta T \\ AE\,\alpha\delta T \end{Bmatrix} \tag{18.8}$$

which is equivalent to

$$\frac{\partial \Lambda^{(e)}}{\partial \{U^{(e)}\}} = [k^{(e)}]\{U^{(e)}\} - \{f^{(e)}\} \tag{18.9}$$

The element stiffness matrix is

$$[k^{(e)}] = \frac{AE}{L}\begin{bmatrix} 1 & -1 \\ -1 & 1 \end{bmatrix} \tag{18.10}$$

and the element force vector is

$$\{f^{(e)}\} = \left\{ \begin{array}{c} -AE\,\alpha\delta T \\ AE\,\alpha\delta T \end{array} \right\} \tag{18.11}$$

The vector $\{U^{(e)}\}$ in (18.9) contains the element displacements, $\{U^{(e)}\}^T = [U_i \quad U_j]$.

The element matrices as given by (18.10) and (18.11) have the same properties as those discussed in Chapter 4. They are easy to program for computer evaluation; it is also easy to determine where each individual coefficient is located in the final system of equations.

The vector $\partial\Pi/\partial\{U\}$ represents the system of equations

$$\frac{\partial\Pi}{\partial\{U\}} = [K]\{U\} - \{F\} - \{P\} = \{0\} \tag{18.12}$$

Equation 18.8 states that the coefficients in the first row of $[k^{(e)}]$ and $\{f^{(e)}\}$ are located in row i of $[K]$ and $\{F\}$ because $\partial\Lambda^{(e)}/\partial U_i$ contributes to the summation associated with $\partial\Pi/\partial U_i$, which is row i in the final system of equations. Also, the coefficients in the second row of $[k^{(e)}]$ and $\{f^{(e)}\}$ are located in row j of $[K]$ and $\{F\}$ because $\partial\Lambda^{(e)}/\partial U_j$ contributes to the summation associated with $\partial\Pi/\partial U_j$. The coefficients of $[k^{(e)}]$ are located in columns i and j of $[K]$ because the coefficients in the first column multiply U_i while those in the second column multiply U_j.

18.2 A SYSTEM OF AXIAL FORCE MEMBERS

The direct stiffness procedure is illustrated by constructing the system of equations for the three axial force members analyzed in Section 17.4. The system and element models are shown in Figure 18.1. The element stiffness matrix is given by (18.10) and the element force vector is given by (18.11). The element data are

e	i	j	$\dfrac{AE}{L}$	$AE\,\alpha\delta T$
1	1	2	$4(10^6)$	33000
2	2	3	$4(10^6)$	33000
3	3	4	$8(10^6)$	52800

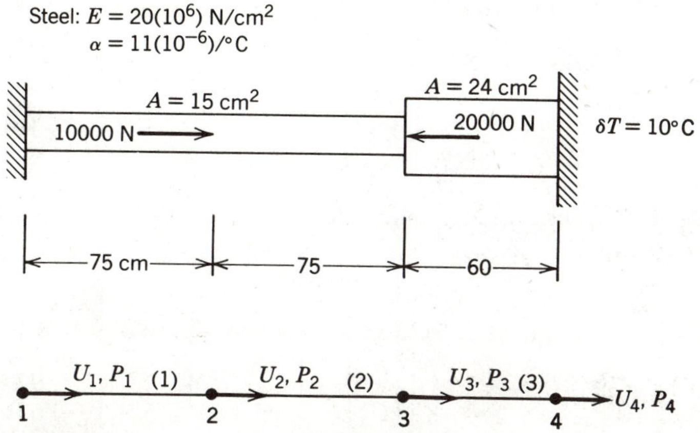

Steel: $E = 20(10^6)$ N/cm^2
$\alpha = 11(10^{-6})/°C$

Figure 18.1. A system of three axial force members.

The element matrices are

$$[k^{(1)}] = 10^6 \begin{matrix} 1 & 2 \\ \begin{bmatrix} 4 & -4 \\ -4 & 4 \end{bmatrix} & \begin{matrix} 1 \\ 2 \end{matrix} \end{matrix}, \qquad \{f^{(1)}\} = \begin{Bmatrix} -33000 \\ 33000 \end{Bmatrix} \begin{matrix} 1 \\ 2 \end{matrix}$$

$$[k^{(2)}] = 10^6 \begin{matrix} 2 & 3 \\ \begin{bmatrix} 4 & -4 \\ -4 & 4 \end{bmatrix} & \begin{matrix} 2 \\ 3 \end{matrix} \end{matrix}, \qquad \{f^{(2)}\} = \begin{Bmatrix} -33000 \\ 33000 \end{Bmatrix} \begin{matrix} 2 \\ 3 \end{matrix}$$

$$[k^{(3)}] = 10^6 \begin{matrix} 3 & 4 \\ \begin{bmatrix} 8 & -8 \\ -8 & 8 \end{bmatrix} & \begin{matrix} 3 \\ 4 \end{matrix} \end{matrix}, \qquad \{f^{(3)}\} = \begin{Bmatrix} -52800 \\ 52800 \end{Bmatrix} \begin{matrix} 3 \\ 4 \end{matrix}$$

Initializing $[K]$ and $\{F\}$ with zeros and adding the coefficients of element one yields

$$[K] = 10^6 \begin{bmatrix} 4 & -4 & 0 & 0 \\ -4 & 4 & 0 & 0 \\ 0 & 0 & 0 & 0 \\ 0 & 0 & 0 & 0 \end{bmatrix}, \qquad \{F\} = \begin{Bmatrix} -33000 \\ 33000 \\ 0 \\ 0 \end{Bmatrix}$$

Adding the values of element two gives

$$[K] = 10^6 \begin{bmatrix} 4 & -4 & 0 & 0 \\ -4 & 8 & -4 & 0 \\ 0 & -4 & 4 & 0 \\ 0 & 0 & 0 & 0 \end{bmatrix}, \qquad \{F\} = \begin{Bmatrix} -33000 \\ 0 \\ 33000 \\ 0 \end{Bmatrix}$$

Adding the values of element three yields

$$[K] = 10^6 \begin{bmatrix} 4 & -4 & 0 & 0 \\ -4 & 8 & -4 & 0 \\ 0 & -4 & 12 & -8 \\ 0 & 0 & -8 & 8 \end{bmatrix}, \quad \{F\} = \begin{Bmatrix} -33000 \\ 0 \\ -19800 \\ 52800 \end{Bmatrix}$$

The addition of element three finishes the summation through the elements. The final system of equations is

$$10^6 \begin{bmatrix} 4 & -4 & 0 & 0 \\ -4 & 8 & -4 & 0 \\ 0 & -4 & 12 & -8 \\ 0 & 0 & -8 & 8 \end{bmatrix} \begin{Bmatrix} U_1 \\ U_2 \\ U_3 \\ U_4 \end{Bmatrix} - \begin{Bmatrix} -33000 \\ 0 \\ -19800 \\ 52800 \end{Bmatrix} - \begin{Bmatrix} P_1 \\ P_2 \\ P_3 \\ P_4 \end{Bmatrix} = \begin{Bmatrix} 0 \\ 0 \\ 0 \\ 0 \end{Bmatrix} \quad (18.13)$$

The final result of the direct stiffness method is a system of four equations. Two of these equations should not be included, however, because U_1 and U_4 have known values; that is, $U_1 = U_4 = 0$. The potential energy can be minimized only with respect to the unknown displacements. Deleting equations one and four gives

$$-4(10^6)U_1 + 8(10^6)U_2 - 4(10^6)U_3 + 0 - P_2 = 0$$
$$-4(10^6)U_2 + 12(10^6)U_3 - 8(10^6)U_4 + 19800 - P_3 = 0 \quad (18.14)$$

Substituting the known values for U_1 and U_4 as well as for P_2 and P_3 yields

$$8(10^6)U_2 - 4(10^6)U_3 + 0 - 10000 = 0$$
$$-4(10^6)U_2 + 12(10^6)U_3 + 19800 + 20000 = 0 \quad (18.15)$$

which is the same as (17.27).

18.3   A GENERAL FORMULATION

The element stiffness matrix and force vector developed for the axial force member were obtained by differentiating the strain energy equation. The strain energy equations for other structural elements are more complicated than (18.6). There is a more efficient way to develop $[k^{(e)}]$ and $\{f^{(e)}\}$.

The general form of the finite element equations for potential energy formulations is

$$\frac{\partial \Pi}{\partial \{U\}} = [K]\{U\} - \{F\} - \{P\} = \{0\} \quad (18.16)$$

The global stiffness matrix $[K]$ and the global force vector $\{F\}$ come from element contributions whereas the force vector $\{P\}$ results from defining a force for every possible displacement. The force vector $\{P\}$ exists for all structural and solid mechanics problems and needs no further discussion. The element contributions to $[K]$ and $\{F\}$ are the items that need to be discussed.

The total potential energy in a structural system consists of the sum of the element contribution minus the work done by the concentrated forces and moments applied at the nodes

$$\Pi = \sum_{e=1}^{n} \Pi^{(e)} - \sum_{i=1}^{r} U_i P_i \qquad (18.17)$$

where r is the total number of displacements and is usually a multiple of the number of nodes. The quantity $\Pi^{(e)}$ consists of the element strain energy minus the work terms that are element-related.

$$\Pi^{(e)} = \Lambda^{(e)} - W^{(e)} \qquad (18.18)$$

Examples of $W^{(e)}$ include the work done by body forces and distributed loads acting within and on the boundary of the element, respectively.

Equation β in the final system is

$$\frac{\partial \Pi}{\partial U_\beta} = \sum_{e=1}^{n} \frac{\partial \Pi^{(e)}}{\partial U_\beta} - P_\beta = 0 \qquad (18.19)$$

where $\partial \Pi^{(e)}/\partial U_\beta = 0$ unless β is one of the element displacements. The element contribution to equation β comes from evaluating $\partial \Pi^{(e)}/\partial U_\beta$. The element contribution to the final system of equations comes from evaluating $\partial \Pi^{(e)}/\partial \{U^{(e)}\}$, where $\{U^{(e)}\}$ contains the element displacements. The element contribution is

$$\frac{\partial \Pi^{(e)}}{\partial \{U^{(e)}\}} = [k^{(e)}]\{U^{(e)}\} - \{f^{(e)}\} \qquad (18.20)$$

The theoretical developments in structural and solid mechanics applications always produce a $\Pi^{(e)}$ that has the matrix form

$$\Pi^{(e)} = \tfrac{1}{2} \{U^{(e)}\}^T [A]\{U^{(e)}\} - \{U^{(e)}\}^T \{C\} \qquad (18.21)$$

where $[A]$ is symmetric and $\{C\}$ is a column vector. The fact to be proved here is that $[A]$ and $\{C\}$ in (18.21) are actually $[k^{(e)}]$ and $\{f^{(e)}\}$.

Differentiating (18.21) with respect to $\{U^{(e)}\}$ and using the rules of matrix calculus developed in Appendix II produces

$$\frac{\partial \Pi^{(e)}}{\partial \{U^{(e)}\}} = \frac{1}{2} [A]^T \{U^{(e)}\} + \frac{1}{2} [A]\{U^{(e)}\} - \{C\} \qquad (18.22)$$

Since $[A] = [A]^T ([A]$ is symmetric),

$$\frac{\partial \Pi^{(e)}}{\partial \{U^{(e)}\}} = [A]\{U^{(e)}\} - \{C\} \qquad (18.23)$$

Equating (18.20) and (18.23) gives

$$[k^{(e)}] = [A] \quad \text{and} \quad \{f^{(e)}\} = \{C\} \qquad (18.24)$$

The result is very useful. The element stiffness matrix and force vector are known as soon as $\Pi^{(e)}$ has been written in a form similar to (18.21). It is relatively easy to write $\Pi^{(e)}$ in this form, much easier than obtaining an explicit equation for $\Pi^{(e)}$

and than differentiating the equation with respect to the element displacements. The strain energy equation for the axial force member, (18.6), has the matrix form

$$
\Lambda^{(e)} = \tfrac{1}{2} \quad [U_i \quad U_j] \begin{bmatrix} \dfrac{AE}{L} & -\dfrac{AE}{L} \\[2mm] -\dfrac{AE}{L} & \dfrac{AE}{L} \end{bmatrix} \left\{ \begin{matrix} U_i \\ U_j \end{matrix} \right\}
$$

$$
- \lfloor U_i \quad U_j \rfloor \left\{ \begin{matrix} -AE\,\alpha\delta T \\ AE\,\alpha\delta T \end{matrix} \right\} \tag{18.25}
$$

The element quantities are easily identified and agree with (18.10) and (18.11).

18.4 INTERNAL FORCES

The internal forces acting at the ends of a structural element are calculated once the element displacements are known. These forces are easily calculated if an energy theorem developed by Castigliano (Langhaar, 1962) is used. This theorem states that

$$
\frac{\partial \Lambda^{(e)}}{\partial U_\beta} = S_\beta \tag{18.26}
$$

where S_β is the force acting in the direction of displacement U_β. If U_β is a rotation, then S_β is a moment. The set of element forces, $\{S^{(e)}\}$, is given by

$$
\{S^{(e)}\} = \frac{\partial \Lambda^{(e)}}{\partial \{U^{(e)}\}} = [k^{(e)}]\{U^{(e)}\} - \{f_{se}^{(e)}\} \tag{18.27}
$$

where $\{f_{se}^{(e)}\}$ is the strain energy's contribution to $\{f^{(e)}\}$. The internal forces at the nodes of an element can be calculated using the element stiffness matrix and a part of the element force vector. The physical meaning of the components of $\{S^{(e)}\}$ changes with the element under consideration.

Applying (18.27) to the axial force element gives

$$
\left\{ \begin{matrix} S_i^{(e)} \\ S_j^{(e)} \end{matrix} \right\} = \frac{AE}{L} \begin{bmatrix} 1 & -1 \\ -1 & 1 \end{bmatrix} \left\{ \begin{matrix} U_i \\ U_j \end{matrix} \right\} - \left\{ \begin{matrix} -AE\,\alpha\delta T \\ AE\,\alpha\delta T \end{matrix} \right\} \tag{18.28}
$$

where $S_i^{(e)}$ and $S_j^{(e)}$ are the axial forces at nodes i and j, respectively. A positive value indicates that the force is in the direction of a positive displacement. The application of (18.28) is shown in the following example.

ILLUSTRATIVE EXAMPLE

Calculate the internal axial force in member two of the axial force system shown in Figure 18.1. The displacements of nodes two and three were calculated in Chapter 17, (17.28), and are

$$
U_2 = -0.0004900 \text{ cm} \quad \text{and} \quad U_3 = -0.003480 \text{ cm}
$$

The physical parameters for member two are

$$\frac{AE}{L}=4(10^6) \qquad \text{and} \qquad AE\,\alpha\delta T = 33000$$

Substituting these values and the calculated displacements in (18.28) gives

$$\begin{Bmatrix} S_2^{(2)} \\ S_3^{(2)} \end{Bmatrix} = (10^6)\begin{bmatrix} 4 & -4 \\ -4 & 4 \end{bmatrix}\begin{Bmatrix} -0.0004900 \\ -0.003480 \end{Bmatrix} - \begin{Bmatrix} -33000 \\ 33000 \end{Bmatrix}$$

$$= \begin{Bmatrix} 11960 \\ -11960 \end{Bmatrix} - \begin{Bmatrix} -33000 \\ 33000 \end{Bmatrix} = \begin{Bmatrix} 44960 \\ -44960 \end{Bmatrix}$$

The positive value for $S_2^{(2)}$ indicates that its direction is the same as that of a positive U_2. The minus sign on $S_3^{(2)}$ indicates that its direction is opposite to that of a positive U_3. The signs lead us to the conclusion that member two is in compression.

PROBLEMS

18.1 Solve Problem 17.1 using the direct stiffness procedure discussed in this chapter.

18.2 Solve Problem 17.2 using the direct stiffness procedure discussed in this chapter.

18.3 Solve Problem 17.3 using the direct stiffness procedure discussed in this chapter.

18.4 Solve Problem 17.4 using the direct stiffness procedure discussed in this chapter.

18.5 Solve Problem 17.5 using the direct stiffness procedure discussed in this chapter.

Chapter 19

THE TRUSS ELEMENT

The efficient implementation of the displacement method of structural analysis utilizes an element stiffness matrix and force vector. A general procedure for determining these quantities was developed in the previous chapter. This chapter is the first of six in which the derivation of specific element matrices is considered. The truss element, a member pinned at each end, is discussed in this chapter. It is an ideal starting point for the discussion of structural elements because the element equations are related to the axial force member discussed in the previous two chapters.

19.1 THE STRUCTURAL MODEL

An axial force member in a truss has an arbitrary orientation and is connected to other members with arbitrary orientations. The joint displacements of a loaded truss are generally neither horizontal nor vertical, although they can be resolved into horizontal and vertical components. The accepted procedure in structural analysis is to use the displacement components as the unknown quantities rather than the resultant displacement and its direction.

The displacement components at an arbitrary joint, i, are shown in Figure 19.1a. Both are denoted by U with a subscript to indicate the difference between them. The horizontal component is always U_{2i-1} whereas the vertical component is U_{2i}. The two displacements carry consecutive numerical subscripts that are calculated using the joint number. For example, if $i=5$, then the joint displacements are U_9 and U_{10} (Figure 19.1b).

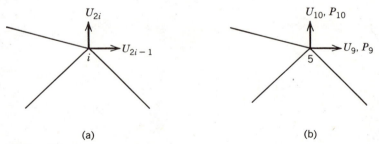

Figure 19.1. Displacement notation for a truss joint. (a) General notation. (b) Notation for joint five.

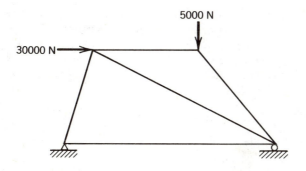

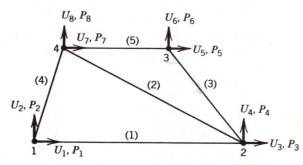

Figure 19.2. Nodal displacements and forces for a specific truss.

The external force acting on a joint can also be resolved into components parallel to the displacements. These components are P_{2i-1} and P_{2i}. They carry the same subscripts as the displacements, and they are positive when they act in the same direction as a positive displacement.

The idealization of a two-dimensional structure parallels the procedure used in the previous chapters. The members and nodes are labeled, and the known support conditions and joint loads are identified. The implementation of this procedure for a pin-connected truss is shown in Figure 19.2. The structure is always separated from its supports during the idealization process. The member numbers are in parentheses.

The support conditions for the truss in Figure 19.2 are $U_1 = U_2 = U_4 = 0$. These are incorporated after the system of displacement equations is developed. The loading conditions are $P_7 = 30000$ N, $P_6 = -50000$ N, and $P_1 = P_2 = P_3 = P_4 = P_8 = P_5 = 0$.

19.2 THE ELEMENT MATRICES

The axial force member under discussion has an arbitrary orientation; therefore, the x, y-coordinate system, which is parallel to the displacements, is generally not parallel with the member. A member coordinate system, $(\bar{x}, \bar{y})$ with the origin at node i and $\bar{x}$ directed along the member is introduced. The general orientation of the truss element and the two coordinate systems is shown in Figure 19.3. The member has a cross-sectional area A, an elastic modulus E, a length L, a thermal coefficient α, and a temperature change δT. It is oriented at an angle θ from the x-axis.

The strain energy in the member coordinate system is

$$\Lambda^{(e)} = \tfrac{1}{2}\{\bar{U}^{(e)}\}^T[\bar{k}^{(e)}]\{\bar{U}^{(e)}\} - \{\bar{U}^{(e)}\}^T\{\bar{f}^{(e)}\} \tag{19.1}$$

where $\{\bar{U}^{(e)}\}$ contains the nodal displacements parallel to the member. The element matrices are denoted as $[\bar{k}^{(e)}]$ and $\{\bar{f}^{(e)}\}$ because they are for the member co-ordinate system. They are the same as those given in (18:10) and (18.11).

The displacement vector for the truss element is $\{U^{(e)}\}$

$$\{U^{(e)}\}^T = [U_{2i-1} \quad U_{2i} \quad U_{2j-1} \quad U_{2j}] \tag{19.2}$$

The two sets of displacements are shown in Figure 19.3. The strain energy equation, (19.1), can be written in terms of $\{U^{(e)}\}$ if a relationship between $\{\bar{U}^{(e)}\}$ and $\{U^{(e)}\}$ is available.

After analyzing Figure 19.3, it becomes clear that

$$\bar{U}_i = U_{2i-1}\cos\theta + U_{2i}\sin\theta$$

and

$$\tag{19.3}$$

$$\bar{U}_j = U_{2j-1}\cos\theta + U_{2j}\sin\theta$$

The two sets of displacements are related by

$$\left\{\begin{matrix}\bar{U}_i\\\bar{U}_j\end{matrix}\right\} = \begin{bmatrix}\cos\theta & \sin\theta & 0 & 0\\0 & 0 & \cos\theta & \sin\theta\end{bmatrix}\left\{\begin{matrix}U_{2i-1}\\U_{2i}\\U_{2j-1}\\U_{2j}\end{matrix}\right\} \tag{19.4}$$

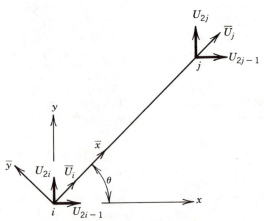

Figure 19.3. The two sets of displacements for a general truss element.

or

$$\{\bar{U}^{(e)}\}=[T]\{U^{(e)}\} \tag{19.5}$$

where $[T]$ is the transformation matrix

$$[T]=\begin{bmatrix} \cos\theta & \sin\theta & 0 & 0 \\ 0 & 0 & \cos\theta & \sin\theta \end{bmatrix} \tag{19.6}$$

Using (19.5) and the transpose property of matrix algebra allows the strain energy to be written as

$$\Lambda^{(e)}=\tfrac{1}{2}\{U^{(e)}\}^T[T]^T[\bar{k}^{(e)}][T]\{U^{(e)}\}-U^{(e)}\}^T[T]^T\{\bar{f}^{(e)}\} \tag{19.7}$$

By using the results of the previous chapter, the element matrices are easily recognized as

$$[k^{(e)}]=[T]^T[\bar{k}^{(e)}][T] \tag{19.8}$$

and

$$\{f^{(e)}\}=[T]^T\{\bar{f}^{(e)}\} \tag{19.9}$$

Substitution of (18.10) for $[\bar{k}^{(e)}]$ and (18.11) for $\{\bar{f}^{(e)}\}$ along with (19.6) and performing the matrix operations give

$$[k^{(e)}]=\frac{AE}{L}\begin{bmatrix} \overset{2i-1}{C^2} & \overset{2i}{CS} & \overset{2j-1}{-C^2} & \overset{2j}{-CS} \\ CS & S^2 & -CS & -S^2 \\ -C^2 & -CS & C^2 & CS \\ -CS & -S^2 & CS & S^2 \end{bmatrix}\begin{matrix} 2i-1 \\ 2i \\ 2j \\ 2j-1 \end{matrix} \tag{19.10}$$

and

$$\{f^{(e)}\}=\begin{Bmatrix} -AE\alpha\delta T\,(C) \\ -AE\alpha\delta T\,(S) \\ AE\alpha\delta T\,(C) \\ AE\alpha\delta T\,(S) \end{Bmatrix}\begin{matrix} 2i-1 \\ 2i \\ 2j-1 \\ 2j \end{matrix} \tag{19.11}$$

where $C=\cos\theta$ and $S=\sin\theta$.

The first row of $[k^{(e)}]$ and $\{f^{(e)}\}$ contributes to the minimization equation $\partial\Pi/\partial U_{2i-1}=0$; therefore, each coefficient in this row is located in row $2i-1$ of $[K]$ and $\{F\}$. For example, if $i=6$, the coefficients in the first row of $[k^{(e)}]$ and $\{f^{(e)}\}$ are located in row 11 of $[K]$ and $\{F\}$. In general, the rows of $[k^{(e)}]$ and $\{f^{(e)}\}$ are associated with rows $2i-1, 2i, 2j-1$, and $2j$ of $[K]$ and $\{F\}$. The columns of $[k^{(e)}]$ are associated with columns $2i-1, 2i, 2j-1$, and $2j$ of $[K]$. The indices for the rows and columns are given with $[k^{(e)}]$ and $\{f^{(e)}\}$ in (19.10) and (19.11). These indices are the same and in the same order as the subscripts on the displacements in $\{U^{(e)}\}$. This is true for all solid and structural mechanics problems.

Equations (19.10) and (19.11) are the element matrices for the truss element.

They are used in the formulation of the global stiffness matrix $[K]$ and global force vector $\{F\}$ as discussed in Chapter 18. They can also be used to calculate the horizontal and vertical force components at each node. The axial force directed along the member can be obtained by resolving S_{2j-1} and S_{2j} into a single force, or it can be calculated using a variation of (18.28) written in the member coordinate system.

$$\left\{ \begin{matrix} \bar{S}_i^{(e)} \\ \bar{S}_j^{(e)} \end{matrix} \right\} = \frac{AE}{L} \begin{bmatrix} 1 & -1 \\ -1 & 1 \end{bmatrix} \left\{ \begin{matrix} \bar{U}_i \\ \bar{U}_j \end{matrix} \right\} - \left\{ \begin{matrix} -AE\,\alpha\delta T \\ AE\,\alpha\delta T \end{matrix} \right\} \tag{19.12}$$

Substituting (19.3) for $\bar{U}_i$ and $\bar{U}_j$ and evaluating the matrix product gives

$$\bar{S}_i^{(e)} = \frac{AE}{L}\left[(U_{2i-1} - U_{2j-1})\cos\theta + (U_{2i} - U_{2j})\sin\theta \right] + AE\,\alpha\delta T \tag{19.13}$$

$$\bar{S}_j^{(e)} = \frac{AE}{L}\left[(U_{2j-1} - U_{2i-1})\cos\theta + (U_{2j} - U_{2i})\sin\theta \right] - AE\,\alpha\delta T \tag{19.14}$$

Since $\bar{S}_i^{(e)} = -\bar{S}_j^{(e)}$, only one equation, (19.13) or (19.14), needs to be evaluated. The force at j, $\bar{S}_j^{(e)}$, is usually selected because the force in the member can be interpreted as tension if $\bar{S}_j^{(e)}$ is positive and compression if $\bar{S}_j^{(e)}$ is negative.

19.3   ANALYSIS OF A PINNED TRUSS

The objective here is to analyze the three-member pinned truss shown in Figure 19.4. The truss is statically determinant, but the procedure applies equally well to statically indeterminate trusses. The truss experiences a temperature change of 5°C. The members are made from steel with the properties $E = 20(10^6)$ N/cm^2 and $\alpha = 11(10^{-6})$°C. A 100-kN load is applied as shown.

The truss is modeled using three line segments to represent the members. The

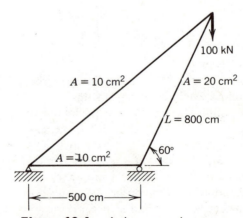

Figure 19.4.   A three-member truss.

nodes and members are numbered counterclockwise, although the direction is not important. The nodal displacements are shown in Figure 19.5. The known displacements are $U_1 = U_2 = U_4 = 0$ because of the pinned support at node one and the roller at node two. The known forces are $P_1 = P_2 = P_3 = P_4 = P_5 = 0$, and $P_6 = -100 \text{ kN}$.

The first step is to calculate the element matrices and merge them to obtain the global stiffness matrix $[K]$ and the global force vector $\{F\}$. The general system of equations is

$$[K]\{U\} - \{F\} - \{P\} = \{0\}$$

where

$$\{P\}^T = [0 \quad 0 \quad 0 \quad 0 \quad 0 \quad -100000]$$

The first step in calculating the element matrices is to define the origin of the local coordinate system. This is the same as defining node i of the element. Member one is defined as going from $i = 1$ to $j = 3$, member two from $i = 1$ to $j = 2$, and member three from $i = 3$ to $j = 2$. The orientation of each member is shown in Figure 19.6.

The pertinent member properties can be tabulated. They are

Member	A	L	$\cos \theta$	$\sin \theta$
1	10	1135	0.7928	0.6095
2	20	500	1.0	0.0
3	20	800	-0.5	-0.8660

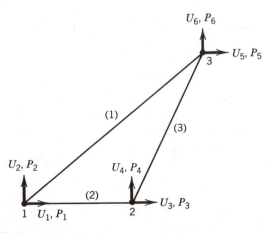

Figure 19.5.   Nodal displacements for the three-member truss.

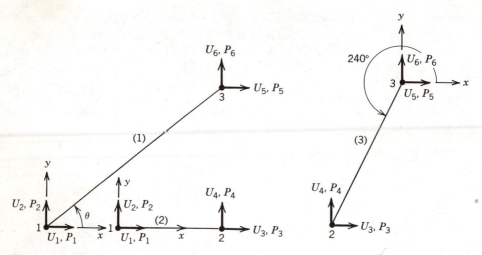

Figure 19.6. The local coordinate system and nodal displacements for each member in the truss example.

A study of the element matrices indicates that there are five quantities that need to be evaluated for each member.

	Member		
	1	2	3
$(AE \cos^2 \theta)/L$	110800	800000	125000
$(AE \cos \theta \sin \theta)/L$	85150	0	216500
$(AE \sin^2\theta)/L$	65460	0	375000
$AE \, \alpha\delta T \cos \theta$	8721	22000	-11000
$AE \, \alpha\delta T \sin \theta$	6705	0	-19050

The element stiffness matrices and their degrees of freedom are

$$
[k^{(1)}] =
\begin{matrix}
& 1 & 2 & 5 & 6 & \\
& \begin{bmatrix}
110800 & 85150 & -110800 & -85150 \\
85150 & 65460 & -85150 & -65460 \\
-110800 & -85150 & 110800 & 85150 \\
-85150 & -65460 & 85150 & 65460
\end{bmatrix} &
\begin{matrix} 1 \\ 2 \\ 5 \\ 6 \end{matrix}
\end{matrix}
$$

$$
[k^{(2)}] =
\begin{matrix}
& 1 & 2 & 3 & 4 & \\
& \begin{bmatrix}
800000 & 0 & -800000 & 0 \\
0 & 0 & 0 & 0 \\
-800000 & 0 & 800000 & 0 \\
0 & 0 & 0 & 0
\end{bmatrix} &
\begin{matrix} 1 \\ 2 \\ 3 \\ 4 \end{matrix}
\end{matrix}
$$

$$[k^{(3)}] = \begin{array}{c} 5 6 3 4 \end{array}$$

$$[k^{(3)}] = \begin{bmatrix} 125000 & 216500 & -125000 & -216500 \\ 216500 & 375000 & -216500 & -375000 \\ -125000 & -216500 & 125000 & 216500 \\ -216500 & -375000 & 216500 & 375000 \end{bmatrix} \begin{array}{c} 5 \\ 6 \\ 3 \\ 4 \end{array}$$

The element force vectors are

$$\{f^{(1)}\} = \begin{Bmatrix} -8721 \\ -6705 \\ 8721 \\ 6705 \end{Bmatrix} \begin{array}{c} 1 \\ 2 \\ 5 \\ 6 \end{array}, \quad \{f^{(2)}\} = \begin{Bmatrix} -22000 \\ 0 \\ 22000 \\ 0 \end{Bmatrix} \begin{array}{c} 1 \\ 2 \\ 3 \\ 4 \end{array}$$

and

$$\{f^{(3)}\} = \begin{Bmatrix} 11000 \\ 19050 \\ -11000 \\ -19050 \end{Bmatrix} \begin{array}{c} 5 \\ 6 \\ 3 \\ 4 \end{array}$$

Adding the element stiffness matrices using the direct stiffness method gives

$$[K] = \begin{bmatrix} 910800 & 85150 & -800000 & 0 & -110800 & -85150 \\ 85150 & 65460 & 0 & 0 & -85150 & -65460 \\ -800000 & 0 & 925000 & 216500 & -125000 & -216500 \\ 0 & 0 & 216500 & 375000 & -216500 & -375000 \\ -110800 & -85150 & -125000 & -216500 & 235800 & 301700 \\ -85150 & -65460 & -216500 & -375000 & 301700 & 440400 \end{bmatrix}$$

Three checks can be made on the accuracy of $[K]$. First, it should be symmetric. Second, each of the diagonal coefficients should be positive. Third, the sum of each row and column should be zero. Each of these checks comes from similar properties that exist for $[k^{(e)}]$. Since $[K]$ as given above satisfies all of these properties, the calculations are reasonably correct. There is always a possibility of errors in the basic calculations.

The coefficients of the global force vector come from the sum of the three element force vectors added in the direct stiffness sense. The vector is

$$\{F\} = \begin{Bmatrix} -30720 \\ -6705 \\ 11000 \\ -19050 \\ 19720 \\ 25760 \end{Bmatrix}$$

The final system of equations is obtained by deleting rows one, two, and four from $[K]$ and $\{F\}$. Deleting these equations, combining the $\{F\}$ and $\{P\}$ vectors, and noting that $U_1 = U_2 = U_4 = 0$ yields

$$
\begin{bmatrix}
92500 & -125000 & -216500 \\
-125000 & 235800 & 301700 \\
-216500 & 301700 & 440500
\end{bmatrix}
\begin{Bmatrix} U_3 \\ U_5 \\ U_6 \end{Bmatrix}
=
\begin{Bmatrix} 11000 \\ 19720 \\ -74240 \end{Bmatrix}
$$

as the system of equations that must be solved for the unknown displacements. Using the Gaussian elimination process gives these displacements as

$$U_3 = -0.1020 \text{ cm} \qquad U_5 = 2.500 \text{ cm} \qquad \text{and} \qquad U_6 = -1.930 \text{ cm}$$

The axial force in each member is calculated using (19.14).

Member One. $i=1, j=3$.

$$\bar{S}_3^{(1)} = \frac{AE}{L}\left[(U_5 - U_1)\cos\theta + (U_6 - U_2)\sin\theta\right] - AE\,\alpha\delta T$$

$$= 176200[2.5(0.7928) + (-1.930)(0.6095)] - 11000$$

$$= 131000 \text{ N}$$

Member Two. $i=1, j=2$.

$$\bar{S}_2^{(2)} = \frac{AE}{L}\left[(U_3 - U_1)\cos\theta + (U_4 - U_2)\sin\theta - AE\,\alpha\delta T\right.$$

$$= 800000[(-0.1020)(1) + (0)(0)] - 22000$$

$$= -103600 \text{ N}$$

Member Three. $i=3, j=2$.

$$\bar{S}_3^{(3)} = \frac{AE}{L}\left[(U_3 - U_5)\cos\theta + (U_4 - U_6)\sin\theta\right] - AE\,\alpha\delta T$$

$$= 500000[(-0.1020 - 2.50)(-0.50) + 0 - (-1.930)(-0.8660)] - 22000$$

$$= -207200 \text{ N}$$

The member forces are shown on the free body diagram in Figure 19.7. A (T) means tension and (C) means compression.

The member forces of a statically determinant truss can be obtained faster using statics, but the displacement method of analysis gives the joint displacements and the member forces. We now know that joint three moves 2.50 cm to the right and downward 1.930 cm. A statics analysis does not give this information, and it is not easily obtained using "strength of materials" techniques.

A comment on the global stiffness matrix $[K]$ closes this chapter. The stiffness matrix is singular and the system of equations cannot be solved until at least three support conditions are incorporated. These conditions must prevent rigid body translations and rotation. If the matrix remains singular after this fixation, the structural system is kinematically unstable and it will collapse under a load.

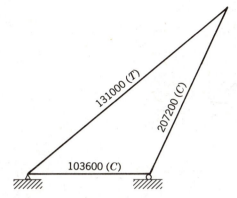

Figure 19.7. The calculated force in each member of the truss example.

PROBLEMS

19.1–19.6 Each joint of the structural systems shown here is a pinned joint. The node numbers and element numbers are given on the sketch. The cross-sectional area of each member in cm² is underlined. Each member is made from steel, $E = 20(10^6)$ N/cm². All lengths are given in centimeters. Calculate the unknown nodal displacements and the axial force in each member, and show that the calculated axial forces produce a system that is in equilibrium. There is no temperature change.

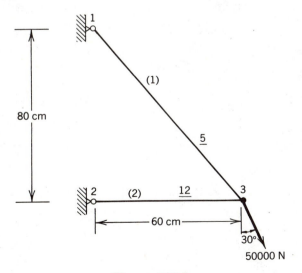

Figure P19.1

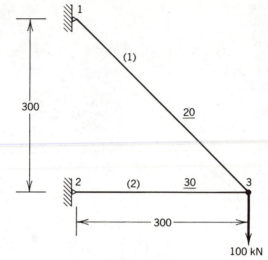

Figure P19.2

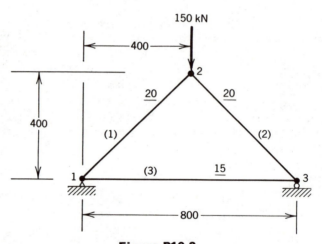

Figure P19.3

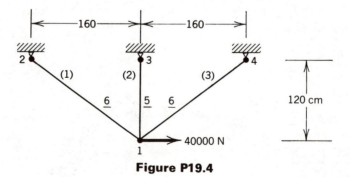

Figure P19.4

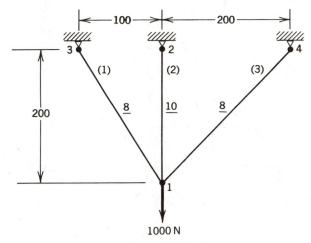

Figure P19.5

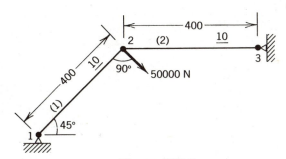

Figure P19.6

19.7 The displacements for the truss shown in Figure P19.7 are

$$
\begin{array}{ll}
U_1 = 0.0 & U_2 = 0.0 \\
U_3 = 1.748 & U_4 = -0.3333 \\
U_5 = -0.5000 & U_6 = 0.0 \\
U_7 = 2.248 & U_8 = -3.690
\end{array}
$$

The area of each member in square centimeters is underlined. The length dimensions are in centimeters and each member is made from steel, $E = 20(10^6)\,\text{N/cm}^2$. Calculate the axial force in each member connected to joint two and show that this joint is in equilibrium. There is no temperature change.

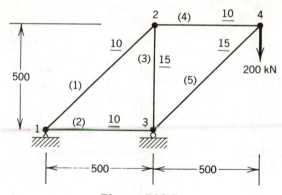

Figure P19.7

19.8 Do Problem 19.7 for joint four.
19.9 Do Problem 19.7 for joint three.
19.10 The displacements for the truss shown in Figure P19.10 are

$$U_1 = -0.3333 \qquad U_2 = -2.780$$
$$U_3 = 0.8700 \qquad U_4 = -1.847$$
$$U_5 = -0.2667 \qquad U_6 = -0.9590$$
$$U_7 = 0.7812 \qquad U_8 = 0.0$$
$$U_9 = 0.0 \qquad U_{10} = 0.0$$

The area of each member in square centimeters is underlined. The length dimensions are in centimeters and each member is made from steel, $E = 20(10^6)\,\text{N/cm}^2$. Calculate the axial force in each member connected to joint one and show that this joint is in equilibrium. There is no temperature change.

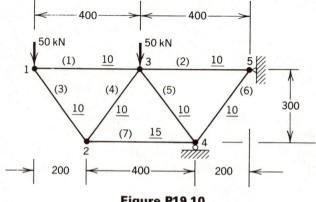

Figure P19.10

19.11 Do Problem 19.10 for joint two.

19.12 Do Problem 19.10 for joint three.

19.13 Verify to your own satisfaction, by evaluating the element stiffness matrices, that it does not make any difference in the final results which node of the element shown in Figure P19.13 is denoted as node *i*.

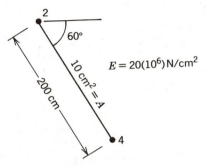

Figure P19.13

19.14 Calculate the axial force in each member of the truss system shown in Figure P19.14 using the computer program **FRAME** discussed in Chapter 25. The truss is made from steel, $E = 20(10^6)\,\text{N/cm}^2$. The cross-sectional area of each member in square centimeters is underlined. All length dimensions are in centimeters. There is no temperature change.

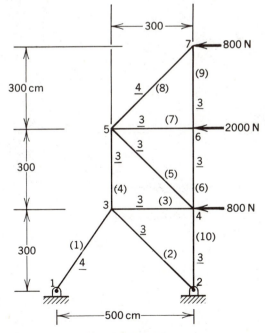

Figure P19.14

19.15 Do Problem 19.14 for the configuration shown in Figure P19.15.

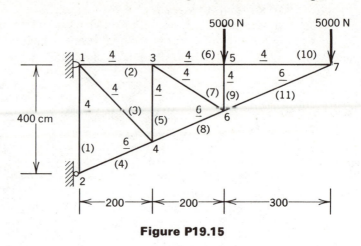

Figure P19.15

19.16 Do Problem 19.14 for the configuration shown in Figure P19.16.

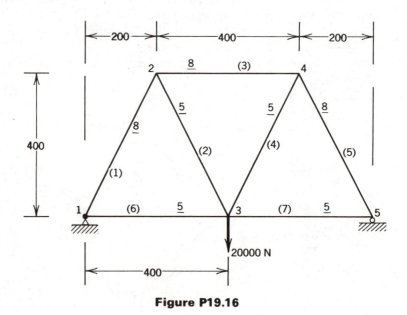

Figure P19.16

Chapter 20

A BEAM ELEMENT

Any member subjected to a transverse load is a beam. The transverse load induces normal and shear stresses within the beam and displacements perpendicular to its longitudinal axis. Beams can be straight or curved. They can have a constant depth or the depth can vary with the length. The cross-sectional shape can have two axes of symmetry, such as the I-beam, one axis of symmetry, such as a T-beam, or no axes of symmetry. The loading can consist of concentrated and/or distributed forces, and the forces can act in one or more planes. Any systematic method of beam analysis must be capable of handling the numerous possible combinations of loads and shapes that can occur.

This chapter contains an introduction to beam analysis. The discussion is limited to straight beams with a uniform depth and at least one axis of symmetry. The discussion is limited to concentrated loads; these loads must lie in the plane of symmetry.

20.1 THE STRUCTURAL MODEL

A straight beam is modeled by a line segment with a vertical displacement and a rotation at each end. The beam has a length L and a section property EI that combines the elastic modulus E with the area moment I (Figure 20.1).

The specific nodal displacements are vertical translations, U_{2i-1} and U_{2j-1}, and rotations U_{2i} and U_{2j}. Positive translations are in the positive y-coordinate direction. Positive rotations are counterclockwise. These positive directions are

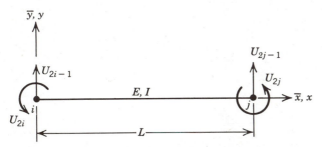

Figure 20.1.   The beam element and its nodal displacements.

shown in Figure 20.1. The element vector of nodal displacements, $\{U^{(e)}\}$, is

$$\{U^{(e)}\}^T = [U_{2i-1} \quad U_{2i} \quad U_{2j-1} \quad U_{2j}] \tag{20.1}$$

where the even-numbered displacements are rotations, whereas the odd-numbered displacements are vertical translations.

The external forces acting on the joints of a beam consist of vertically directed concentrated loads and concentrated moments. The direction of a positive force coincides with that of a positive displacement. The word *force* is used here in a generalized sense. It denotes either a concentrated force or a concentrated moment.

The guidelines for locating the nodes when modeling a continuous beam include placing nodes

1. At each external support.
2. At each end of the beam.
3. Wherever the section property EI changes.
4. Wherever there is a concentrated moment.
5. Wherever there is a concentrated force.
6. Wherever the value of the deflection is needed.

A beam subjected to two concentrated forces and a concentrated moment (Figure 20.2) is modeled using three elements. Nodes are located at both concentrated forces as well as the two supports. The elements and nodes are numbered in the usual manner. There are eight nodal displacements: four vertical and four rotational displacements. The vertical displacements, U_1 and U_5, are zero as well as the rotation U_2. The nonzero concentrated forces are $P_3 = -15000$ N and $P_7 = 10000$ N. The nonzero concentrated moment is $P_4 = 20000$ N · cm. The other joint forces and moments, P_1, P_2, P_5, P_6, and P_8, are zero.

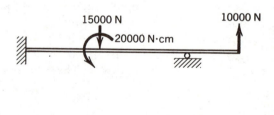

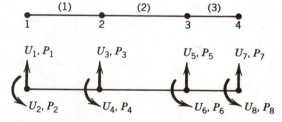

Figure 20.2.   The elements and nodal displacements for a specific beam.

20.2   THE STRAIN ENERGY EQUATION

The total potential energy for a continuous beam with n elements is

$$\Pi = \sum_{e=1}^{n} \Lambda^{(e)} - \{U\}^T \{P\} \tag{20.2}$$

where $\{U\}$ is the vector containing all of the possible nodal values and $\{P\}$ is the vector of joint forces.

In order to utilize the principle of minimum potential energy, $\Lambda^{(e)}$ must be written in terms of the joint displacements. This is a two-part process. An equation for the strain energy in a beam must be developed; then an equation for the beam deflection must be specified. The strain energy equation is developed in this section and the deflection equation in the next.

The strain energy in a beam is

$$\Lambda^{(e)} = \frac{E}{2} \int_V \varepsilon_{xx}^2 \, dV \tag{20.3}$$

assuming that the contribution from the shear stress τ_{xy} is negligible. The total strain e_{xx} is related to the beam deflection, v, by

$$e_{xx} = -y \frac{d^2v}{dx^2} \tag{20.4}$$

where a negative strain occurs at the top of the beam for positive curvature. Equation (20.4) is derived in all basic books covering the mechanics of solids, (Popov, 1976 or Higdon, et al., 1976). In our case, $e_{xx} = \varepsilon_{xx}$, since we will not include a thermal change.

Substitution of (20.4) into (20.3) gives

$$\Lambda^{(e)} = \frac{E}{2} \int_V \left(-y \frac{d^2v}{dx^2}\right)^2 dV = \frac{E}{2} \int_V y^2 \left(\frac{d^2v}{dx^2}\right)^2 dV \tag{20.5}$$

If we note that $dV = dA \, dx$, (20.5) can be written as

$$\Lambda^{(e)} = \frac{E}{2} \int_0^L \left(\frac{d^2v}{dx^2}\right)^2 \left(\int_A y^2 \, dA\right) dx \tag{20.6}$$

The area integral is the definition of the area moment of inertia, I; therefore,

$$\Lambda^{(e)} = \frac{EI}{2} \int_0^L \left(\frac{d^2v}{dx^2}\right)^2 dx \tag{20.7}$$

Equation (20.7) gives the strain energy in terms of the beam deflection, v. This integral cannot be evaluated until an equation, $v = f(x)$, which accurately describes the beam deflection, has been specified.

20.3   THE DISPLACEMENT EQUATION

The selection of a displacement equation to describe the action of a beam subjected to different types of loadings is relatively easy because there are several guidelines. First, there is the governing differential equation for beam deflections

$$\frac{d^4v}{dx^4} = w(x) \tag{20.8}$$

where $w(x)$ is the distributed load (Popov, 1976). Second, there are four boundary conditions to impose

$$v(0) = U_{2i-1} \qquad v(L) = U_{2j-1}$$

$$\frac{dv}{dx}(0) = U_{2i} \qquad \frac{dv}{dx}(L) = U_{2j} \tag{20.9}$$

These boundary conditions require that the equation have four undetermined coefficients. A third but less obvious condition is that the equation must have a continuous first derivative; otherwise, the integral in (20.7) is not defined.

The nodal conditions of (20.9) are compatible with the differential equation (20.8), since it is a fourth-order equation. If we assume that no distributed loads, $w(x) = 0$, the differential equation becomes

$$\frac{d^4v}{dx^4} = 0 \tag{20.10}$$

which has the general solution

$$v = a_1 + a_2 x + a_3 x^2 + a_4 x^3 \tag{20.11}$$

The solution of beam problems is built around this equation.

Note that x as used in (20.10) and (20.11) is a local coordinate system rather than a global system. The x-coordinate as used here is identical to the s-coordinate discussed in Chapter 6.

Application of the boundary conditions (20.9) to (20.11) generates the system of equations

$$U_{2i-1} = a_1$$

$$U_{2i} = a_2$$

$$U_{2j-1} = a_1 + a_2 L + a_3 L^2 + a_4 L^3$$

$$U_{2j} = a_2 L + 2a_3 L + 3a_4 L^2 \tag{20.12}$$

which can be solved for a_1, a_2, a_3, and a_4 yielding

$$a_1 = U_{2i-1}$$

$$a_2 = U_{2i}$$

$$a_3 = \frac{3}{L^2}(U_{2j-1} - U_{2i-1}) - \frac{1}{L}(2U_{2i} + U_{2j})$$

$$a_4 = \frac{2}{L^3}(U_{2i-1} - U_{2j-1}) + \frac{1}{L^2}(U_{2i} + U_{2j})$$

Substituting these results into (20.10) and rearranging gives

$$v = N_{2i-1}U_{2i-1} + N_{2i}U_{2i} + N_{2j-1}U_{2j-1} + N_{2j}U_{2j} \tag{20.14}$$

where

$$N_{2i-1} = 1 - \frac{3x^2}{L^2} + \frac{2x^3}{L^3}, \qquad N_{2i} = x - \frac{2x^2}{L} + \frac{x^3}{L^2}$$

$$N_{2j-1} = \frac{3x^2}{L^2} - \frac{2x^3}{L^3}, \qquad N_{2j} = -\frac{x^2}{L} + \frac{x^3}{L^2} \tag{20.15}$$

Equation (20.14) can also be written in the familiar form

$$v = [N_{2i-1} \quad N_{2i} \quad N_{2j-1} \quad N_{2j}]\{U^{(e)}\} = [N]\{U^{(e)}\} \tag{20.16}$$

The shape functions in (20.15) belong to the class of interpolating polynomials known as the Hermite polynomials (Conte and deBoor, 1981). These polynomials are used to construct interpolation functions for which the first derivative must be continuous. They possess a different set of properties from those observed for the linear shape functions of Chapters 2 and 5. The following relationships hold at nodes i and j.

At node i:

$$N_{2i-1} = 1, \qquad N_{2i} = N_{2j-1} = N_{2j} = 0$$

$$\frac{dN_{2i}}{dx} = 1, \qquad \frac{dN_{2i-1}}{dx} = \frac{dN_{2j-1}}{dx} = \frac{dN_{2j}}{dx} = 0 \tag{20.17}$$

At node j:

$$N_{2j-1} = 1, \qquad N_{2i-1} = N_{2i} = N_{2j} = 0$$

$$\frac{dN_{2j}}{dx} = 1, \qquad \frac{dN_{2i-1}}{dx} = \frac{dN_{2i}}{dx} = \frac{dN_{2j-1}}{dx} = 0$$

Also

$$N_{2i-1} + N_{2j-1} = 1 \tag{20.19}$$

for all values of x.

20.4 THE ELEMENT STIFFNESS MATRIX

The element matrices emerge after the displacement relationship (20.16) has been substituted into the strain energy equation (20.7). Differentiating the displacement equation twice gives

$$\frac{d^2v}{dx^2} = \frac{d^2[N]}{dx^2}\{U^{(e)}\} = \left[\frac{d^2N_{2i-1}}{dx^2} \quad \frac{d^2N_{2i}}{dx^2} \quad \frac{d^2N_{2j-1}}{dx^2} \quad \frac{d^2N_{2j}}{dx^2}\right]\{U^{(e)}\}$$

$$= [B_1 \quad B_2 \quad B_3 \quad B_4]\{U^{(e)}\} = [B]\{U^{(e)}\} \quad (20.20)$$

where

$$B_1 = \frac{d^2N_{2i-1}}{dx^2}, \qquad B_2 = \frac{d^2N_{2i}}{dx^2}$$

$$B_3 = \frac{d^2N_{2j-1}}{dx^2}, \qquad B_4 = \frac{d^2N_{2j}}{dx^2} \qquad (20.21)$$

The square of d^2v/dx^2 is

$$\left(\frac{d^2v}{dx^2}\right)^2 = \{U^{(e)}\}^T[B]^T[B]\{U^{(e)}\} \qquad (20.22)$$

Substituting this into (20.7) produces

$$\Lambda^{(e)} = \frac{EI}{2}\int_0^L \{U^{(e)}\}^T[B]^T[B]\{U^{(e)}\}\,dx$$

$$= \frac{1}{2}\{U^{(e)}\}^T\left(EI\int_0^L [B]^T[B]\,dx\right)\{U^{(e)}\} \qquad (20.23)$$

The product $[B]^T[B]$ is symmetric; therefore, (20.23) has the general form given in (18.21) and $[k^{(e)}]$ is identified as

$$[k^{(e)}] = EI\int_0^L [B]^T[B]\,dx \qquad (20.24)$$

The element force vector $\{f^{(e)}\}$ is zero because a temperature gradient across the beam depth was not included in the analysis.

Evaluation of the integral in (20.24) is straightforward but somewhat tedious. The matrix product to be integrated is

$$[k^{(e)}] = \int_0^L \frac{EI}{L^3}\begin{bmatrix} B_1^2 & B_1B_2 & B_1B_3 & B_1B_4 \\ B_1B_2 & B_2^2 & B_2B_3 & B_2B_4 \\ B_1B_3 & B_2B_3 & B_3^2 & B_3B_4 \\ B_1B_4 & B_2B_4 & B_3B_4 & B_4^2 \end{bmatrix}dx \qquad (20.25)$$

Consider the B_1B_2 product

$$\int_0^L B_1B_2\cdot dx = \int_0^L \left(\frac{d^2N_{2i-1}}{dx^2}\right)\left(\frac{d^2N_{2i}}{dx^2}\right)dx \qquad (20.26)$$

where

$$\frac{d^2N_{2i-1}}{dx^2} = -\frac{6}{L} + \frac{12x}{L^3} \quad \text{and} \quad \frac{d^2N_{2i}}{dx^2} = -\frac{4}{L} + \frac{6x}{L^2} \qquad (20.27)$$

thus

$$\int_0^L B_1 B_2 \, dx = 12 \int_0^L \left(\frac{2}{L^3} - \frac{7x}{L^4} + \frac{6x^2}{L^5} \right) dx = \frac{6}{L^2} \tag{20.28}$$

Evaluation of the other nine coefficients gives the element stiffness matrix

$$[k^{(e)}] = \frac{EI}{L^3} \begin{array}{cccc} 2i-1 & 2i & 2j-1 & 2j \\ \begin{bmatrix} 12 & 6L & -12 & 6L \\ 6L & 4L^2 & -6L & 2L^2 \\ -12 & -6L & 12 & -6L \\ 6L & 2L^2 & -6L & 4L^2 \end{bmatrix} & \begin{array}{c} 2i-1 \\ 2i \\ 2j-1 \\ 2j \end{array} \end{array} \tag{20.29}$$

The indices to use with the direct stiffness procedure are given in (20.29). These indices have the same values and ordering as the subscripts of $\{U^{(e)}\}$.

The four internal forces, two at each node, are calculated using (18.30). Since $\{f^{(e)}\}$ is zero, the calculation is

$$\{S^{(e)}\} = \begin{Bmatrix} Sh_i^{(e)} \\ M_i^{(e)} \\ Sh_j^{(e)} \\ M_j^{(e)} \end{Bmatrix} = [k^{(e)}]\{U^{(e)}\} \tag{20.30}$$

The shear forces, $Sh_i^{(e)}$ and $Sh_j^{(e)}$, and moments, $M_i^{(e)}$ and $M_j^{(e)}$, acting at the two nodes are shown in Figure 20.3. The shear forces and moments are positive as shown.

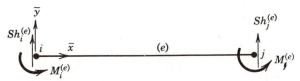

Figure 20.3. The internal shear forces and bending moments acting at each node.

20.5   ANALYSIS OF A STATICALLY INDETERMINATE BEAM

A beam, fixed at one end and supported by a roller at the other end, has a 20000 N concentrated load applied at the center of the span (Figure 20.4). Calculate the deflection under the load and construct the shear force and bending moment diagrams for the beam.

A study of the beam reveals that $U_1 = U_2 = U_5 = 0$, $P_3 = -20000$ N, and $P_1 = P_2 = P_4 = P_5 = P_6 = 0$.

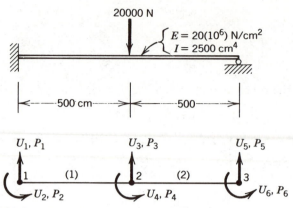

Figure 20.4. A single span beam supporting a concentrated load.

The element stiffness matrix (20.29) has four different parameters whose numerical values are

$$\frac{EI}{L^3} = \frac{2500(20)10^6}{500^3} = 400 \text{ N/cm}$$

$$6L = 6(500) = 3000 \text{ cm}$$
$$4L^2 = 4(500)^2 = 1000000 \text{ cm}^2$$
$$2L^2 = 2(500)^2 = 500000 \text{ cm}^2$$

The mixture of units occurs because the displacements have different units. The translations are in centimeters; the rotations are in radians. The values in $\{P\}$ have units of N or N·cm depending on whether they are a force or a moment.

If we use the parameters calculated above and the displacement subscripts given on the element grid (Figure 20.4), the element matrices are

$$[k^{(1)}] = 400 \begin{array}{c} \\ \begin{array}{cccc} 1 & 2 & 3 & 4 \end{array} \\ \left[\begin{array}{cccc} 12 & 3000 & -12 & 3000 \\ 3000 & 1000000 & -3000 & 500000 \\ -12 & -3000 & 12 & -3000 \\ 3000 & 500000 & -3000 & 1000000 \end{array} \right] \begin{array}{c} 1 \\ 2 \\ 3 \\ 4 \end{array} \end{array} \qquad (20.31)$$

$$[k^{(2)}] = 400 \begin{array}{c} \\ \begin{array}{cccc} 3 & 4 & 5 & 6 \end{array} \\ \left[\begin{array}{cccc} 12 & 3000 & -12 & 3000 \\ 3000 & 1000000 & -3000 & 500000 \\ -12 & -3000 & 12 & -3000 \\ 3000 & 500000 & -3000 & 1000000 \end{array} \right] \begin{array}{c} 3 \\ 4 \\ 5 \\ 6 \end{array} \end{array} \qquad (20.32)$$

Combining $[k^{(1)}]$ and $[k^{(2)}]$ using the direct stiffness method, and modifying the resulting six equations to incorporate the known displacements $U_1 = U_2 =$

$U_5 = 0$, yields the system of equations

$$400 \begin{bmatrix} 24 & 0 & 3000 \\ 0 & 2000000 & 500000 \\ 3000 & 500000 & 1000000 \end{bmatrix} \begin{Bmatrix} U_3 \\ U_4 \\ U_6 \end{Bmatrix} = \begin{Bmatrix} -20000 \\ 0 \\ 0 \end{Bmatrix}$$

The solution of this system gives

$$U_3 = -3.646 \text{ cm}, \qquad U_4 = -0.003125 \text{ rad}, \qquad \text{and} \qquad U_6 = 0.01250 \text{ rad}$$

The complete displacement vector for the beam is

$$\{U\}^T = \begin{bmatrix} 0 & 0 & -3.646 & -0.003125 & 0 & 0.01250 \end{bmatrix}$$

The internal forces at each end of an individual member are obtained by using (20.30). The calculations for member one are

$$\{S^{(1)}\} = \begin{Bmatrix} Sh_1^{(1)} \\ M_1^{(1)} \\ Sh_2^{(1)} \\ M_2^{(1)} \end{Bmatrix}, \qquad \{U^{(1)}\} = \begin{Bmatrix} 0 \\ 0 \\ -3.646 \\ -0.003125 \end{Bmatrix} \tag{20.33}$$

because $i=1$ and $j=2$ for element one. Combining these with $[k^{(1)}]$, (20.31), we find that

$$\{S^{(1)}\} = 400 \begin{bmatrix} 12 & 3000 & -12 & 3000 \\ 3000 & 1000000 & -3000 & 500000 \\ -12 & -3000 & 12 & -3000 \\ -3000 & 500000 & -3000 & 1000000 \end{bmatrix} \begin{Bmatrix} 0 \\ 0 \\ -3.646 \\ -0.003125 \end{Bmatrix}$$

and

$$\{S^{(1)}\} = \begin{Bmatrix} 13750 \text{ N} \\ 3750000 \text{ N} \cdot \text{cm} \\ -13750 \text{ N} \\ 3125000 \text{ N} \cdot \text{cm} \end{Bmatrix} \tag{20.34}$$

The forces on member one are shown in Figure 20.5a
The displacement vector $\{U^{(2)}\}$ is

$$\{U^{(2)}\}^T = \begin{bmatrix} -3.646 & -0.003125 & 0 & 0.01250 \end{bmatrix} \tag{20.35}$$

Using (20.32) for $[k^{(2)}]$ gives

$$\{S^{(2)}\} = \begin{Bmatrix} Sh_2^{(2)} \\ M_2^{(2)} \\ Sh_3^{(2)} \\ M_3^{(2)} \end{Bmatrix} = \begin{Bmatrix} -6251 \text{ N} \\ -3125000 \text{ N} \cdot \text{cm} \\ 6251 \text{ N} \\ 0 \text{ N} \cdot \text{cm} \end{Bmatrix} \tag{20.36}$$

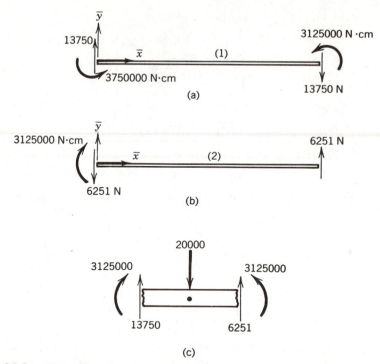

Figure 20.5. The internal nodal forces for each member of the beam example and a free body diagram of joint two.

These forces are illustrated in Figure 20.5b. Note that $M_3^{(2)}=0$ as expected at the roller support.

The accuracy of the calculations can be checked by performing an equilibrium analysis of joint two. The forces acting on this joint are shown in Figure 20.5c. Applying the equations of statics gives

$$+\uparrow \Sigma F_y = 13750 + 6251 - 20000 = 1 \approx 0$$
$$+\, \backslash \Sigma M = 3125000 - 3125000 = 0$$

The joint is in equilibrium; thus the calculations are correct.

The shear force and bending moment diagrams are shown beneath the members in Figure 20.6. Note that there are two different sign conventions involved: (1) the positive shear and moment definitions that are used in the analysis and in the interpretation of the calculated results; and (2) an entirely different sign convention used to construct the shear force and bending moment diagrams. The latter sign convention is shown in Figure 20.6. These two sign conventions can be a source of confusion. The sign convention for shear force and bending moment diagrams is not to be used until the calculated values have been interpreted using the analysis sign convention.

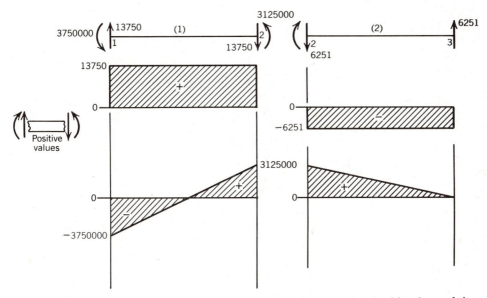

Figure 20.6.   Shear force and bending moment diagrams for the Members of the example problem.

PROBLEMS

20.1–20.12   Calculate the nodal displacements and the internal member forces for each of the beam loadings shown. Construct the shear force and bending moment diagram for each member. Check the accuracy of your solution by considering the equilibrium of each member. Use the element and nodes shown under each figure. Use $E = 20(10^6)\,\text{N/cm}^2$ and $I = 8000\,\text{cm}^2$ in each problem

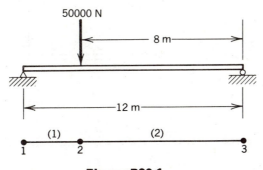

Figure P20.1

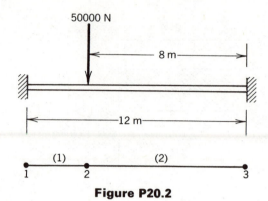

Figure P20.2

50000 N

8 m

12 m

(1)   (2)

1   2   3

50000 N

8 m

12 m

(1)   (2)

1   2   3

Figure P20.3

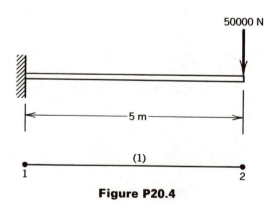

50000 N

5 m

(1)

1   2

Figure P20.4

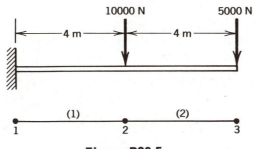

Figure P20.5

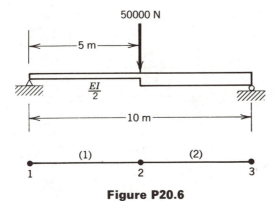

Figure P20.6

50000 N

5 m

50000 N

10 m

(1)   (2)
1   2   3

Figure P20.7

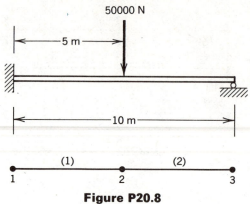

Figure P20.8

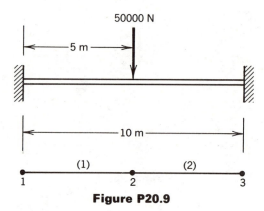

Figure P20.9

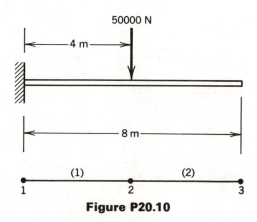

Figure P20.10

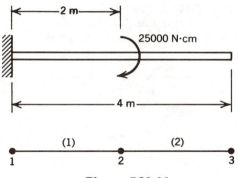

Figure P20.11

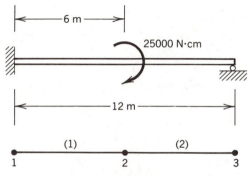

Figure P20.12

20.13 Evaluate the coefficient in $[k^{(e)}]$ resulting from

$$(a)=EI \int_0^L B_2 B_3 \, dx \qquad (b)=EI \int_0^L B_1 B_4 \, dx$$

$$(c)=EI \int_0^L B_3 B_4 \, dx \qquad (d)=EI \int_0^L B_2^2 \, dx$$

20.14 Show that the strain energy equation is

$$\Lambda^{(e)} = \frac{EI}{2} \int_0^L \left(\frac{d^2 v}{dx^2}\right)^2 dx + \frac{EI \, \alpha \delta T}{h} \int_0^L \left(\frac{d^2 v}{dx^2}\right) dx$$

when there is a linear thermal gradient across the depth of the beam. In this case,

$$e_{xx} = \varepsilon_{xx} + \alpha \left(T_a + \frac{\delta T}{h} y\right)$$

where T_a is the average temperature, δT is the temperature difference between the top and bottom of the beam, and h is the beam depth.

20.15 Use the equation for $\Lambda^{(e)}$ given in Problem 20.14 along with (20.16) and show that

$$\{f^{(e)}\} = -\frac{EI\,\alpha\delta T}{h} \int_0^L [B]^T\,dx = \frac{EI\,\alpha\delta T}{h} \begin{Bmatrix} 0 \\ 1 \\ 0 \\ -1 \end{Bmatrix}$$

20.16 Solve one or more of the beam deflection problems in 20.1 through 20.12 using the **FRAME** program discussed in Chapter 25.

Chapter 21

A PLANE FRAME ELEMENT

Bolted or welded joints are very common in machinery and building frameworks. All welded joints and any joint with two or more bolts behave as a rigid joint. All members connecting at a rigid joint translate and rotate the same amount. The members connecting to a rigid joint must support axial and shear forces as well as a bending moment. The plane frame element discussed in this chapter is used to model two-dimensional members that connect at a rigid joint.

21.1 THE STRUCTURAL MODEL

The plane frame element is a combination of the two-dimensional axial force element and the beam element. It has a horizontal and vertical displacement at each node plus a rotation (Figure 21.1). The generalized notation places consecutive subscripts on the horizontal and vertical displacements and rotation, respectively. For example, if $i=5$, then the generalized displacements are U_{13}, U_{14}, and U_{15}. The external forces applied at joint five are P_{13}, P_{14}, and P_{15}. Positive forces have the same direction as positive displacements.

The important element parameters are the elastic modulus, E; the cross-sectional area, A; the area moment, I; and the length L. A temperature change is not included in this discussion.

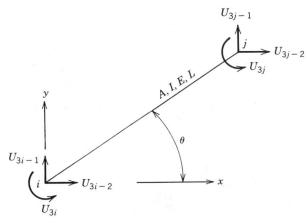

Figure 21.1.   The frame element and its nodal displacements.

21.2   THE ELEMENT STIFFNESS MATRIX

The element stiffness matrix has six rows and columns because there are six element displacements. The most convenient procedure for developing the element stiffness matrix is to start with a horizontal member and a member coordinate system and add the results for the axial force member and the beam element. Once $[\bar{k}^{(e)}]$ is known, $[k^{(e)}]$ can be obtained for an arbitrary orientation using a coordinate transformation.

Consider the horizontal member in Figure 21.2 with the nodal displacements in the following ordering

$$\{\bar{U}^{(e)}\}^T = [\bar{u}_i \quad \bar{v}_i \quad \bar{\phi}_i \quad \bar{u}_j \quad \bar{v}_j \quad \bar{\phi}_j] \tag{21.1}$$

The bars indicate that the displacements are for a member coordinate system. The strain energy in this member is the sum of the strain energy due to the axial displacement, $\Lambda_A^{(e)}$, and that resulting from bending, $\Lambda_B^{(e)}$,

$$\Lambda^{(e)} = \Lambda_A^{(e)} + \Lambda_B^{(e)} \tag{21.2}$$

The two components, $\Lambda_A^{(e)}$ and $\Lambda_B^{(e)}$, can be written in terms of the element nodal displacements given in (21.1).

The axial strain energy given by (18.25) can be written in terms of the new displacements by expanding (18.10) and assigning zero values for the coefficients that multiply $\bar{v}_i$, $\bar{\phi}_i$, $\bar{v}_j$, and $\bar{\phi}_j$. The result is

$$\Lambda_A^{(e)} = \tfrac{1}{2}\{\bar{U}^{(e)}\}^T[\bar{k}_A^{(e)}]\{\bar{U}^{(e)}\} \tag{21.3}$$

where

$$[\bar{k}_A^{(e)}] = \begin{bmatrix} \dfrac{AE}{L} & 0 & 0 & -\dfrac{AE}{L} & 0 & 0 \\ 0 & 0 & 0 & 0 & 0 & 0 \\ 0 & 0 & 0 & 0 & 0 & 0 \\ -\dfrac{AE}{L} & 0 & 0 & 0 & \dfrac{AE}{L} & 0 \\ 0 & 0 & 0 & 0 & 0 & 0 \\ 0 & 0 & 0 & 0 & 0 & 0 \end{bmatrix} \tag{21.4}$$

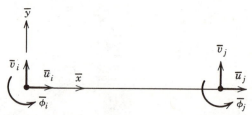

Figure 21.2.   The nodal displacements for a frame element in the member coordinate system.

The strain energy for bending, (20.23), can be written in terms of the six nodal displacements as

$$\Lambda_B^{(e)} = \tfrac{1}{2}\{\bar{U}^{(e)}\}^T[\bar{k}_B^{(e)}]\{\bar{U}^{(e)}\} \tag{21.5}$$

where

$$[\bar{k}^{(e)}] = \frac{EI}{L^3}\begin{bmatrix} 0 & 0 & 0 & 0 & 0 & 0 \\ 0 & 12 & 6L & 0 & -12 & 6L \\ 0 & 6L & 4L^2 & 0 & -6L & 2L^2 \\ 0 & 0 & 0 & 0 & 0 & 0 \\ 0 & -12 & -6L & 0 & 12 & -6L \\ 0 & 6L & 2L^2 & 0 & -6L & 4L^2 \end{bmatrix} \tag{21.6}$$

Adding $[\bar{k}_A^{(e)}]$ and $[\bar{k}_B^{(e)}]$ gives the total strain energy in the member coordinate-system as

$$\Lambda^{(e)} = \tfrac{1}{2}\{\bar{U}^{(e)}\}^T[\bar{k}^{(e)}]\{\bar{U}^{(e)}\} \tag{21.7}$$

where $[\bar{k}^{(e)}]$ is given by

$$[k^{(e)}] = \frac{EI}{L^3}\begin{bmatrix} \dfrac{AL^2}{I} & 0 & 0 & -\dfrac{AL^2}{I} & 0 & 0 \\ 0 & 12 & 6L & 0 & -12 & 6L \\ 0 & 6L & 4L^2 & 0 & -6L & 2L^2 \\ -\dfrac{AL^2}{I} & 0 & 0 & \dfrac{AL^2}{I} & 0 & 0 \\ 0 & -12 & -6L & 0 & 12 & -6L \\ 0 & 6L & 2L^2 & 0 & -6L & 4L^2 \end{bmatrix} \tag{21.8}$$

The next step is to write $\Lambda^{(e)}$ in terms of the generalized displacements. The generalized element displacement vector, $\{U^{(e)}\}$, is

$$\{U^{(e)}\}^T = [U_{3i-2} \quad U_{3i-1} \quad U_{3i} \quad U_{3j-2} \quad U_{3j-1} \quad U_{3j}] \tag{21.9}$$

The two sets of element displacements for node i are shown in Figure 21.3. A similar set exists for node j. The two displacement vectors are related by

$$\{\bar{U}^{(e)}\} = [T]\{U^{(e)}\} \tag{21.10}$$

where $[T]$ is the coordinate transformation matrix

$$[T] = \begin{bmatrix} \cos\theta & \sin\theta & 0 & 0 & 0 & 0 \\ -\sin\theta & \cos\theta & 0 & 0 & 0 & 0 \\ 0 & 0 & 1 & 0 & 0 & 0 \\ 0 & 0 & 0 & \cos\theta & \sin\theta & 0 \\ 0 & 0 & 0 & -\sin\theta & \cos\theta & 0 \\ 0 & 0 & 0 & 0 & 0 & 1 \end{bmatrix} \tag{21.11}$$

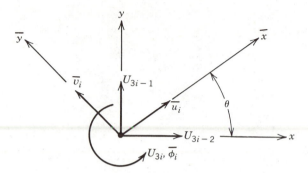

Figure 21.3.   The member and global displacements at node i.

The first, second, fourth, and fifth rows in (21.11) come from equations similar to (19.3); the third and sixth row state that $\bar\phi_i = U_{3i}$ and $\phi_j = U_{3j}$. The transformation matrix in (21.11) is similar to, but not the same as, the one used with the truss element, (19.6).

Substituting (21.10) into (21.7) produces

$$\Lambda^{(e)} = \tfrac{1}{2}\{U^{(e)}\}^T[T]^T[\bar{k}^{(e)}][T]\{U^{(e)}\} \tag{21.12}$$

The element stiffness matrix is immediately recognized as

$$[k^{(e)}] = [T]^T[\bar{k}^{(e)}][T] \tag{21.13}$$

where $[T]$ is defined by (21.11) and $[\bar{k}^{(e)}]$ is defined by (21.8). The element stiffness matrix is defined by a sequence of matrix products. This is the best form in which to leave it. Program the computer to calculate $[T]$ and $[\bar{k}^{(e)}]$ and the matrix multiplications.

21.3   THE INTERNAL FORCES

The internal nodal forces for the plane frame element are given by

$$\{\bar S^{(e)}\} = \frac{\partial \Lambda^{(e)}}{\partial\{\bar U^{(e)}\}} = [\bar{k}^{(e)}]\{\bar U^{(e)}\} \tag{21.14}$$

The internal force values are needed in the member coordinate system, where the axial and shear forces are parallel and perpendicular to the member. This orientation makes it easier to construct the shear force and axial force diagrams needed by the analyst. The nodal forces can be written in terms of the calculated nodal displacements using (21.10)

$$\{\bar S^{(e)}\} = [\bar{k}^{(e)}][T]\{U^{(e)}\} \tag{21.15}$$

The components of $\{\bar S^{(e)}\}$ are

$$\{\bar S^{(e)}\}^T = [Ax_i \quad Sh_i \quad M_i \quad Ax_j \quad Sh_j \quad M_j] \tag{21.16}$$

These components are shown in Figure 21.4.

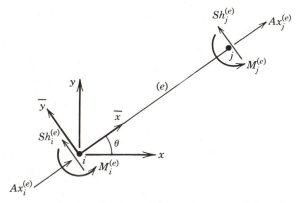

Figure 21.4. The internal axial force, shear force, and bending moment acting at each node.

ILLUSTRATIVE EXAMPLE

The displacement vector for the two-member rigid frame in Figure 21.5 is

$$\{U\}^T=[0 \quad 0 \quad 0 \quad 0.06619 \quad -0.1989 \quad 0.0001776 \quad 0 \quad 0 \quad 0]$$

Calculate the nodal forces for member one.

The nodal forces are given by (21.15); thus numerical values are needed for the coefficients in $[\bar{k}^{(1)}]$ and $[T^{(1)}]$. The parameters in $[\bar{k}^{(1)}]$ are

$$\frac{EI}{L^3}=\frac{20(10^6)(400)}{400^3}=125 \text{ N/cm}$$

$$\frac{AL^2}{I}=\frac{30(400^2)}{400}=12000$$

$$2L^2=2(400^2)=320000 \text{ cm}^2$$

$$4L^2=640000 \text{ cm}^2$$

$$6L=24000 \text{ cm}$$

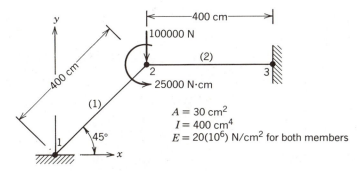

Figure 21.5. A two-member rigid frame.

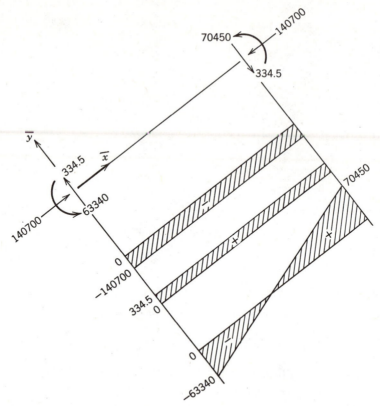

Figure 21.6. The nodal forces and moments and appropriate diagrams for member one.

Substituting these calculated values into (21.8) yields

$$
[\bar{k}^{(1)}] = 125
\begin{bmatrix}
12000 & 0 & 0 & -12000 & 0 & 0 \\
0 & 12 & 2400 & 0 & -12 & 2400 \\
0 & 2400 & 640000 & 0 & -2400 & 320000 \\
-12000 & 0 & 0 & 12000 & 0 & 0 \\
0 & -12 & -2400 & 0 & 12 & -2400 \\
0 & 2400 & 320000 & 0 & -2400 & 640000
\end{bmatrix}
$$

The transformation matrix $[T^{(1)}]$ defined by (21.11) is

$$
[T^{(1)}] =
\begin{bmatrix}
0.7071 & 0.7071 & 0 & 0 & 0 & 0 \\
-0.7071 & 0.7071 & 0 & 0 & 0 & 0 \\
0 & 0 & 1 & 0 & 0 & 0 \\
0 & 0 & 0 & 0.7071 & 0.7071 & 0 \\
0 & 0 & 0 & -0.7071 & 0.7071 & 0 \\
0 & 0 & 0 & 0 & 0 & 1
\end{bmatrix}
$$

The element nodal forces are

$$\{\bar{S}^{(1)}\} = [\bar{k}^{(1)}][T^{(1)}]\{U^{(1)}\}$$

where $\{U^{(1)}\}$ is

$$\{U^{(1)}\}^T = [0 \quad 0 \quad 0 \quad 0.06619 \quad -0.1989 \quad 0.0001776]$$

The numerical values are

$$\{\bar{S}(1)\}^T = [140700 \quad 334.5 \quad 63340 \quad -140700 \quad -334.5 \quad 70450] \quad (21.17)$$

These force components are shown in Figure 21.6. The axial force is compressive, and the 63340 N · cm moment at node one is negative relative to the construction of the bending moment diagram even though it is positive in (21.17).

PROBLEMS

21.1 Calculate the nodal forces (internal) for member two of the configuration shown in Figure 21.5.

21.2 Calculate the displacement at the free end of the inclined beam shown in Figure P21.2.

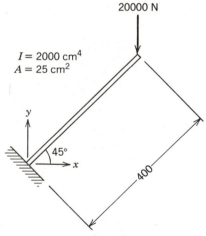

20000 N

$I = 2000$ cm^4
$A = 25$ cm^2

y

$45°$

x

400

Figure P21.2

21.3 Do Problem 21.2 when the applied load is parallel to the x-axis and has a magnitude of 15000 N.

21.4 Write a computer subroutine that will evaluate $[k^{(e)}]$ given A, EI, L, and θ.

21.5–21.10 Use the computer program **FRAME** Chapter 25, to analyze the rigid frames shown in the respective figures and construct shear force and bending moment diagrams for the individual members. Each frame is made from steel, $E = 20(10^6)$ N/cm^2. There is no thermal change. Every joint where two or more members are connected is a rigid joint. All length values are in centimeters.

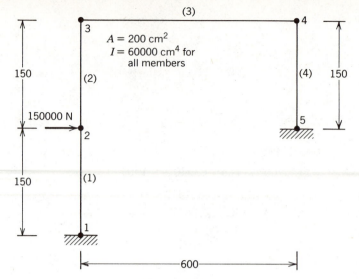

Figure P21.5

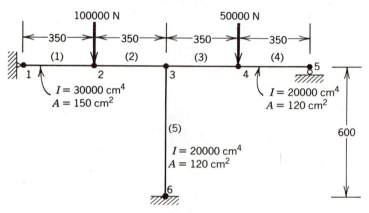

Figure P21.6

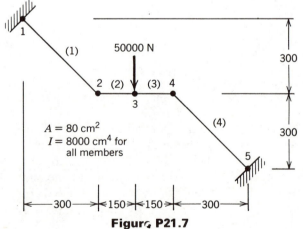

Figure P21.7

284

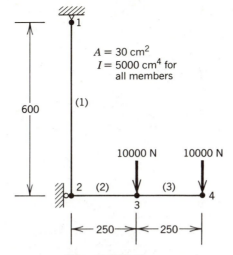

Figure P21.8

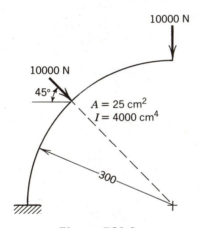

Figure P21.9

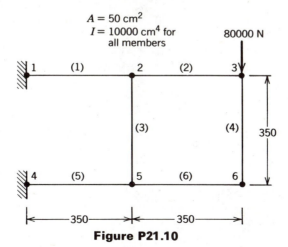

Figure P21.10

Chapter **22**

THEORY OF ELASTICITY

Applications of the finite element method to solid mechanics problems are extensive. These applications include elasticity problems, the analysis of plates and shells, the buckling of structures, continuum vibrations, elastic-plastic behavior, and viscoelastic analysis. The discussion in this book is confined to elasticity problems and is divided into three chapters. A general derivation of the equations for the element matrices is presented in this chapter. A discussion of two-dimensional elasticity follows in Chapter 23 and the analysis of axisymmetric configurations is in Chapter 24.

The derivation of the equations for $[k^{(e)}]$ and $\{f^{(e)}\}$ given in this chapter and the derivations for other solid mechanics problems are very similar. A thorough understanding of this derivation forms a good basis for understanding the derivations for plate or shell structures. Some of the definitions change in these other topic areas, but the general procedure remains the same.

22.1 STRESS, STRAIN, AND HOOKE'S LAW

The theory of elasticity involves several concepts that were not used in the previous chapters. These concepts are briefly reviewed here.

The state of stress at a point is defined by the six stress components shown in Figure 22.1. These stress components are produced by internal forces that counteract the externally applied forces. The stress components are positive when the outward normal is in a positive coordinate direction and the stress component is in a positive coordinate direction. There is a similar definition for the components acting on a negative face. These six components are placed in the column vector $\{\sigma\}$.

$$\{\sigma\}^T = [\sigma_{xx} \quad \sigma_{yy} \quad \sigma_{zz} \quad \sigma_{xy} \quad \sigma_{xz} \quad \sigma_{yz}] \tag{22.1}$$

The application of forces and/or heat to a solid body causes the body to deform. Each point in the body moves to a new location. The resultant displacement has three components u, v, and w parallel to the x-, y-, and z-axes, respectively. Six strain components are defined to assist in the study of how a body deforms. Since the deformation of the body can result from applied loads and/or thermal changes, the strain components are separated into elastic (load-produced) and thermal strains. The three sets of strain components are the total strain, $\{e\}$, elastic

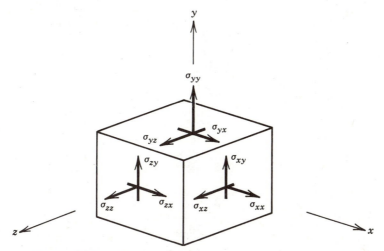

Figure 22.1. The six stress components acting at a point.

strain, $\{\varepsilon\}$, and thermal strain, $\{\varepsilon_T\}$. The entries in each vector are

$$\{e\}^T = [e_{xx} \quad e_{yy} \quad e_{zz} \quad e_{xy} \quad e_{xz} \quad e_{yz}] \tag{22.2}$$

$$\{\varepsilon\}^T = [\varepsilon_{xx} \quad \varepsilon_{yy} \quad \varepsilon_{zz} \quad \varepsilon_{xy} \quad \varepsilon_{xz} \quad \varepsilon_{yz}] \tag{22.3}$$

and

$$\{\varepsilon_T\}^T = [\alpha\delta T \quad \alpha\delta T \quad \alpha\delta T \quad 0 \quad 0 \quad 0] \tag{22.4}$$

where α is the coefficient of thermal expansion and δT is the temperature change. The three strain vectors are related by

$$\{e\} = \{\varepsilon\} + \{\varepsilon_T\} \tag{22.5}$$

The stress and elastic strain components are related by a set of coefficients known as the generalized Hooke's law. The coefficients are established by subjecting a piece of material to loadings that produce known stress distributions. The strains are calculated from the measured deformations.

The generalized Hooke's law can be written as

$$\{\varepsilon\} = [C]\{\sigma\} \quad \text{or} \quad \{\sigma\} = [D]\{\varepsilon\} \tag{22.6}$$

The coefficients in $[C]$ are

$$[C] = \frac{1}{E}\begin{bmatrix} 1 & -\mu & -\mu & 0 & 0 & 0 \\ -\mu & 1 & -\mu & 0 & 0 & 0 \\ -\mu & -\mu & 1 & 0 & 0 & 0 \\ 0 & 0 & 0 & a & 0 & 0 \\ 0 & 0 & 0 & 0 & a & 0 \\ 0 & 0 & 0 & 0 & 0 & a \end{bmatrix} \tag{22.7}$$

where E is the elastic modulus, μ is Poisson's ratio, and $a = 2(1 + \mu)$.

The coefficients in $[D]$ are

$$[D] = \frac{E}{1+\mu} \begin{bmatrix} d & b & b & 0 & 0 & 0 \\ b & d & b & 0 & 0 & 0 \\ b & b & d & 0 & 0 & 0 \\ 0 & 0 & 0 & c & 0 & 0 \\ 0 & 0 & 0 & 0 & c & 0 \\ 0 & 0 & 0 & 0 & 0 & c \end{bmatrix} \tag{22.8}$$

where

$$d = \frac{1-\mu}{1-2\mu}, \qquad b = \frac{\mu}{1-2\mu} \qquad \text{and} \qquad c = \frac{1}{2} \tag{22.9}$$

Note that $[D][C] = [C][D] = [I]$, where $[I]$ is the identity matrix. Poisson's ratio must be less than $\frac{1}{2}$ for (22.7) and (22.8) to be valid.

22.2 THE STRAIN-DISPLACEMENTS EQUATIONS

Each displacement component of each point is a function of the three coordinate directions, that is,

$$u = f(x, y, z), \qquad v = g(x, y, z) \qquad \text{and} \qquad w = h(x, y, z) \tag{22.10}$$

The objective of every analytical and finite element analysis is to determine the equations corresponding to $f(x, y, z)$, $g(x, y, z)$, and $h(x, y, z)$. The finite element approximations for these functions are continuous, piecewise smooth equations defined over the individual elements. The element equations depend on the type of element used to solve the problem. Since there are several elements to choose from when solving elasticity problems, the displacement equations are left in the general form

$$\begin{Bmatrix} u \\ v \\ w \end{Bmatrix} = [N]\{U^{(e)}\} \tag{22.11}$$

where $\{U^{(e)}\}$ is a column vector containing the element nodal displacements. The matrix $[N]$ contains the element shape functions. It has three rows and as many columns as there are components in $\{U^{(e)}\}$.

The strain components in $\{e\}$ and the displacements are related. These relationships are called the strain-displacement equations and are derived in all elasticity books (Fung, 1965). The set consists of

$$e_{xx} = \frac{\partial u}{\partial x}, \qquad e_{yy} = \frac{\partial v}{\partial y}, \qquad e_{zz} = \frac{\partial w}{\partial z}$$
$$e_{xy} = \frac{\partial u}{\partial y} + \frac{\partial v}{\partial x}, \qquad e_{xz} = \frac{\partial u}{\partial z} + \frac{\partial w}{\partial x}, \qquad e_{yz} = \frac{\partial v}{\partial z} + \frac{\partial w}{\partial y} \tag{22.12}$$

It is important to realize that these equations relate the total strains and the displacements. Most books consider the equations as relationships between the elastic strains and displacements. This is correct only when the thermal strain is zero.

The equations in (22.12) are used to obtain the strain energy equation in the next section. A general matrix $[B]$ is defined to assist with this evaluation.

$$\{e\} = [B]\{U^{(e)}\} \tag{22.13}$$

The matrix $[B]$ has six rows and as many columns as there are rows in $\{U^{(e)}\}$. The first row of $[B]$ is obtained by differentiating the displacement equation for u with respect to x, that is, $\partial u / \partial x$. The second row contains $\partial v / \partial y$, and so on. Each equation of (22.12) is used to generate one row of $[B]$.

22.3 THE ELEMENT MATRICES

The element stiffness matrix and the element force vector are the element's contribution to the system of equations that result when the potential energy is minimized. The potential energy consists of the strain energy in the system minus the work done by the forces acting on the system. This derivation is divided into two parts: a discussion of the strain energy term and a discussion of the work terms.

22.3.1 The Strain Energy Equation

The strain energy in a three-dimensional elastic body is

$$\Lambda^{(e)} = \tfrac{1}{2} \int_V (\sigma_{xx}\varepsilon_{xx} + \sigma_{yy}\varepsilon_{yy} + \sigma_{zz}\varepsilon_{zz} + \sigma_{xy}\varepsilon_{xy} + \sigma_{xz}\varepsilon_{xz} + \sigma_{yz}\varepsilon_{yz})\, dV \tag{22.14}$$

which can be neatly written as

$$\Lambda^{(e)} = \tfrac{1}{2} \int_V \{\sigma\}^T \{\varepsilon\}\, dV \tag{22.15}$$

The stress components can be replaced using the second form of (22.6). This substitution produces

$$\Lambda^{(e)} = \tfrac{1}{2} \int_V \{\varepsilon\}^T [D]\{\varepsilon\}\, dV \tag{22.16}$$

because $[D]$ is symmetric.

The strain energy, $\Lambda^{(e)}$, must be written in terms of the displacements. The nodal displacements, however, are related to the total strain components not the elastic strain components. Solving (22.5) for $\{\varepsilon\}$ gives

$$\{\varepsilon\} = \{e\} - \{\varepsilon_T\} \tag{22.17}$$

Substitution for $\{\varepsilon\}$ in (22.16) yields

$$\Lambda^{(e)} = \tfrac{1}{2} \int_V (\{e\}^T - \{\varepsilon_T\}^T)[D](\{e\} - \{\varepsilon_T\}) \, dV \tag{22.18}$$

Expanding gives

$$\Lambda^{(e)} = \tfrac{1}{2} \int_V \{e\}^T [D]\{e\} \, dV - \tfrac{1}{2} \int_V \{e\}^T [D]\{\varepsilon_T\} \, dV$$

$$- \tfrac{1}{2} \int_V \{\varepsilon_T\}^T [D]\{e\} \, dV + \tfrac{1}{2} \int_V \{\varepsilon_T\}^T [D]\{\varepsilon_T\} \, dV \tag{22.19}$$

The product $[D]\{\varepsilon_T\}$ is a column vector; therefore,

$$\{e\}^T([D]\{\varepsilon_T\}) = ([D]\{\varepsilon_T\})^T\{e\}$$
$$= \{\varepsilon_T\}^T [D]^T\{e\} = \{\varepsilon_T\}^T [D]\{e\} \tag{22.20}$$

The matrix operations in the second and third integrals of (22.19) are identical. Using this property and the fact that the last integral is independent of the displacements (and can be discarded), (22.19) can be simplified to

$$\Lambda^{(e)} = \tfrac{1}{2} \int_V \{e\}^T [D]\{e\} \, dV - \int_V \{e\}^T [D]\{\varepsilon_T\} \, dV \tag{22.21}$$

The last step is to substitute (22.13) for $\{e\}$. The strain energy equation written in terms of the element nodal displacements is

$$\Lambda^{(e)} = \tfrac{1}{2} \int_V \{U^{(e)}\}^T [B]^T [D][B]\{U^{(e)}\} \, dV - \int_V \{U^{(e)}\}^T [B]^T [D]\{\varepsilon_T\} \, dV \tag{22.22}$$

22.3.2 The Work Terms

The work done by the applied loads can be separated into three distinct parts: that due to the concentrated loads, that resulting from the stress components acting on the outside surface, W_p, and that done by the body forces, W_{bf}.

The work done by the concentrated forces is the $\{U\}^T\{P\}$ product observed with the structural applications. Minimization produces the $\{P\}$ vector in the final system of equations.

The work done by the body forces, $\mathscr{X}$, $\mathscr{Y}$, $\mathscr{Z}$, is given by

$$W_{bf}^{(e)} = \int_V (u\mathscr{X} + v\mathscr{Y} + w\mathscr{Z}) \, dV \tag{22.23}$$

where u, v, and w are the x-, y-, and z-components of the displacement within the element. The integral is necessary because u, v, and w along with $\mathscr{X}$, $\mathscr{Y}$, and $\mathscr{Z}$ can vary within the element. Using (22.11) allows (22.23) to be written as

$$W_{bf}^{(e)} = \int_V \{U^{(e)}\}^T [N]^T \begin{Bmatrix} \mathscr{X} \\ \mathscr{Y} \\ \mathscr{Z} \end{Bmatrix} dV \tag{22.24}$$

The work done by the distributed loads that act on the surface is

$$W_p^{(e)} = \int_{\Gamma} (up_x + vp_y + wp_z)\, d\Gamma \tag{22.25}$$

where u, v, and w are the displacement components and p_x, p_y, and p_z are the stress components parallel to the x-, y-, and z-coordinate directions. A comparison of (22.25) and (22.23) indicates that they are identical in form; therefore,

$$W_p^{(e)} = \int_{\Gamma} \{U^{(e)}\}^T [N]^T \begin{Bmatrix} p_x \\ p_y \\ p_z \end{Bmatrix} d\Gamma \tag{22.26}$$

22.3.3   The Element Matrices

The total potential energy in a continuous three-dimensional elastic system is

$$\Pi = \sum_{e=1}^{n} \Pi^{(e)} - \{U\}^T \{P\} \tag{22.27}$$

where

$$\Pi^{(e)} = \Lambda^{(e)} - W_{bf}^{(e)} - W_p^{(e)} \tag{22.28}$$

Substituting (22.22), (22.24), and (22.26) for the three terms in (22.28) gives

$$\Pi^{(e)} = \tfrac{1}{2}\{U^{(e)}\}^T \left(\int_V [B]^T [D][B]\, dV \right) \{U^{(e)}\}$$

$$- \{U^{(e)}\}^T \left(\int_V [B]^T [D]\{\varepsilon_T\}\, dV + \int_V [N]^T \begin{Bmatrix} \mathscr{X} \\ \mathscr{Y} \\ \mathscr{Z} \end{Bmatrix} dV \right.$$

$$\left. + \int_{\Gamma} [N]^T \begin{Bmatrix} p_x \\ p_y \\ p_z \end{Bmatrix} d\Gamma \right) \tag{22.29}$$

which is in the same form as (18.21). Thus the element stiffness matrix is

$$[k^{(e)}] = \int_V [B]^T [D][B]\, dV \tag{22.30}$$

and the element force vector is

$$\{f^{(e)}\} = \int_V [B]^T [D]\{\varepsilon_T\}\, dV + \int_V [N]^T \begin{Bmatrix} \mathscr{X} \\ \mathscr{Y} \\ \mathscr{Z} \end{Bmatrix} dV + \int_{\Gamma} [N]^T \begin{Bmatrix} p_x \\ p_y \\ p_z \end{Bmatrix} d\Gamma \tag{22.31}$$

22.4 THE STRESS COMPONENTS

The design criteria for structural and machine components involves the values of the stress components or the maximum displacement. The nodal displacements are available once the system of finite element equations has been solved. The element stress components are calculated once the element nodal displacements are known.

The general equation for calculating the stress components emerges after several of the equations discussed in the first two sections of this chapter are combined. Starting with the second equation of (22.6), we obtain

$$\{\sigma\} = [D]\{\varepsilon\} \tag{22.32}$$

and replacing $\{\varepsilon\}$ with (22.17),

$$\{\sigma\} = [D](\{e\} - \{\varepsilon_T\}) \tag{22.33}$$

This equation can be written in terms of the element nodal displacements using (22.13). The final result gives the stress components as a function of the element nodal displacements and thermal strain vector

$$\{\sigma^{(e)}\} = [D]([B]\{U^{(e)}\} - \{\varepsilon_T\}) \tag{22.34}$$

Chapter **23**

TWO-DIMENSIONAL ELASTICITY

The derivation presented in the previous chapter assumed a three-dimensional state of stress within an elastic body. Special cases of this theory are now considered. Two-dimensional problems are discussed in this chapter and axisymmetric problems in Chapter 24. The discussion is confined to the triangular element because the integrals can be evaluated relatively easily. The element matrices for the rectangular element and all higher-order elements are evaluated using the numerical integration techniques discussed in the last chapters of this book.

23.1 PLANE STRESS AND PLANE STRAIN

The reduction of a three-dimensional problem to a two-dimensional problem can occur in two ways. These choices are called plane stress and plane strain. The resulting equations for each option are developed in this section.

23.1.1 Plane Stress

A state of plane stress is said to exist when the elastic body is very thin and there are no loads applied in the coordinate direction parallel to the thickness. The stress components associated with the thickness direction, σ_{zz}, σ_{zx}, and σ_{zy} are very small and assumed to be zero, when the applied loads lie in the x-y plane (Figure 23.1).

Substituting the zero values for σ_{zz}, σ_{zx}, and σ_{zy} into (22.1) indicates that the non-zero stress components are σ_{xx}, σ_{yy}, and σ_{xy}. The stress vector can be written as

$$\{\sigma\}^T = [\sigma_{xx} \quad \sigma_{yy} \quad \sigma_{xy}] \tag{23.1}$$

Substitution of the zero stress values into Hooke's law $\{\varepsilon\} = [C]\{\sigma\}$ using (22.7) for $[C]$ shows that

$$\varepsilon_{zz} = -\frac{\mu}{E}(\sigma_{xx} + \sigma_{yy})$$

$$\varepsilon_{xz} = \varepsilon_{yz} = 0 \tag{23.2}$$

The normal strain ε_{zz} is not zero but it can be calculated once σ_{xx} and σ_{yy} are known.

293

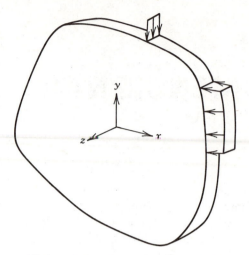

Figure 23.1. A thin body in a state of plane stress.

The two shear strains associated with the z-axis are zero. The elastic strain vector is

$$\{\varepsilon\}^T = [\varepsilon_{xx} \quad \varepsilon_{yy} \quad \varepsilon_{xy}] \tag{23.3}$$

The total strain vector reduces to

$$\{e\}^T = [e_{xx} \quad e_{yy} \quad e_{xy}] \tag{23.4}$$

while

$$\{\varepsilon_T\}^T = [\alpha\delta T \quad \alpha\delta T \quad 0] \tag{23.5}$$

The matrix $[D]$, which relates $\{\sigma\}$ and $\{\varepsilon\}$ for plane stress, is obtained by deleting rows and columns three, five, and six from (22.7) and inverting the remaining 3×3 matrix. The final relationship is

$$[D] = \frac{E}{1-\mu^2} \begin{bmatrix} 1 & \mu & 0 \\ \mu & 1 & 0 \\ 0 & 0 & \dfrac{1-\mu}{2} \end{bmatrix} \tag{23.6}$$

23.1.2 Plane Strain

The state of plane strain occurs in members that are not free to expand in the direction perpendicular to the plane of the applied loads. If we assume that the applied loads lie in the x-y plane, then w, the displacement in the z-direction is zero and the displacements u and v are functions of only x and y. This set of displacements makes e_{zz}, e_{xz}, and e_{yz} each zero.

Substitution for the zero values reduces the strain vectors to

$$\{e\}^T = [e_{xx} \quad e_{yy} \quad e_{xy}] \tag{23.7}$$

$$\{\varepsilon\}^T = [\varepsilon_{xx} \quad \varepsilon_{yy} \quad \varepsilon_{xy}] \tag{23.8}$$

and

$$\{\varepsilon_T\}^T = [\alpha\delta T \quad \alpha\delta T \quad 0] \tag{23.9}$$

while

$$\{\sigma\}^T = [\sigma_{xx} \quad \sigma_{yy} \quad \sigma_{xy}] \tag{23.10}$$

The above three vectors are identical to those which occur in a state of plane stress. The stress vector, (23.10), is obtained by substituting $\varepsilon_{zz} = \varepsilon_{xz} = \varepsilon_{yz} = 0$ into Hooke's law $\{\sigma\} = [D]\{\varepsilon\}$, using (22.8) for $[D]$ and noting which stress components are unknown. In this case, σ_{xz} and σ_{yz} are zero while

$$\sigma_{zz} = \frac{E}{1+\mu}\left[\frac{\mu}{1-2\mu}(\sigma_{xx} + \sigma_{yy}) - \left(\frac{1-\mu}{1-2\mu}\right)\alpha\delta T\right] \tag{23.11}$$

The Hooke's law relationship for plane strain is obtained by deleting rows and columns three, five, and six from (22.8). The resulting matrix is

$$[D] = \frac{E}{1+\mu}\begin{bmatrix} d & b & 0 \\ b & d & 0 \\ 0 & 0 & \frac{1}{2} \end{bmatrix} \tag{23.12}$$

where

$$d = \frac{1-\mu}{1-2\mu} \quad \text{and} \quad b = \frac{\mu}{1-2\mu}$$

23.2   THE DISPLACEMENT EQUATIONS

There are two unknown displacements in a two-dimensional elasticity problem, u and v. The displacement parallel to the z-axis, w, is zero when plane strain exists and is related to u and v when plane stress exists.

The u and v displacements are modeled in a continuum element by defining two displacement components at each node (Figure 23.2). The notation used here is identical to that used for the truss element.

The simplest model for u and v is to use a linear variation for each displacement within the element. The horizontal displacement u is approximated using

$$u(x, y) = N_i U_{2i-1} + N_j U_{2j-1} + N_k U_{2k-1} \tag{23.13}$$

while the vertical component v is represented by

$$v(x, y) = N_i U_{2i} + N_j U_{2j} + N_k U_{2k} \tag{23.14}$$

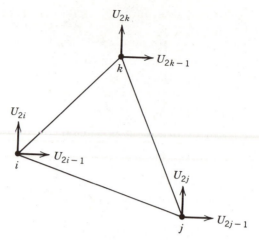

Figure 23.2. The nodal displacements for a triangular elasticity element.

In each equation, N_i, N_j, and N_k are the linear shape functions developed in Chapter 5 and given by (5.8), (5.9), and (5.10).

Equations (23.13) and (23.14) can be written in terms of all the nodal values by adding zeros that multiply the missing displacements. The displacement equations become

$$u = N_i U_{2i-1} + 0 U_{2i} + N_j U_{2j-1} + 0 U_{2j} + N_k U_{2k-1} + 0 U_{2k}$$

$$v = 0 U_{2i-1} + N_i U_{2j} + 0 U_{2j-1} + N_j U_{2j} + 0 U_{2k-1} + N_k U_{2k} \tag{23.15}$$

Utilizing matrix notation yields

$$
\left\{ \begin{matrix} u(x,y) \\ v(x,y) \end{matrix} \right\} = \begin{bmatrix} N_i & 0 & N_j & 0 & N_k & 0 \\ 0 & N_i & 0 & N_j & 0 & N_k \end{bmatrix} \left\{ \begin{matrix} U_{2i-1} \\ U_{2i} \\ U_{2j-1} \\ U_{2j} \\ U_{2k-1} \\ U_{2k} \end{matrix} \right\} \tag{23.16}
$$

or

$$\left\{ \begin{matrix} u(x,y) \\ v(x,y) \end{matrix} \right\} = [N]\{U^{(e)}\} \tag{23.17}$$

where $[N]$ is the 2×6 matrix containing the element shape functions and $\{U^{(e)}\}$ contains the element nodal displacements.

The three-dimensional strain-displacement relationships, (22.12), reduce to

$$e_{xx} = \frac{\partial u}{\partial x}, \qquad e_{yy} = \frac{\partial v}{\partial y}, \qquad \text{and} \qquad e_{xy} = \frac{\partial u}{\partial y} + \frac{\partial v}{\partial x} \tag{23.18}$$

because w is zero and u and v are not functions of z. Applying the operations defined in (23.18) to (23.13), we find that

$$e_{xx} = \frac{1}{2A}(b_i U_{2i-1} + b_j U_{2j-1} + b_k U_{2k-1})$$

$$e_{yy} = \frac{1}{2A}(c_i U_{2i} + c_j U_{2j} + c_k U_{2k}) \tag{23.19}$$

$$e_{xy} = \frac{1}{2A}(c_i U_{2i-1} + b_i U_{2i} + c_j U_{2j-1} + b_j U_{2j} + c_k U_{2k-1} + b_k U_{2k}$$

where the b and c coefficients are defined in Chapter 5. The equations in (23.19) have the matrix form

$$\begin{Bmatrix} e_{xx} \\ e_{yy} \\ e_{xy} \end{Bmatrix} = \frac{1}{2A} \begin{bmatrix} b_i & 0 & b_j & 0 & b_k & 0 \\ 0 & c_i & 0 & c_j & 0 & c_k \\ c_i & b_i & c_j & b_j & c_k & b_k \end{bmatrix} \begin{Bmatrix} U_{2i-1} \\ U_{2i} \\ U_{2j-1} \\ U_{2j} \\ U_{2k-1} \\ U_{2k} \end{Bmatrix} \tag{23.20}$$

or

$$\{e\} = [B]\{U^{(e)}\} \tag{23.21}$$

Equation (23.21) defines the gradient matrix $[B]$ for the triangular element. It is a 3×6 matrix. The number of rows exceeds the dimension of the problem because there are three unknown strain components in the two-dimensional problem.

23.3 THE ELEMENT MATRICES

The element stiffness matrix is given by (22.30)

$$[k^{(e)}] = \int_V [B]^T[D][B]\, dV \tag{23.22}$$

where $[B]$ is defined by (23.21) and $[D]$ is either (23.6) or (23.12). This integral is readily evaluated because $[B]$ and $[D]$ consist entirely of constant terms. The result is

$$[k^{(e)}] = [B]^T[D][B]tA \tag{23.23}$$

where t is the element thickness and A is the element area. The matrix product $[B]^T[D][B]$ is not evaluated because the final result is quite complicated when written in equation form. The best approach is to evaluate $[B]$ and $[D]$ and let the computer perform the multiplications.

The value of t used in (23.23) is the actual thickness of the body for plane stress and unity for plane strain.

The element force vector is given by (22.31) after neglecting the body force $\mathscr{Z}$ and the surface stress p_z. Neither of these components exist in the two-dimensional problem. The resulting equation is

$$\{f^{(e)}\} = \int_V [B]^T[D]\{\varepsilon_T\}\, dV + \int_V [N]^T \left\{\begin{matrix} \mathscr{X} \\ \mathscr{Y} \end{matrix}\right\} dV \int_\Gamma [N]^T \left\{\begin{matrix} p_x \\ p_y \end{matrix}\right\} d\Gamma \quad (23.24)$$

where $[N]$ is defined by (23.17).

The first integral in (23.24) is easily evaluated because the matrices contain constant coefficients. The integral yields

$$\int_V [B]^T[D]\{\varepsilon_T\}\, dV = [B]^T[D]\{\varepsilon_T\} tA \quad (23.25)$$

The matrix product is relatively easy to evaluate, but the best procedure is to let the computer perform the products. Note that the $[B]^T[D]$ product also occurs in $[k^{(e)}]$, (23.23). If this is evaluated first, both (23.23) and (23.25) can be evaluated in the same DO-loop.

The volume integral involving the body forces is easy to evaluate if the shape functions are replaced by their equivalent area coordinate. The body force integral is

$$\int_V [N]^T \left\{\begin{matrix} \mathscr{X} \\ \mathscr{Y} \end{matrix}\right\} dV = \int_V \begin{bmatrix} N_i & 0 \\ 0 & N_i \\ N_j & 0 \\ 0 & N_j \\ N_k & 0 \\ 0 & N_k \end{bmatrix} \left\{\begin{matrix} \mathscr{X} \\ \mathscr{Y} \end{matrix}\right\} dV$$

$$= \int_V \left\{\begin{matrix} N_i\,\mathscr{X} \\ N_i\,\mathscr{Y} \\ N_j\,\mathscr{X} \\ N_j\,\mathscr{Y} \\ N_k\,\mathscr{X} \\ N_k\,\mathscr{Y} \end{matrix}\right\} dV = t \int_A \left\{\begin{matrix} L_1\,\mathscr{X} \\ L_1\,\mathscr{Y} \\ L_2\,\mathscr{X} \\ L_2\,\mathscr{Y} \\ L_3\,\mathscr{X} \\ L_3\,\mathscr{Y} \end{matrix}\right\} dA = \frac{tA}{3} \left\{\begin{matrix} \mathscr{X} \\ \mathscr{Y} \\ \mathscr{X} \\ \mathscr{Y} \\ \mathscr{X} \\ \mathscr{Y} \end{matrix}\right\} \quad (23.26)$$

The integral is dividing each body force component $tA\mathscr{X}$ or $tA\mathscr{Y}$ equally among the three nodes.

The integral in (23.24) involving the surface stresses p_x and p_y must be integrated along the edge of the element. Noting that $d\Gamma = t\, d\ell_2$, we find that the integral is

$$\int_\Gamma [N]^T \left\{\begin{matrix} p_x \\ p_y \end{matrix}\right\} d\Gamma = tL \int_0^1 [N]^T \left\{\begin{matrix} p_x \\ p_y \end{matrix}\right\} d\ell_2 \quad (23.27)$$

where L is the length of a side. The integral in (23.27) has three different values, one for each side of the element. By assuming that the surface stresses act on the side ij, (23.27) becomes

$$tL_{ij} \int_0^1 [N]^T \begin{Bmatrix} p_x \\ p_y \end{Bmatrix} d\ell_2 = tL_{ij} \int_0^1 \begin{bmatrix} N_i & 0 \\ 0 & N_i \\ N_j & 0 \\ 0 & N_j \\ N_k & 0 \\ 0 & N_k \end{bmatrix} \begin{Bmatrix} p_x \\ p_y \end{Bmatrix} d\ell_2 \qquad (23.28)$$

However, N_k is zero along side ij. Using this fact and substituting the area coordinates for the shape functions yields

$$tL_{ij} \int_0^1 [N]^T \begin{Bmatrix} p_x \\ p_y \end{Bmatrix} d\ell_2 = tL_{ij} \int_0^1 \begin{Bmatrix} \ell_1 p_x \\ \ell_1 p_y \\ \ell_2 p_x \\ \ell_2 p_y \\ 0 \\ 0 \end{Bmatrix} d\ell_2 = \frac{tL_{ij}}{2} \begin{Bmatrix} p_x \\ p_y \\ p_x \\ p_y \\ 0 \\ 0 \end{Bmatrix} \qquad (23.29)$$

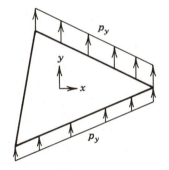

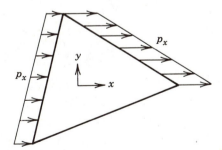

Figure 23.3.   The direction of positive surface stresses.

The resulting equation (23.29) can be interpreted as follows. The quantities $p_x t L_{ij}$ and $p_y t L_{ij}$ represent the force components acting on side ij. The integral has allotted one-half of each force component to each node on side ij.

Evaluation of the surface integral gives similar results for the other sides. The results are

$$
\int_\Gamma [N]^T \begin{Bmatrix} p_x \\ p_y \end{Bmatrix} d\Gamma = \frac{t L_{jk}}{2} \begin{Bmatrix} 0 \\ 0 \\ p_x \\ p_y \\ p_x \\ p_y \end{Bmatrix} ; \qquad \frac{t L_{ik}}{2} \begin{Bmatrix} p_x \\ p_y \\ 0 \\ 0 \\ p_x \\ p_y \end{Bmatrix} \tag{23.30}
$$

for sides jk and ik, respectively.

It is important to realize that p_x and p_y are positive when they are directed in the positive coordinate directions. These positive directions are shown in Figure 23.3.

ILLUSTRATIVE EXAMPLE

Calculate the element stiffness matrix and the thermal force vector for the plane stress element shown in Figure 23.4. The element experiences a 10°C increase in temperature.

The element stiffness matrix is given by (23.23) as $[k^{(e)}] = tA[B]^T[D][B]$. The gradient matrix $[B]$ is

$$
[B] = \frac{1}{2A} \begin{bmatrix} b_i & 0 & b_j & 0 & b_k & 0 \\ 0 & c_i & 0 & c_j & 0 & c_k \\ c_i & b_i & c_j & b_j & c_k & b_k \end{bmatrix}
$$

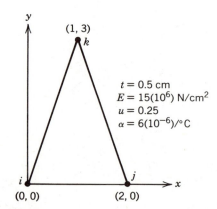

Figure 23.4.   A triangular elasticity element.

where $A = (3)(2)/2 = 3 \text{ cm}^2$ and

$$
\begin{aligned}
b_i &= Y_j - Y_k = -3 & c_i &= X_k - X_j = -1 \\
b_j &= Y_k - Y_i = 3 & c_j &= X_i - X_k = -1 \\
b_k &= Y_i - Y_j = 0 & c_k &= X_j - X_i = 2
\end{aligned}
$$

Substitution gives

$$
[B] = \frac{1}{6}
\begin{bmatrix}
-3 & 0 & 3 & 0 & 0 & 0 \\
0 & -1 & 0 & -1 & 0 & 2 \\
-1 & -3 & -1 & 3 & 2 & 0
\end{bmatrix}
$$

The Hooke's law matrix $[D]$ given by (23.6) is

$$
[D] = \frac{E}{1-\mu^2}
\begin{bmatrix}
1 & \mu & 0 \\
\mu & 1 & 0 \\
0 & 0 & \dfrac{1-\mu}{2}
\end{bmatrix}
$$

$$
= \frac{15(10^6)}{1 - 0.25^2}
\begin{bmatrix}
1 & \frac{1}{4} & 0 \\
\frac{1}{4} & 1 & 0 \\
0 & 0 & \frac{3}{8}
\end{bmatrix}
= 10^6
\begin{bmatrix}
16 & 4 & 0 \\
4 & 16 & 0 \\
0 & 0 & 6
\end{bmatrix}
$$

The evaluation of $[k^{(e)}]$ starts by evaluating $[B]^T[D]$ because this product also occurs in the evaluation of the thermal force vector.

$$
[B]^T[D] = \frac{10^6}{6}
\begin{bmatrix}
-3 & 0 & -1 \\
0 & -1 & -3 \\
3 & 0 & -1 \\
0 & -1 & 3 \\
0 & 0 & 2 \\
0 & 2 & 0
\end{bmatrix}
\begin{bmatrix}
16 & 4 & 0 \\
4 & 16 & 0 \\
0 & 0 & 6
\end{bmatrix}
$$

$$
[B]^T[D] = \frac{10^6}{6}
\begin{bmatrix}
-48 & -12 & -6 \\
-4 & -16 & -18 \\
48 & 12 & -6 \\
-4 & -16 & 18 \\
0 & 0 & 12 \\
8 & 32 & 0
\end{bmatrix}
$$

$$
[k^{(e)}] = \frac{tA10^6}{36}
\begin{bmatrix}
-48 & -12 & -6 \\
-4 & -16 & -18 \\
48 & 12 & -6 \\
-4 & -16 & 18 \\
0 & 0 & 12 \\
8 & 32 & 0
\end{bmatrix}
\begin{bmatrix}
-3 & 0 & 3 & 0 & 0 & 0 \\
0 & -1 & 0 & -1 & 0 & 2 \\
-1 & -3 & -1 & 3 & 2 & 0
\end{bmatrix}
$$

$$[k^{(e)}] = 41667 \begin{bmatrix} 150 & 30 & -138 & -6 & -12 & -24 \\ 30 & 70 & 6 & -38 & -36 & -32 \\ -138 & 6 & 150 & -30 & -12 & 24 \\ -6 & -38 & -30 & 70 & 36 & -32 \\ -12 & -36 & -12 & 36 & 24 & 0 \\ -24 & -32 & 24 & -32 & 0 & 64 \end{bmatrix}$$

The thermal force vector given by (23.25) is $\{f_t^{(e)}\} = [B]^T[D]\{\varepsilon_T\} t A$. The strain vector is

$$\{\varepsilon_T\} = \begin{Bmatrix} \alpha \delta T \\ \alpha \delta T \\ 0 \end{Bmatrix} = 10^{-6} \begin{Bmatrix} 60 \\ 60 \\ 0 \end{Bmatrix}.$$

Using the $[B]^T[D]$ product evaluated earlier, we obtain

$$\{f_t^{(e)}\} = \frac{(10^{-6})(10^6)}{6} \begin{bmatrix} -48 & -12 & -6 \\ -4 & -16 & -18 \\ 48 & 12 & -6 \\ -4 & -16 & 18 \\ 0 & 0 & 12 \\ 8 & 32 & 0 \end{bmatrix} \begin{Bmatrix} 60 \\ 60 \\ 0 \end{Bmatrix} = \begin{Bmatrix} -600 \\ -200 \\ 600 \\ -200 \\ 0 \\ 400 \end{Bmatrix}$$

ILLUSTRATIVE EXAMPLE

Calculate the equivalent set of concentrated forces acting at the nodes 10, 12, and 18 (Figure 23.5). The length of each element perpendicular to the surface stress is given.

We could use one of the surface integrals in (23.29) or (23.30) to evaluate the force components, but an easier and faster method is to use the physical interpretation of these equations. We calculate the total force acting on the side of the element and then divide it equally between the two modes.

Starting with element one, we find that

$$\text{force}^{(1)} = p_n A = p_n t L$$
$$= 5000(2.5)(1.5) = 18750 \text{ N}$$

while

$$\text{force}^{(2)} = p_n A = p_n t L$$
$$= 5000(2.5)(2) = 25000 \text{ N}$$

Alloting the forces to the nodes produces the situation shown in Figure 23.6a. Adding the two forces at node 12 gives the three forces shown in Figure 23.6b. Resolution of the forces in Figure 23.6b into their x- and y-components produces the set of forces in Figure 23.6c. The forces in Figure 23.6c can be handled in the

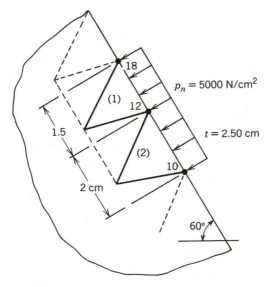

Figure 23.5. Surface stress and elements along an exterior boundary.

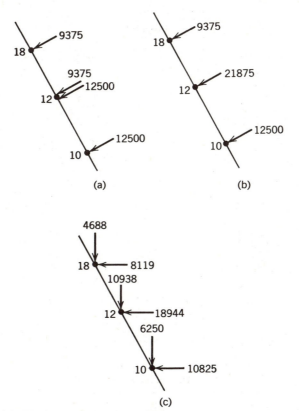

Figure 23.6. (a) The element contributions to the nodal forces. (b) The nodal forces. (c) The nodal forces resolved into components.

same manner as any other concentrated force. This set of forces is equivalent to

$$P_{19} = -10825 \text{ N}, \qquad P_{23} = -18944 \text{ N}, \qquad P_{35} = -8119 \text{ N}$$
$$P_{20} = -6250 \text{ N}, \qquad P_{24} = -10938 \text{ N}, \qquad P_{36} = -4688 \text{ N}$$

23.4 ELEMENT STRESSES

The desired results in any elasticity problem are the stress components acting within the body. The stress components within an element can be calculated once the element nodal displacements are known. The stress components are given by (22.34) using the vectors $\{\sigma\}$, $\{\varepsilon\}$, and $\{\varepsilon_T\}$ as defined at the beginning of this chapter.

ILLUSTRATIVE EXAMPLE

Calculate the stress components for the element in Figure 23.5 when the nodal displacements are

$$U_{2i-1} = 0, \qquad U_{2j-1} = 0.0001 \text{ cm}, \qquad U_{2k-1} = 0.0004 \text{ cm}$$
$$U_{2i} = 0, \qquad U_{2j} = 0.0006 \text{ cm}, \qquad U_{2k} = -0.0010 \text{ cm}$$

The stress components are given by $\{\sigma\} = [D]([B]\{U^{(e)}\} - \{\varepsilon_T\})$. Using for $[B]$, $[D]$, and $\{\varepsilon_T\}$ the matrices developed for the example associated with Figure 23.5, we obtain

$$[B]\{U^{(e)}\} = \frac{1}{6}\begin{bmatrix} -3 & 0 & 3 & 0 & 0 & 0 \\ 0 & -1 & 0 & -1 & 0 & 2 \\ -1 & -3 & -1 & 3 & 2 & 0 \end{bmatrix} \begin{Bmatrix} 0 \\ 0 \\ 0.0001 \\ 0.0006 \\ 0.0004 \\ -0.0010 \end{Bmatrix}$$

$$= 10^{-6} \begin{Bmatrix} 50 \\ 433 \\ 267 \end{Bmatrix}$$

and

$$[B]\{U^{(e)}\} - \{\varepsilon_T\} = 10^{-6} \begin{Bmatrix} 50 \\ 433 \\ 267 \end{Bmatrix} - 10^{-6} \begin{Bmatrix} 60 \\ 60 \\ 0 \end{Bmatrix}$$

$$= 10^{-6} \begin{Bmatrix} -10 \\ 373 \\ 267 \end{Bmatrix}$$

The stress components are

$$\{\sigma\} = 10^6 \begin{bmatrix} 16 & 4 & 0 \\ 4 & 16 & 0 \\ 0 & 0 & 8 \end{bmatrix} \begin{Bmatrix} -10(10^{-6}) \\ 373(10^{-6}) \\ 267(10^{-6}) \end{Bmatrix} = \begin{Bmatrix} 1332 \\ 5928 \\ 2136 \end{Bmatrix}$$

The stress components are constant within the linear triangular element. They are generally assumed to be the values at the centroid of the triangle. The set of constant values is a major disadvantage of the linear element.

23.5 DISCUSSION OF A COMPUTER EXAMPLE

The thin plate subjected to a distributed load in the middle third of one side as shown in Figure 23.7 is analyzed in Chapter 25 in conjunction with the discussion of the computer program **STRESS**. The bottom surface of the plate is free to move horizontally relative to the support surface. The finite element grid used to solve for the stress distribution is shown in Figure 23.8. The horizontal displacements at nodes 1, 5, 9, and 13 are zero as well as the vertical displacements at nodes 1, 2, 3, and 4. The complete computer output is given in Chapter 25. The three stress components in each element are summarized in Table 23.1.

The objective of this section is to illustrate an equilibrium analysis of the plate as a check on the calculated stress values. The stress components are constant within the triangular element and the values are usually assumed to act at the centroid of the element. The stress values acting on two cutting planes, $A-A$ and $B-B$ (Figure 23.9) are shown in free body diagrams in Figures 23.10, 23.11, and 23.12. The free body analysis of each diagram is discussed in the following paragraphs.

The cutting plane $A-A$ passes through elements (1), (2), (3), (4), (5), and (6) (Figure 23.10). The exposed area of each element is $(5)(3) = 15$ cm^2. A summation

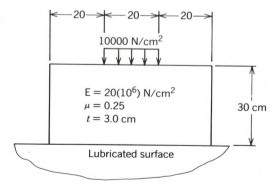

Figure 23.7. A thin plate with a distributed load over a part of one side.

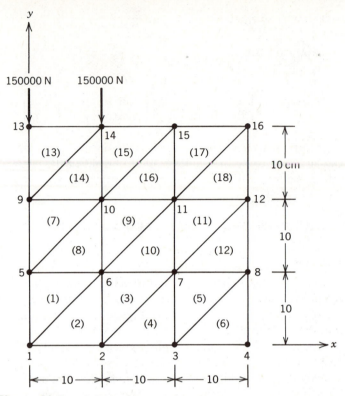

Figure 23.8. A finite element grid for the partially loaded plate.

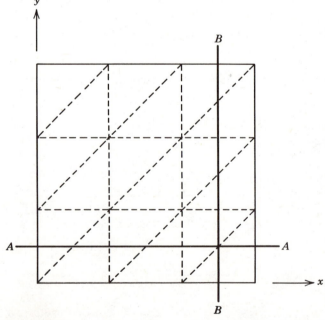

Figure 23.9. Location of the cutting planes *A-A* and *B-B*.

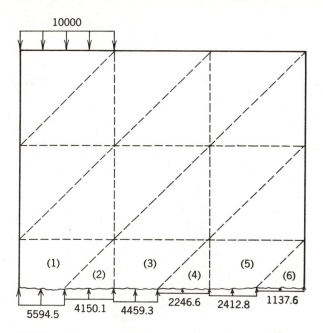

Figure 23.10. The stress component σ_{yy} on the cutting plane A-A.

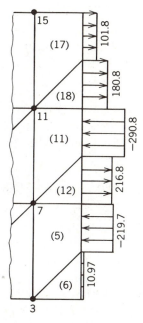

Figure 23.11. The stress component σ_{xx} on the cutting plane B-B.

of forces in the vertical direction gives

$$+\uparrow\Sigma F_y = (5594.5 + 4150.1 + 4459.3 + 2246.6 + 2412.8 + 1137.6)(15) - (10000)(10)(3)$$
$$= (200000.9)(15) - (10000)(30) = 13.5 \text{ N} \approx 0$$

Table 23.1   Element Stress Components

Element	σ_{xx}	σ_{yy}	σ_{xy}
1	710.9	− 5594.5	529.8
2	1190.6	− 4150.1	− 47.4
3	− 46.2	− 4459.3	788.5
4	449.3	− 2246.6	− 24.3
5	− 219.7	− 2512.8	462.7
6	11.0	− 1137.6	11.0
7	− 69.1	− 6425.1	1272.4
8	1036.5	− 4292.1	300.8
9	− 573.9	− 4694.7	1627.5
10	606.5	− 1848.5	419.5
11	− 290.8	− 2072.8	582.2
12	216.8	− 666.8	8.1
13	− 3206.2	− 7858.5	2141.5
14	337.2	− 4800.0	160.9
15	− 784.0	− 5080.3	2261.2
16	348.2	− 1006.6	699.4
17	101.8	− 1068.2	186.4
18	180.8	− 101.8	− 101.8

There is a small imbalance in the forces, but the 13.5-N force is close enough to zero to assume that the calculated values of σ_{yy} comprise a set of stresses that are in equilibrium with the applied load.

The distribution of σ_{xx} on section $B-B$ is shown in Figure 23.11. This cutting plane passes through elements (5), (6), (11), (12), (17), and (18) and the exposed area for each element is 15 cm^2. The force produced by the σ_{xx} distribution must be zero because there are no external forces applied to the body in the x-direction. The summation of forces yields

$$+\rightarrow\Sigma F_x = (101.8 + 180.8 - 290.8 + 216.8 - 219.7 + 10.97)(15)$$
$$= (-0.13)(15) = -1.95 \text{ N} \approx 0$$

Equilibrium in the x-direction is satisfied.

The cutting planes, $A-A$ and $B-B$, have two stress components acting on the exposed surface, a normal stress and a shear stress. Only the normal stresses were shown in Figures 23.10 and 23.11. A free body diagram of the section to the right of $B-B$ with the shear stress σ_{xy} is shown in Figure 23.12. The normal stresses σ_{xx} are the same (opposite direction) as those in Figure 23.11 and are not shown. The normal stress acting upward on element (6) is the σ_{yy} value for this element. A

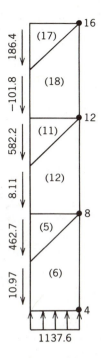

Figure 23.12.   The stress component $\sigma_x{}^y$ on the cutting plane B-B.

summation of forces produces

$$+ \uparrow \Sigma \ F_y = (-186.4 + 101.8 - 582.2 - 8.11 - 462.7 - 10.97)(15) + (1137.6)(15)$$
$$= 164.7 \ \text{N}$$

The summation is not as close to zero as the previous two because the $\sigma_{yy} = 1136.6$ value applied to the exterior boundary of element (6) is not correct. A free body analysis that places $\sigma_{yy} = 1137.6$ at the centroid of element (6) and eliminates the 10.97 shear stress value yields a net vertical for of zero newtons.

The three free body analyses discussed here verify that the stress distributions satisfy the equilibrium equation. More accurate stress values can be obtained by refining the grid.

PROBLEMS

23.1–23.8   The nodal coordinates and the nodal displacements for some plane stress elasticity elements are listed in the following problems. Calculate the element stiffness matrix, $[k^{(e)}]$, and the element stress vector $\{\sigma\}$. The coordinates and displacements are given in centimeters. The element thickness is 2 cm. The material matrix is

$$[D] = 10^6 \begin{bmatrix} 16 & 4 & 0 \\ 4 & 16 & 0 \\ 0 & 0 & 6 \end{bmatrix}$$

23.1

$X_i = 2$	$X_j = 1$	$X_k = 1$
$Y_i = 0$	$Y_j = 1$	$Y_k = 0$
$U_{2i-1} = 0.003$	$U_{2j-1} = 0.001$	$U_{2k-1} = 0.0015$
$U_{2i} = 0.0$	$U_{2j} = -0.0003$	$U_{2k} = 0.0$

23.2

$X_i = 4$	$X_j = 2$	$X_k = 2$
$Y_i = 3$	$Y_j = 4$	$Y_k = 3$
$U_{2i-1} = 0.004$	$U_{2j-1} = 0.0026$	$U_{2k-1} = 0.003$
$U_{2j-1} = -0.001$	$U_{2j} = -0.0013$	$U_{2k} = -0.001$

23.3

$X_i = 6.6$	$X_j = 7.9$	$X_k = 6.0$
$Y_i = 3.0$	$Y_j = 4.0$	$Y_k = 4.0$
$U_{2i-1} = 0.007$	$U_{2j-1} = 0.006$	$U_{2k-1} = 0.006$
$U_{2i} = -0.001$	$U_{2j} = -0.0015$	$U_{2k} = -0.0018$

23.4

$X_i = 9.9$	$X_j = 8.5$	$X_k = 8.0$
$Y_i = 0.0$	$Y_j = 1.0$	$Y_k = 0.0$
$U_{2i-1} = 0.012$	$U_{2j-1} = 0.010$	$U_{2k-1} = 0.010$
$U_{2i} = 0.0$	$U_{2j} = -0.003$	$U_{2k} = 0.0$

23.5

$X_i = 11.0$	$X_j = 10.0$	$X_k = 10.0$
$Y_i = 2.0$	$Y_j = 1.8$	$Y_k = 1.5$
$U_{2i-1} = 0.016$	$U_{2j-1} = 0.015$	$U_{2k-1} = 0.014$
$U_{2i} = -0.0008$	$U_{2j} = -0.0006$	$U_{2k} = -0.0004$

23.6

$X_i = 0.0$	$X_j = 10.0$	$X_k = 0.0$
$Y_i = 0.0$	$Y_j = 10.0$	$Y_k = 10.0$
$U_{2i-1} = 0.0$	$U_{2j-1} = 0.001$	$U_{2k-1} = 0.0$
$U_{2i} = 0.0$	$U_{2j} = -0.0022$	$U_{2k} = -0.0028$

23.7

$X_i = 10.0$	$X_j = 20.0$	$X_k = 10.0$
$Y_i = 0.0$	$Y_j = 10.0$	$Y_k = 10.0$
$U_{2i-1} = 0.0011$	$U_{2j-1} = 0.0016$	$U_{2k-1} = 0.0011$
$U_{2i} = 0.0$	$U_{2j} = -0.0012$	$U_{2k} = -0.0022$

23.8

$X_i = 0.0$	$X_j = 10.0$	$X_k = 0.0$
$Y_i = 20.0$	$Y_j = 30.0$	$Y_k = 30.0$
$U_{2i-1} = 0.0$	$U_{2j-1} = -0.0006$	$U_{2k-1} = 0.0$
$U_{2i} = -0.006$	$U_{2j} = 0.007$	$U_{2k} = -0.0096$

23.9 Evaluate the contribution of a thermal change of $\delta T = 10°C$ to $\{f^{(e)}\}$ for the element in Problem 23.1 using $\alpha = 12(10^{-6})/°C$.

23.10 Do Problem 23.9 for the element in Problem 23.2.

23.11 Evaluate the surface integral

$$\int_{\Gamma} [N]^T \begin{Bmatrix} p_x \\ p_y \end{Bmatrix} d\Gamma$$

for the surface loading shown in Figure P23.11.

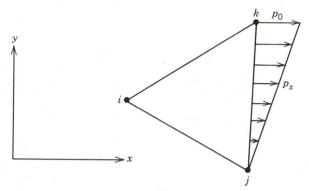

Figure P23.11

23.12 Evaluate the surface integral in Problem 23.11 for the surface loading shown in Figure P23.12.

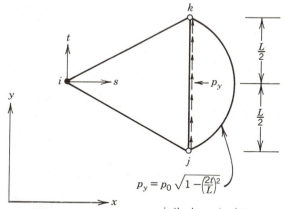

p_0 is the largest value

Figure P23.12

23.13 Solve the plane stress problem shown in Figure 23.7 using the computer program **STRESS** discussed in Chapter 25 and a minimum of 30 elements.

23.14–23.18 Calculate the stress distribution in one of the bodies shown below using the computer program **STRESS** discussed in Chapter 25. Make use of symmetry when possible.

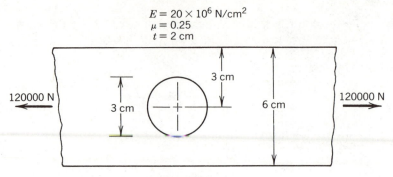

Figure P23.14

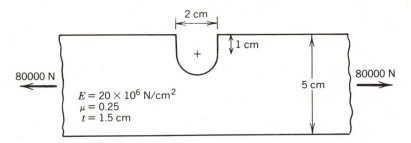

Figure   P23.15

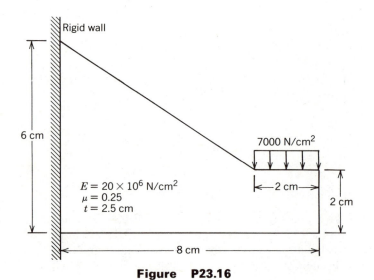

Figure   P23.16

312

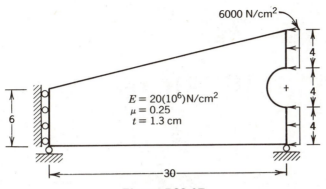

Figure P23.17

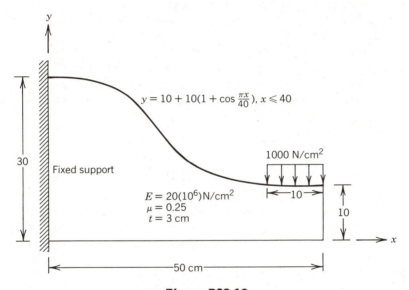

Figure P23.18

Chapter 24

AXISYMMETRIC ELASTICITY

An important group of elasticity problems involves a body of revolution with an axisymmetric loading. The body is three-dimensional, but the geometry of the body and loading are not a function of the circumferential direction. The body can be analyzed using two-dimensional techniques. The axisymmetric triangular element is obtained by revolving the linear triangular element through 360 degrees to form a triangular torus.

The element integrals for the axisymmetric problem look very similar to those evaluated in the previous chapter. The evaluation process and results, however, are quite different. Our objective here is to discuss axisymmetric elasticity problems and to highlight those areas where it differs significantly from the two-dimensional problems.

24.1 DEFINITIONS IN CYLINDRICAL COORDINATES

The first and most obvious change from two-dimensional to axisymmetric problems is the change in coordinate systems. Axisymmetric problems are solved using a cylindrical coordinate system with coordinates r, θ, z (Figure 24.1). The elasticity quantities in this coordinate system are discussed by Fung (1965) and summarized here.

The vector of stress components is

$$\{\sigma\}^T = [\sigma_{rr} \quad \sigma_{\theta\theta} \quad \sigma_{zz} \quad \sigma_{r\theta} \quad \sigma_{rz} \quad \sigma_{\theta z}] \tag{24.1}$$

The elastic strain components are

$$\{\varepsilon\} = [\varepsilon_{rr} \quad \varepsilon_{\theta\theta} \quad \varepsilon_{zz} \quad \varepsilon_{r\theta} \quad \varepsilon_{rz} \quad \varepsilon_{\theta z}] \tag{24.2}$$

and the total strain components are

$$\{e\}^T = [e_{rr} \quad e_{\theta\theta} \quad e_{zz} \quad e_{r\theta} \quad e_{rz} \quad e_{\theta z}] \tag{24.3}$$

The thermal strain vector remains the same as (22.4) whereas $[D]$ in Hooke's law, $\{\sigma\} = [D]\{\varepsilon\}$, is the matrix defined in (22.7). The relationship (22.5), $\{e\} = \{\varepsilon\} + \{\varepsilon_T\}$ is also valid. The body force components in (22.23) and the surface stress com-

314

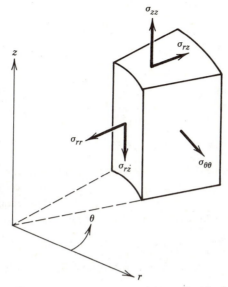

Figure   24.1.   The stress components in a cylindrical coordinate system.

ponents are

$$\begin{Bmatrix} \mathcal{R} \\ \mathcal{T} \\ \mathcal{L} \end{Bmatrix} \qquad \text{and} \qquad \begin{Bmatrix} p_r \\ p_\theta \\ p_z \end{Bmatrix}$$

respectively.

The equations that define the relationship between the total strains and the three displacements have a new look. These equations are

$$e_{rr} = \frac{\partial u}{\partial r} \qquad\qquad e_{r\theta} = \frac{1}{r}\frac{\partial u}{\partial \theta} + \frac{\partial v}{\partial r} - \frac{v}{r}$$

$$e_{\theta\theta} = \frac{u}{r} + \frac{1}{r}\frac{\partial v}{\partial \theta} \qquad\qquad e_{rz} = \frac{\partial u}{\partial z} + \frac{\partial w}{\partial r}$$

$$e_{zz} = \frac{\partial w}{\partial z} \qquad\qquad e_{\theta z} = \frac{\partial v}{\partial z} + \frac{1}{r}\frac{\partial w}{\partial \theta} \qquad\qquad (24.4)$$

where u, v, and w are the displacements in the r-, θ-, and z-directions, respectively.

The derivation of the equations for the element matrices in cylindrical coordinates proceeds exactly as given in Section 22.3. In fact, the end result is symbolically identical,

$$[k^{(e)}] = \int_V [B]^T [D][B]\, dV \qquad\qquad (24.5)$$

and

$$\{f^{(e)}\} = \int_V [B]^T[D]\{\varepsilon_T\}\,dV + \int_V [N]^T \begin{Bmatrix} \mathcal{R} \\ \mathcal{T} \\ \mathcal{Z} \end{Bmatrix} dV + \int_\Gamma [N]^T \begin{Bmatrix} p_r \\ p_\theta \\ p_z \end{Bmatrix} d\Gamma \qquad (24.6)$$

The actual coefficients in $[B]$, however, are different because the strain-displacement equations, (24.4), are not the same as those defined by (22.11).

24.2   AXISYMMETRIC ELASTICITY

An axisymmetric problem exists when neither the geometry nor the surface loading are a function of the circumferential direction θ. This situation makes the circumferential displacement v zero and leaves the u and w displacements as functions of r and z only. In a general form, the displacements equations are

$$u = f(r, z), \qquad v = 0, \qquad \text{and} \qquad w = g(r, z) \qquad (24.7)$$

Knowing that $v = 0$ and that u and w do not vary with θ reduces the strain-displacement equations (24.4) to

$$e_{rr} = \frac{\partial u}{\partial r}, \qquad e_{\theta\theta} = \frac{u}{r}, \qquad e_{zz} = \frac{\partial w}{\partial z}$$

$$e_{r\theta} = 0, \qquad e_{rz} = \frac{\partial u}{\partial z} + \frac{\partial w}{\partial r}, \qquad e_{\theta z} = 0 \qquad (24.8)$$

There are four nonzero total strain components. The zero values for $e_{r\theta}$ and $e_{\theta z}$ imply that $\varepsilon_{r\theta} = \varepsilon_{rz} = 0$ and there are four nonzero elastic strain components. The strain vectors for axisymmetric elasticity are

$$\{e\}^T = \begin{bmatrix} e_{rr} & e_{\theta\theta} & e_{zz} & e_{rz} \end{bmatrix} \qquad (24.9)$$
$$\{\varepsilon\}^T = \begin{bmatrix} \varepsilon_{rr} & \varepsilon_{\theta\theta} & \varepsilon_{zz} & \varepsilon_{rz} \end{bmatrix} \qquad (24.10)$$

and

$$\{\varepsilon_T\}^T = \begin{bmatrix} \alpha\delta T & \alpha\delta T & \alpha\delta T & 0 \end{bmatrix} \qquad (24.11)$$

Substitution of $\varepsilon_{r\theta} = \varepsilon_{\theta z} = 0$ into Hooke's law, $\{\sigma\} = [D]\{\varepsilon\}$, reveals that $\sigma_{r\theta} = \sigma_{\theta z} = 0$ and there are four nonzero stress components. The vector of stress components is

$$\{\sigma\}^T = \begin{bmatrix} \sigma_{rr} & \sigma_{\theta\theta} & \sigma_{zz} & \sigma_{rz} \end{bmatrix} \qquad (24.12)$$

The stress components are shown in Figure 24.1. The materials matrix $[D]$ in Hooke's law reduces to

$$[D] = \frac{E}{1+\mu} \begin{bmatrix} d & b & b & 0 \\ b & d & b & 0 \\ b & b & d & 0 \\ 0 & 0 & 0 & \frac{1}{2} \end{bmatrix} \qquad (24.13)$$

where

$$d = \frac{1-\mu}{1-2\mu} \quad \text{and} \quad b = \frac{\mu}{1-2\mu}$$

The body force vector and the surface stress vectors reduce to

$$\begin{Bmatrix} \mathscr{R} \\ \mathscr{L} \end{Bmatrix} \quad \text{and} \quad \begin{Bmatrix} p_r \\ p_z \end{Bmatrix} \tag{24.14}$$

respectively.

24.3 ELEMENT MATRICES

The unknown displacements in an axisymmetric problem are u and w. These displacements can be written in terms of the element nodal values similar to (23.15). The displacement equations are

$$\begin{Bmatrix} u(r,z) \\ w(r,z) \end{Bmatrix} = \begin{bmatrix} N_i & 0 & N_j & 0 & N_k & 0 \\ 0 & N_i & 0 & N_j & 0 & N_k \end{bmatrix} \begin{Bmatrix} U_{2i-1} \\ U_{2i} \\ U_{2j-1} \\ U_{2j} \\ U_{2k-1} \\ U_{2k} \end{Bmatrix} \tag{24.15}$$

or

$$\begin{Bmatrix} u(r,z) \\ v(r,z) \end{Bmatrix} = [N]\{U^{(e)}\} \tag{24.16}$$

The shape functions are identical to those given in (5.8) through (5.10) with x replaced by r and y replaced by z. The equation for N_i is

$$N_i = \frac{1}{2A}(a_i + b_i r + c_i z) \tag{24.17}$$

with $a_i = R_j Z_k - R_k Z_j$, $b_i = Z_j - Z_k$, and $c_i = R_k - R_j$.

The differentiation of (24.15) using the strain-displacement relationships in (24.8) yields

$$\{e\} = \frac{1}{2A} \begin{bmatrix} b_i & 0 & b_j & 0 & b_k & 0 \\ \dfrac{2AN_i}{r} & 0 & \dfrac{2AN_j}{r} & 0 & \dfrac{2AN_k}{r} & 0 \\ 0 & c_i & 0 & c_j & 0 & c_k \\ c_i & b_i & c_j & b_j & c_k & b_k \end{bmatrix} \{U^{(e)}\} \tag{24.18}$$

The coefficient matrix in (24.18) is $[B]$, since $\{e\} = [B]\{U^{(e)}\}$.

The evaluation of the element integrals is no longer the simple procedure that occurred for the two-dimensional problems. The $[B]$ matrix contains terms that are a function of the coordinates and cannot be removed from under the integral sign. One popular procedure for evaluating the integrals is to make $[B]$ a constant by evaluating the $2AN_\beta/r$ terms using $\bar{r}$ and $\bar{z}$. This procedure allows $[B]$ to be removed from under the integral sign, or

$$[k^{(e)}] = [\bar{B}]^T[D][\bar{B}] \int_V dV \tag{24.19}$$

The integral is the element volume that is $V = 2\pi\bar{r}A$; thus the final equation for $[k^{(e)}]$ is

$$[k^{(e)}] = 2\pi A\bar{r}[\bar{B}]^T[D][\bar{B}] \tag{24.20}$$

The bar on $[\bar{B}]$ indicates an approximate value. Equation 24.20 yields acceptable results if a subdivision consistent with the final stress distribution is used, that is, small elements in the regions of high stress gradients.

The column vector associated with the thermal change is handled in the same manner, since $[B]$ occurs in the integral. The approximate solution is

$$\int_V [B][D]\{\varepsilon_T\}\, dV = 2\pi\bar{r}A[\bar{B}]^T[D]\{\varepsilon_T\} \tag{24.21}$$

The volume integral involving the body forces can be integrated exactly by using area coordinates. The integral written in terms of area coordinates is

$$\{f_b^{(e)}\} = \int_A \begin{bmatrix} rL_1 & 0 \\ 0 & rL_1 \\ rL_2 & 0 \\ 0 & rL_2 \\ rL_3 & 0 \\ 0 & rL_3 \end{bmatrix} \begin{Bmatrix} \mathscr{R} \\ \mathscr{L} \end{Bmatrix} 2\pi\, dA \tag{24.22}$$

where $2\pi r\, dA$ has been substituted for dV. The radial distance r can also be written in terms of area coordinates

$$r = R_iL_1 + R_jL_2 + R_kL_3 \tag{24.23}$$

and substituted into (24.22). This substitution produces L_1L_2- and L_1^2-type products, which are evaluated using (6.29). The final result is

$$\{f_b^{(e)}\} = \frac{2\pi A}{12} \begin{Bmatrix} (R_i + 3\bar{r})\mathscr{R} \\ (R_i + 3\bar{r})\mathscr{L} \\ (R_j + 3\bar{r})\mathscr{R} \\ (R_j + 3\bar{r})\mathscr{L} \\ (R_k + 3\bar{r})\mathscr{R} \\ (R_k + 3\bar{r})\mathscr{L} \end{Bmatrix} \tag{24.24}$$

where $3\bar{r} = R_i + R_j + R_k$. Equation (24.24) does not distribute $\mathscr{R}$ or $\mathscr{L}$ equally between the three nodes. The nodes farthest from the axis of revolution receive a larger share of the body forces.

The integral involving the surface stresses is also evaluated using area coordinates. The integral is

$$\{f_p^{(e)}\} = \int_\Gamma [N]^T \begin{Bmatrix} p_r \\ p_z \end{Bmatrix} d\Gamma \tag{24.25}$$

where p_r and p_z are the surface stresses in the r- and z- directions. Confining our discussion to the side, between nodes i and j means that $N_k=0$ and

$$\{f_p^{(e)}\} = L_{ij} \int_0^1 2\pi \begin{bmatrix} rL_1 & 0 \\ 0 & rL_1 \\ rL_2 & 0 \\ 0 & rL_2 \\ 0 & 0 \\ 0 & 0 \end{bmatrix} \begin{Bmatrix} p_r \\ p_z \end{Bmatrix} d\ell_2 \tag{24.26}$$

where $d\Gamma = 2\pi r d\ell_2$. This integral is evaluated using (6.29) after substituting (24.23). The final result is

$$\{f_p^{(e)}\} = \frac{2\pi L_{ij}}{6} \begin{Bmatrix} (2R_i + R_j)p_r \\ (2R_i + R_j)p_z \\ (R_i + 2R_j)p_r \\ (R_i + 2R_j)p_z \\ 0 \\ 0 \end{Bmatrix} \tag{24.27}$$

Equation (24.27) is applicable to any surface. For a vertical surface we find that $R_i = R_j = R$, and

$$\{f^{(e)}\} = \frac{2\pi R L_{ij}}{2} \begin{Bmatrix} p_r \\ p_z \\ p_r \\ p_z \\ 0 \\ 0 \end{Bmatrix} \tag{24.28}$$

The stress components are converted to forces and distributed equally between the nodes. This is identical to the results obtained for the two-dimensional problem. On the other hand, if we are considering a horizontal surface, $R_i \neq R_j$, and a larger proportion of the load is given to the node farthest from the axis of revolution.

The other two solutions to (24.25) are

$$\frac{2\pi L_{jk}}{6} \begin{Bmatrix} 0 \\ 0 \\ (2R_j+R_k)p_r \\ (2R_j+R_k)p_z \\ (R_j+2R_k)p_r \\ (R_j+2R_k)p_z \end{Bmatrix}, \quad \frac{2\pi L_{ik}}{6} \begin{Bmatrix} (2R_i+R_k)p_r \\ (2R_i+R_k)p_z \\ 0 \\ 0 \\ (R_i+2R_k)p_r \\ (R_i+2R_k)p_z \end{Bmatrix} \quad (24.29)$$

on sides L_{jk} and L_{ik}, respectively.

24.4 SURFACE LOADS

The surface stresses p_r and p_z cannot be calculated in an intuitive way when the surface is not vertical. We illustrate this by considering the simple example of a cylinder subjected to an axial compression load.

Consider the cylinder shown in Figure 24.2a with the element subdivision shown in Figure 24.2b. Only the elements near the load are shown. Assume that side jk of each element is subjected to the load; then $\{f_p^{(e)}\}$ is given by the first vector of (24.29)

$$\{f^{(e)}\} = \frac{2\pi L_{jk}}{6} \begin{Bmatrix} 0 \\ 0 \\ 0 \\ (2R_j+R_k)p_z \\ 0 \\ (R_j+2R_k)p_z \end{Bmatrix}$$

The extra zeros result because $p_r = 0$.

The data required for the computations are summarized in the following table.

e	R_j	R_k	$(2R_j+R_k)$	(R_j+2R_k)
1	2	0	4	2
2	4	2	10	8
3	6	4	16	14

The forces acting at the element nodes are

$$\text{Node } j: \quad P_j = \frac{2\pi L_{jk}}{6}(2R_j+R_k)p_z = 400\pi(2R_j+R_k)$$

$$\text{Node } k: \quad P_k = \frac{2\pi L_{jk}}{6}(R_j+2R_k)p_z = 400\pi(R_j+2R_k)$$

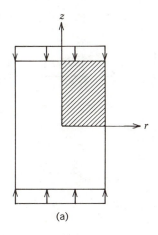

(a)

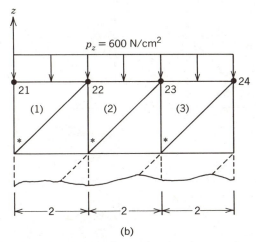

(b)

Figure 24.2. A uniformly distributed load acting on the ends of a cylinder.

Substituting for R_j and R_k in each element gives

	Element		
Node	(1)	(2)	(3)
j	1600π	4000π	6400π
k	800π	3200π	5600π

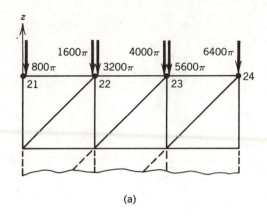

(a)

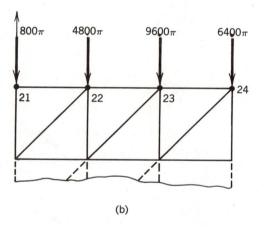

(b)

Figure 24.3. (a) Element contributions to the nodal forces. (b) Calculated nodal forces.

These forces are shown in Figure 24.3a. The final set of nodal forces is shown in Figure 24.3b. This set of forces is equivalent to

$$P_{42} = -800\pi, \qquad P_{44} = -4800\pi, \qquad P_{46} = -9600\pi, \qquad P_{48} = -6400\pi$$

It is clear from Figure 24.3b that the concentrated forces are not equal.

PROBLEMS

24.1–24.5   The nodal coordinates for five axisymmetric elasticity elements are given below. Evaluate $[\bar{B}]$ for the element. The coordinates are in centimeters.

Element	Problem Number				
Quantity	24.1	24.2	24.3	24.4	24.5
R_i	1.0	6.0	2.0	10.0	18.0
Z_i	1.0	7.0	4.0	2.0	16.0
R_j	3.0	8.0	4.0	14.0	22.0
Z_j	1.0	7.0	4.0	2.0	18.0
R_k	3.0	9.0	3.0	14.0	18.0
Z_k	4.0	10.0	6.0	4.0	18.0

24.6   Evaluate $\{f_T^{(e)}\}$ given by (24.21) for the element in Problem 24.1 when $\alpha = 10(10^{-6})/°C$ and $\delta T = 15°C$. Use $E = 20(10^6)\ N/cm^2$ and $\mu = 1/4$.

24.7   Do Problem 24.6 for the element in Problem 24.2.

24.8   A body force of $\mathscr{Z} = 10\ N/cm^3$ exists in the element in Problem 24.3. Evaluate $\{f_b^{(e)}\}$ given by (24.24). The body force $\mathscr{R}$ is zero.

24.9   A body force of $\mathscr{R} = 20\ N/cm^3$ exists in the element in Problem 24.4. Evaluate $\{f_b^{(e)}\}$ given by (24.24). The body force $\mathscr{Z}$ is zero.

24.10   Calculate the equivalent concentrated forces at nodes one, two, three, and four for the distributed surface stress shown in Figure P24.10. The surface stress p_r is zero. The dimensions are in centimeters.

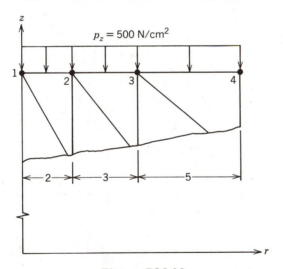

Figure P24.10

24.11   Calculate the equivalent concentrated forces at nodes two, three, and four for the distributed surface stress shown in Figure P24.11. The surface stress p_r is zero. The dimensions are in centimeters.

24.12   Evaluate the surface integral in Figure P24.12 for the variable surface loading acting on the horizontal surface of an axisymmetric element.

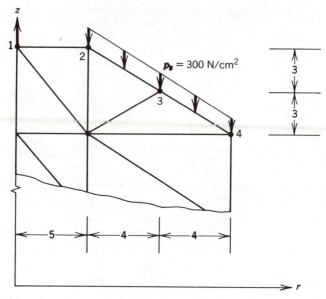

Figure P24.11

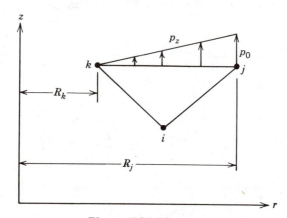

Figure P24.12

24.13 Show that the stress vector $\{\sigma\}$ defined by (24.12) is related to the nodal displacements by

$$\{\sigma\} = [D][\bar{B}]\{U^{(e)}\} - [D]\{\varepsilon_T\}$$

24.14 Calculate the components of $\{\sigma\}$ as defined by the equation in Problem 24.13 for the element in Problem 24.1 given that $\delta T = 0$ and

$$U_{2i-1} = 0.0 \qquad U_{2j-1} = 0.005 \qquad U_{2k-1} = 0.001$$
$$U_{2i} = 0.0 \qquad U_{2j} = 0.0002 \qquad U_{2k} = 0.008$$

24.15 Calculate the components of $\{\sigma\}$ as defined by the equation in Problem 24.13, for the element in Problem 24.2 given that $\delta T = 0$ and

$$U_{2i-1} = 0.005 \qquad U_{2j-1} = 0.003 \qquad U_{2k-1} = 0.006$$
$$U_{2i} = 0.0 \qquad U_{2j} = 0.002 \qquad U_{2k} = 0.001$$

24.16 Evaluate $[B]$ for an axisymmetric rectangular element. Remember that the shape function equations must be written relative to the origin of the global coordinate system. See equation (13.10).

24.17 Evaluate $\{f_p^{(e)}\}$ for the surface stress p_z shown acting on the rectangular element in Figure P24.17. See equation (13.10).

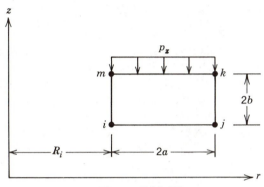

Figure P24.17

Chapter 25

COMPUTER PROGRAMS FOR STRUCTURAL AND SOLID MECHANICS

The matrix analysis of structures and the finite element solution of elasticity problems have a limited usefulness if a digital computer is not available. Computer programs that can be used to analyze plane structures and plane stress elasticity problems are presented in this chapter. Both programs are written for educational use rather than commercial problem solving. The elasticity program, in particular, contains several diagnostic checks to locate errors that are common to the beginning user.

25.1 PROGRAM FRAME

The computer program **FRAME** is used to analyze one- and two-dimensional plane structures. The program uses the element stiffness matrix for the plane frame member, (21.13), in an expanded form and obtains the element stiffness matrix for the axial and truss elements by setting the area moment, I, to zero. The element stiffness matrix for the beam element is obtained by keeping the plane frame element horizontal.

The program **FRAME** has five subroutines; **MODFRM, DCMPBD, SLVBD, CONVERT** and **INVERT**. A brief discussion of each program is given in the following paragraphs.

Subroutine MODFRM. The subroutine **MODFRM** incorporates the specified nodal displacements into the system of equations using the method of deletion of rows and columns (see Appendix II). The subroutine is also used to add concentrated force and moment values directly to $\{F\}$. It is important that the user realize that this subroutine has READ statements and that these statements must be terminated by specifying a numerical value before execution is returned to the main program.

The subroutine **MODFRM** is nearly identical to the subroutine **MODIFY** used with the programs **TDFIELD** (Chapter 16) and **STRESS**. The major difference is that **MODFRM** calls the subroutines **CONVERT** and **INVERT** whereas **MODIFY** does not.

326

Subroutine DCMPBD. The subroutine **DCMPBD** decomposes the global stiffness matrix $[K]$ into an upper triangular form using the method of Gaussian elimination (Conte and deBoor, 1980). This subroutine assumes that $[K]$ is symmetric and only those elements within the bandwidth and on or above the main diagonal are stored. The programming logic is not easy to follow because the coefficients of $[K]$ are stored in a vector rather than a two-dimensional array.

Subroutine SLVBD. The subroutine **SLVBD** is a companion program to **DCMPBD**. This subroutine decomposes the global force vector, $\{F\}$, and solves the system of equations using back substitution. The solution of the system of equations is separated into two subroutines so that they can be used to solve time-dependent problems, where each new time step requires the decomposition of $\{F\}$ but may not require that $[K]$ be converted into an upper triangular form.

Subroutine CONVERT. The subroutine **CONVERT** makes the program **FRAME** look like an axial force, truss, beam, or frame program to the user. The element stiffness matrix in **FRAME** has three displacements at each node, but **CONVERT** makes the program appear as though it has one, two, or three displacements at a node depending on the numerical value of **ITYP** (line 30 of **FRAME**).

Subroutine INVERT. The subroutine **INVERT** is a companion program to **CONVERT**. The subroutine takes results calculated using three displacements at a node and presents them so the program appears as though it has one, two, or three displacements at a node depending on the numerical value of **ITYP**.

```
                    PROGRAM FRAME(INPUT,OUTPUT,TAPE60=INPUT,TAPE61=OUTPUT)
                    DIMENSION NS(6),ESM(6,6),EF(6),F(6),XC(100),YC(100)
                    DIMENSION ELMOD(2),PROP(50,2),NEL(50,2),ISECT(50)
                    DIMENSION ICK(100)
       5            COMMON/TLE/TITLE(20)
                    COMMON/AV/A(5000),JGF,JGSM,NP,NBW
                    COMMON/CT/ITYP,IB
                    DATA ELMOD /20000000.,0./
                    DATA IN/60/,IO/61/
      10            DATA YC/100*0.0/
              C**************
              C**************
              C
              C  DEFINITION OF THE INPUT VARIABLES FOR THIS PROGRAM
      15      C     THIS LIST DOES NOT INCLUDE VALUES
              C     READ BY THE SUBROUTINE MODFRM
              C
              C**************
              C**************
      20      C
              C  TITLE AND PARAMETERS
              C
              C          TITLE - DESCRIPTIVE STATEMENT OF THE PROBLEM
              C                    BEING SOLVED
      25      C     NN - NUMBER OF NODES
              C     NE - NUMBER OF MEMBERS
              C     NSP - NUMBER OF SETS OF SECTION PROPERTIES
              C     IEM - ELASTIC MODULUS INDEX
              C            1 - STEEL, SI UNITS, NEWTONS/SQ. CM.
```

```
30      C                         2 - EXTERNALLY DEFINED VALUE
        C              ITYP - TYPE OF PROBLEM
        C                         1 - AXIAL FORCE MEMBERS
        C                         2 - TRUSS MEMBERS
        C                         3 - BEAM MEMBERS
35      C                         4 - FRAME MEMBERS
        C
        C   SECTION PROPERTIES
        C
        C              NSEC - SECTION NUMBER
40      C              PROP(NSEC,1) - CROSS SECTIONAL AREA
        C              PROP(NSEC,2) - AREA MOMENT
        C
        C   EXTERNALLY DEFINED ELASTIC MODULUS VALUE
        C
45      C              ELMOD(2) - READ WHEN IEM=2, SKIPPED WHEN IEM=1
        C
        C   NODAL COORDINATES
        C
        C              XC(I) - X COORDINATES OF THE NODES IN NUMERICAL SEQUENCE
50      C              YC(I) - Y COORDINATES OF THE NODES IN NUMERICAL SEQUENCE
        C                        Y COORDINATES NEEDED ONLY FOR TRUSS AND FRAME
        C                        PROBLEMS
        C
        C   ELEMENT DATA
55      C
        C              N - ELEMENT NUMBER
        C              ISECT(N) - SECTION NUMBER OF ELEMENT N
        C              NEL(N,I) - NODE NUMBER OF NODE I
        C              NEL(N,J) - NODE NUMBER OF NODE J
60      C
        C***************
        C***************
        C
        C   INPUT SECTION OF THE PROGRAM
65      C
        C***************
        C***************
        C
        C   INPUT OF THE TITLE CARD AND THE CONTROL PARAMETERS
70      C
              READ(IN,1) TITLE
        1     FORMAT(20A4)
              READ(IN,*) NN,NE,NSP,IEM,ITYP
              WRITE(IO,3) TITLE
75      3     FORMAT(1H1////10X,20A4)
        C
        C   COMPARISON OF NN, NE, NSP, IEM, AND ITYP WITH
        C      THE DIMENSIONED VALUES
        C
80            ISTOP=0
        C   CHECK NN
              IF(NN.LE.100) GOTO32
              WRITE(IO,31)
        31    FORMAT(10X,27HNUMBER OF NODES EXCEEDS 100/
85      +10X,20HEXECUTION TERMINATED)
              ISTOP=1
```

```fortran
      C   CHECK NE
      32    IF (NE.LE.50) GOTO34
            WRITE (IO,33)
      33    FORMAT (IOX,29HNUMBER OF ELEMENTS EXCEEDS 50/
           +IOX,2OHEXECUTION TERMINATED)
            ISTOP=1
      C   CHECK NSP
      34    IF (NSP.LE.50) GOTO36
            WRITE (IO,35)
      35    FORMAT (IOX,29HNUMBER OF SECTIONS EXCEEDS 50/
           +IOX,2OHEXECUTION TERMINATED)
            ISTOP=1
      C   CHECK IEM
      36    IF (IEM.LE.2) GOTO38
            WRITE (IO,37)
      37    FORMAT (IOX,34HNUMBER OF MODULUS VALUES EXCEEDS 2/
           +IOX,2OHEXECUTION TERMINATED)
            ISTOP=1
      C   CHECK ITYP
      38    IF (ITYP.LE.4) GOTO40
            WRITE (IO,39)
      39    FORMAT (IOX,21HTYPE NUMBER EXCEEDS 4/
           +IOX,2OHEXECUTION TERMINATED)
            ISTOP=1
      40    IF (ISTOP.EQ.1) STOP
            WRITE (IO,5)
      5     FORMAT (//IOX,18HSECTION PROPERTIES/IOX,7HSECTION,
           +6X,4HAREA,9X,11HAREA MOMENT)
      C
      C   INPUT OF THE SECTION PROPERTIES
      C
            DO4I=1,NSP
            READ (IN,*) NSEC,PROP (NSEC,1),PROP (NSEC,2)
            WRITE (IO,8) NSEC,PROP (NSEC,1),PROP (NSEC,2)
      8     FORMAT (13X,I3,2E15.5)
            IF (ITYP.EQ.1.OR.ITYP.EQ.2) PROP (NSEC,2)=0.
      4     IF (ITYP.EQ.3) PROP (NSEC,1)=0.
      C
      C   INPUT OF A DIFFERENT ELASTIC MODULUS VALUE
      C
            IF (IEM.EQ.2) READ (IN,*) ELMOD (2)
      6     WRITE (IO,7) ELMOD (IEM)
      7     FORMAT (/IOX,17HELASTIC MODULUS =,E15.5)
      C
      C   INPUT AND ECHO PRINT OF THE NODAL COORDINATES
      C
            WRITE (IO,12)
      12    FORMAT (///IOX,17HNODAL COORDINATES/IOX,
           +4HNODE,9X,1HX,14X,1HY)
            READ (IN,*) (XC (I),I=1,NN)
            IF (ITYP.EQ.2.OR.ITYP.EQ.4) READ (IN,*) (YC (I),I=1,NN)
            DO13I=1,NN
      13    WRITE (IO,14) I,XC (I),YC (I)
      14    FORMAT (IOX,I4,F12.2,3X,F12.2)
      C
      C   INPUT AND ECHO PRINT OF ELEMENT DATA
      C
```

329

```
                    WRITE(IO,21)
145        21       FORMAT(//IOX,12HELEMENT DATA/32X,4HNODE/
                   +IOX,7HELEMENT,3X,7HSECTION,3X,7HNUMBERS)
                    DO231=1,NE
                    READ(IN,*) N,ISECT(N),NEL(N,1),NEL(N,2)
           23       WRITE(IO,24) N,ISECT(N),NEL(N,1),NEL(N,2)
150        24       FORMAT(12X,13,8X,11,5X,214)
           C*********
           C*********
           C
           C  ANALYSIS OF THE NODE NUMBERS
155        C
           C*********
           C*********
           C
           C  INITIALIZATION OF A CHECK VECTOR
160        C
                    DO5001=1,NN
           500      ICK(I)=0
           C
           C  CHECK TO SEE IF ANY NODE NUMBER EXCEEDS NP
165        C
                    DO5011=1,NE
                    DO502J=1,2
                    K=NEL(I,J)
                    ICK(K)=1
170        502      IF(K.GT.NN) WRITE(IO,503) J,I,NN
           503      FORMAT(/IOX,4HNODE,14,11H OF ELEMENT,14,
                   +13H EXCEEDS NN =,14)
           501      CONTINUE
           C
175        C  CHECK TO SEE IF ALL NODE NUMBERS THROUGH
           C     NN ARE INCLUDED
           C
                    DO5051=1,NN
           505      IF(ICK(I).EQ.0) WRITE(IO,506) I
180        506      FORMAT(/IOX,4HNODE,14,15H DOES NOT EXIST)
           C***************
           C***************
           C
           C  INITIALIZATION OF THE A VECTOR
185        C
           C***************
           C***************
           C
           C  DETERMINATION OF THE BANDWIDTH, NBW
190        C       NBW - BANDWIDTH OF THE SYSTEM OF EQUATIONS
           C
                    NBW=0
                    DO521=1,NE
                    IJ=IABS(NEL(I,1)-NEL(I,2))
195        52       IF(IJ.GT.NBW) NBW=IJ
                    NBW=(NBW+1)*3
           C
           C  INITIALIZATION OF THE COLUMN VECTOR A( )
           C
200                 NP=NN*3
```

330

```
                  JGF=NP
                  JGSM=JGF+NP
                  JEND=JGSM+NP*NBW
                  DO611=1,JEND
205            61 A(I)=0.0
                  IF(JEND.GT.2500) GOTO62
                  GOTO69
               62 WRITE(IO,63)
               63 FORMAT(//10X,26HMEMORY REQUIREMENTS EXCEED
210              +30H THE DIMENSION OF THE A VECTOR)
                  STOP
C*************
C************
C
215 C  GENERATION OF THE SYSTEM OF EQUATIONS
    C
C*************
C************
    C
220 69     IFORCE=0
                  EM=ELMOD(IEM)
                  KK=1
    C
    C  RETRIEVAL OF THE SECTION PROPERTIES
225 C
    70     J=ISECT(KK)
                  AREA=PROP(J,1)
                  RI=PROP(J,2)
    C
230 C  CALCULATION OF THE GEOMETRIC DATA
    C
                  J1=NEL(KK,2)
                  J2=NEL(KK,1)
                  XL=XC(J1)-XC(J2)
235              YL=YC(J1)-YC(J2)
                  EL=SQRT(XL*XL+YL*YL)
                  CS=XL/EL
                  SN=YL/EL
    C
240 C  CALCULATION OF THE ELEMENT DEGREES OF FREEDOM
    C
                  NS(1)=J2*3-2
                  NS(2)=J2*3-1
                  NS(3)=J2*3
245              NS(4)=J1*3-2
                  NS(5)=J1*3-1
                  NS(6)=J1*3
    C
    C  CALCULATION OF THE ELEMENT STIFFNESS MATRIX
250 C
                  RM=EM/EL
                  ESM(1,1)=(AREA*CS**2 +12.*RI*SN**2 /EL**2 )*RM
                  ESM(2,1)=(AREA-(12.*RI/EL**2 ))*CS*SN*RM
                  ESM(2,2)=(AREA*SN**2 +(12.*RI/EL**2 )*CS**2 )*RM
255              ESM(3,1)=-(6.*RI/EL)*SN*RM
                  ESM(3,2)=RM*CS*6.*RI/EL
                  ESM(3,3)=4.*RI*RM
```

```
                      ESM(4,1)=-ESM(1,1)
                      ESM(4,2)=-ESM(2,1)
260                   ESM(4,3)=-ESM(3,1)
                      ESM(4,4)=ESM(1,1)
                      ESM(5,1)=+ESM(4,2)
                      ESM(5,2)=-ESM(2,2)
                      ESM(5,3)=-ESM(3,2)
265                   ESM(5,4)=+ESM(2,1)
                      ESM(5,5)=+ESM(2,2)
                      ESM(6,1)=+ESM(3,1)
                      ESM(6,2)=+ESM(3,2)
                      ESM(6,3)=+2.*RI*RM
270                   ESM(6,4)=-ESM(3,1)
                      ESM(6,5)=-ESM(3,2)
                      ESM(6,6)=+ESM(3,3)
                      DO73I=1,5
                      K=I+1
275                   DO73J=K,6
                   73 ESM(I,J)=ESM(J,I)
                      IF(IFORCE.EQ.1) GOTO85
       C
       C    DIRECT STIFFNESS PROCEDURE
280    C
                      DO76I=1,6
                      II=NS(I)
                      DO75J=1,6
                      JJ=NS(J)-NS(I)+1
285                   IF(JJ.LE.0) GOTO75
                      J1=JGSM+(JJ-1)*NP+II-(JJ-1)*(JJ-2)/2
                      A(J1)=A(J1)+ESM(I,J)
                   75 CONTINUE
                   76 CONTINUE
290    C
       C    ANOTHER ELEMENT?
       C
                      KK=KK+1
                      IF(KK.LE.NE) GOTO70
295    C*************
       C*************
       C
       C    MODIFICATION AND SOLUTION OF THE SYSTEM OF EQUATIONS
       C         DATA IS CALLED BY MODFRM
300    C
       C*************
       C*************
       C
       C    SETTING ZERO DIAGIONAL VALUES TO ONE
305    C
                      DO101I=1,NP
                      IF(A(JGSM+I).GT.0.01) GOTO101
                      A(JGSM+I)=1.0
                  101 CONTINUE
310    C
       C    THE SOLUTION SUBROUTINES
       C
                      CALL MODFRM
                      CALL DCMPBD
```

```
315              CALL SLVBD
      C
      C   OUTPUT OF THE CALCULATED DISPLACEMENT VALUES
      C
                 WRITE(10,110) TITLE
320   110        FORMAT(1H1///10X,20A4///10X,25HNODAL DISPLACEMENT VALUES)
                 IF(ITYP.EQ.1) WRITE(10,111)
      111        FORMAT(//10X,4HNODE,6X,12HX DEFLECTION)
                 IF(ITYP.EQ.2) WRITE(10,112)
      112        FORMAT(//10X,4HNODE,6X,12HX DEFLECTION,6X,12HY DEFLECTION)
325              IF(ITYP.EQ.3) WRITE(10,113)
      113        FORMAT(//10X,4HNODE,6X,12HY DEFLECTION,8X,10HZ ROTATION)
                 IF(ITYP.EQ.4) WRITE(10,114)
      114        FORMAT(//10X,4HNODE,6X,12HX DEFLECTION,6X,12HY DEFLECTION,
                +8X,10HZ ROTATION)
330              DO1021=1,NN
                 IF(ITYP.EQ.1) WRITE(10,103) I, A(I*3-2)
                 IF(ITYP.EQ.2) WRITE(10,103) I,A(I*3-2),A(I*3-1)
                 IF(ITYP.EQ.3) WRITE(10,103) I,A(I*3-1),A(I*3)
                 IF(ITYP.EQ.4) WRITE(10,103) I,A(I*3-2),A(I*3-1),A(I*3)
335   102        CONTINUE
      103        FORMAT(11X,I3,3X,E15.6,3X,E15.6,3X,E15.6)
      C************
      C************
      C
340   C   CALCULATION OF ELEMENT NODAL FORCES
      C
      C************
      C************
      C
345   C   OUTPUT OF HEADINGS
      C
                 WRITE(10,80)
       80 FORMAT(////10X,20A4//10X,20HELEMENT NODAL FORCES )
                 IF(ITYP.EQ.1.OR..ITYP.EQ.2) WRITE(10,82)
350   82         FORMAT(//10X,7HELEMENT,3X,4HNODE,4X,11HAXIAL FORCE)
                 IF(ITYP.EQ.3) WRITE(10,83)
      83         FORMAT(//10X,7HELEMENT,3X,4HNODE,4X,11HSHEAR FORCE,
                +3X,14HBENDING MOMENT)
                 IF(ITYP.EQ.4) WRITE(10,81)
355   81         FORMAT(//10X,7HELEMENT,3X,4HNODE,4X,11HAXIAL FORCE,
                +4X,11HSHEAR FORCE,3X,14HBENDING MOMENT)
      C
      C   EVALUATION OF THE ELEMENT STIFFNESS MATRIX
      C
360              IFORCE=1
                 EM=ELMOD(IEM)
                 KK=1
                 GOTO70
      C
365   C   CALCULATION OF THE ELEMENT FORCES IN
      C       THE LOCAL COORDINATE SYSTEM
      C
      85         DO861=1,6
                 EF(I)=0.0
370              DO86J=1,6
                 K=NS(J)
```

```
        86      EF(I)=EF(I)+ESM(I,J)*A(K)
        C
        C   TRANSFORMATION OF THE ELEMENT FORCES
375     C       TO THE MEMBER COORDINATE SYSTEM
        C
                F(1)=CS*EF(1)+SN*EF(2)
                F(2)=-SN*EF(1)+CS*EF(2)
                F(3)=EF(3)
380             F(4)=CS*EF(4)+SN*EF(5)
                F(5)=-SN*EF(4)+CS*EF(5)
                F(6)=EF(6)
        C
        C   OUTPUT OF THE ELEMENT NODAL FORCES
385     C
                IF(ITYP.EQ.1.OR.ITYP.EQ.2) WRITE(10,95) KK,NEL(KK,1),F(1),
               +NEL(KK,2),F(4)
        95      FORMAT(/13X,I2,5X,I3,1X,E15.6/20X,I3,1X,E15.6)
                IF(ITYP.EQ.3) WRITE(10,96) KK,NEL(KK,1),F(2),F(3),
390            +NEL(KK,2),F(5),F(6)
        96      FORMAT(/13X,I2,5X,I3,1X,2E15.6/20X,I3,1X,2E15.6)
                IF(ITYP.EQ.4) WRITE(10,93) KK,NEL(KK,1),F(1),F(2),F(3),
               +NEL(KK,2),F(4),F(5),F(6)
               93 FORMAT(/13X,I2,5X,I3,1X,3E15.6/20X,I3,1X,3E15.6)
395     C
        C   ANOTHER ELEMENT?
        C
                KK=KK+1
                IF(KK.LE.NE) GOTO70
400             END


                SUBROUTINE MODFRM
                COMMON/AV/A(5000),JGF,JGSM,NP,NBW
                COMMON/CT/ITYP,IB
                DATA IN/60/,IO/61/
5       C*********
        C*********
        C
        C   INPUT OF THE NODAL FORCE VALUES
        C       IB - DEGREE OF FREEDOM OF THE FORCE
10      C       BV - VALUE OF THE NODAL FORCE
        C   INPUT OF IB AND BV IS TERMINATED BY
        C       INPUTTING A ZERO VALUE FOR IB
        C
        C*********
15      C*********
                NIW =0
        202     READ(IN,*) IB
                IF(IB.LE.0) GOTO216
                IF(NIW.EQ.0) WRITE(IO,201)
20      201     FORMAT(//10X,31HCONCENTRATED FORCES AND MOMENTS)
                NIW=1
                CALL CONVERT
                READ(IN,*) BV
                A(JGF+IB)=A(JGF+IB)+BV
25              CALL INVERT
                WRITE(IO,203) IB,BV
```

```
203    FORMAT(10X,I3,E15.5)
       GOTO202
C********
C********
C
C   INPUT OF THE PRESCRIBED NODAL VALUES
C        IB - DEGREE OF FREEDOM OF THE KNOWN NODAL VALUE
C        BV - KNOWN NODAL VALUE
C    INPUT OF IB AND BV IS TERMINATED BY INPUTTING
C        A ZERO VAUE FOR IB
C
C********
C********
216    NIW=0
209    READ(IN,*) IB
       IF(IB.LE.0) RETURN
       IF(NIW.EQ.0) WRITE(IO,208)
208    FORMAT(//10X,25HKNOWN DISPLACEMENT VALUES)
       NIW=1
       CALL CONVERT
       READ(IN,*) BV
C
C   MODIFICATION OF THE GLOBAL STIFFNESS MATRIX AND
C        THE GLOBAL FORCE VECTOR USING THE METHOD
C        OF DELETION OF ROWS AND COLUMNS
C
       K=IB-1
       DO211J=2,NBW
       M=IB+J-1
       IF(M.GT.NP) GOTO210
       IJ=JGSM+(J-1)*NP+IB-(J-1)*(J-2)/2
       A(JGF+M)=A(JGF+M)-A(IJ)*BV
       A(IJ)=0.0
210    IF(K.LE.0) GOTO 211
       KJ=JGSM+(J-1)*NP+K-(J-1)*(J-2)/2
       A(JGF+K)=A(JGF+K)-A(KJ)*BV
       A(KJ)=0.0
       K=K-1
211    CONTINUE
       A(JGF+IB)=A(JGSM+IB)*BV
221    CONTINUE
       CALL INVERT
       WRITE(IO,203) IB,BV
       GOTO209
       END


       SUBROUTINE DCMPBD
       COMMON/AV/A(5000),JGF,JGSM,NP,NBW
       IO=61
C*********
C*********
C
C   DECOMPOSITION OF A BANDED MATRIX INTO AN UPPER
C        TRIANGULAR FORM USING GAUSSIAN ELIMINATION
C
C*********
```

```
      C*********
            NP1=NP-1
            DO226I=1,NP1
            MJ=I+NBW-1
15          IF(MJ.GT.NP) MJ=NP
            NJ=I+1
            MK=NBW
            IF((NP-I+1).LT.NBW) MK=NP-I+1
            ND=0
20          DO225J=NJ,MJ
            MK=MK-1
            ND=ND+1
            NL=ND+1
            DO225K=1,MK
25          NK=ND+K
            JK=JGSM+(K-1)*NP+J-(K-1)*(K-2)/2
            INL=JGSM+(NL-1)*NP+I-(NL-1)*(NL-2)/2
            INK=JGSM+(NK-1)*NP+I-(NK-1)*(NK-2)/2
            II=JGSM+I
30    225   A(JK)=A(JK)-A(INL)*A(INK)/A(II)
      226   CONTINUE
            RETURN
            END



            SUBROUTINE SLVBD
            COMMON/AV/A(5000),JGF,JGSM,NP,NBW
            NP1=NP-1
      C*********
5     C*********
      C
      C  DECOMPOSITION OF THE GLOBAL FORCE VECTOR
      C
      C*********
10    C*********
            DO250I=1,NP1
            MJ=I+NBW-1
            IF(MJ.GT.NP) MJ=NP
            NJ=I+1
15          L=1
            DO250J=NJ,MJ
            L=L+1
            IL=JGSM+(L-1)*NP+I-(L-1)*(L-2)/2
      250   A(JGF+J)=A(JGF+J)-A(IL)*A(JGF+I)/A(JGSM+I)
20    C*********
      C*********
      C
      C  BACKWARD SUBSTITUTION FOR DETERMINATION OF
      C     THE NODAL VALUES
25    C
      C*********
      C*********
            A(NP)=A(JGF+NP)/A(JGSM+NP)
            DO252K=1,NP1
30          I=NP-K
            MJ=NBW
            IF((I+NBW-1).GT.NP) MJ=NP-I+1
```

```
                SUM=0.0
                DO251J=2,MJ
35              N=I+J-1
                IJ=JGSM+(J-1)*NP+I-(J-1)*(J-2)/2
     251        SUM=SUM+A(IJ)*A(N)
     252        A(I)=(A(JGF+I)-SUM)/A(JGSM+I)
                RETURN
40              END



                SUBROUTINE CONVERT
                COMMON/CT/ITYP,IB
                IF(ITYP.EQ.4) RETURN
                IF(ITYP.GT.1) GOTO1
5        C
         C  AXIAL FORCE MEMBER
         C
                IB=3*IB-2
                RETURN
10       1      IND=(IB+1)/2
                ICK=(IB/2)*2
                IF(ITYP.GT.2) GOTO2
         C
         C  TRUSS MEMBERS
15       C
                IF(ICK.NE.IB) IB=3*IND-2
                IF(ICK.EQ.IB) IB=3*IND-1
                RETURN
         C
20       C  BEAM MEMBERS
         C
         2      IF(ICK.NE.IB) IB=3*IND-1
                IF(ICK.EQ.IB) IB=3*IND
                RETURN
25              END



                SUBROUTINE INVERT
                COMMON/CT/ITYP,IB
                IF(ITYP.EQ.4) RETURN
                IF(ITYP.GT.1) GOTO1
5        C
         C  AXIAL FORCE MEMBER
         C
                IB=(IB+2)/3
                RETURN
10       1      ICK=(IB-2)/3*3+2
                IF(ITYP.GT.2) GOTO3
         C
         C  TRUSS MEMBERS
         C
15              IF(ICK.EQ.IB) GOTO2
                IND=(IB+2)/3
                IB=2*IND-1
                RETURN
         2      IND=(IB+1)/3
```

```
20              IB=2*IND
                RETURN
        C
        C   BEAM MEMBERS
        C
25      3       IF(ICK.EQ.IB) GOTO4
                IND=IB/3
                IB=2*IND
                RETURN
        4       IND=(IB+1)/3
30              IB=2*IND-1
                RETURN
                END
```

25.2   AN EXAMPLE PROBLEM FOR FRAME

The input data and the computer output for the two-member rigid frame shown in Figure 25.1 are discussed here. The section and elastic properties of the frame are given in the figure.

The input data for the **FRAME** program are given in Table 25.1. The data are to the right of the solid vertical line. The titles on the left indicate the nature of the data. Each line represents a card of input. The computer output follows Figure 25.1.

The interpretation of the member forces is given in Figure 25.2. Remember that the calculated member forces must be interpreted relative to the member coordinate system, $(\bar{x}, \bar{y})$, and that the origin of this system is always at node i with $\bar{x}$ directed along the member. The numerical values given contain four significant digits.

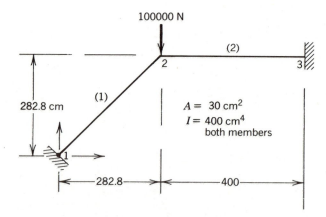

Figure 25.1.   A two-member rigid frame.

FRAME EXAMPLE PROBLEM

SECTION PROPERTIES
SECTION   AREA   AREA MOMENT
    1   .30000E+02   .40000E+03

ELASTIC MODULUS =   .20000E+08

NODAL COORDINATES
NODE   X   Y
   1   0.00   0.00
   2   282.80   282.80
   3   682.80   282.80

ELEMENT DATA
                        NODE
ELEMENT   SECTION   NUMBERS
   1   1   1   2
   2   1   2   3

CONCENTRATED FORCES AND MOMENTS
   5   -.10000E+06

KNOWN DISPLACEMENT VALUES
   1   0.
   2   0.
   7   0.
   8   0.
   9   0.

FRAME EXAMPLE PROBLEM

NODAL DISPLACEMENT VALUES

NODE   X DEFLECTION   Y DEFLECTION   Z ROTATION
   1   0.   0.   -.817600E-03
   2   .664141E-01   -.199321E+00   .225716E-03
   3   0.   0.   0.

ELEMENT   NODE   AXIAL FORCE   SHEAR FORCE   BENDING MOMENT

   1   1   .140990E+06   .104363E+03   .582077E-09
            2   -.140990E+06   -.104363E+03   .417389E+05

   2   2   .996211E+05   -.231266E+03   -.417389E+05
            3   -.996211E+05   .231266E+03   -.507676E+05

Table 25.1   Computer Data for FRAME

Title	TEXTBOOK FRAME EXAMPLE PROBLEM										
Parameters	3	2	1	1	4						
Section properties	1	30.	400.								
X-coordinates	0.	282.8	682.8								
Y-coordinates	0.	282.8	282.8								
Element data	1	1	1	2							
	2	1	2	3							
Known force	5	−10000.	0								
Known displacements	1	0.	2	0.	7	0.	8	0.	9	0.	0

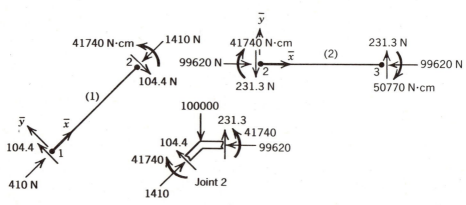

Figure 25.2.   The calculated axial and shear forces and bending moments in each member of a two-member rigid frame.

25.3 **PROGRAM** STRESS

The computer program **STRESS** is used to analyze two-dimensional, plane stress elasticity problems. The program does not include the possibility of body forces, thermal changes, or composite construction. The program is written for educational purposes only. The input data are relatively simple and the program contains several diagnostic checks to detect common errors made by new users. The program uses only the three-node triangular element.

```
      PROGRAM STRESS (INPUT,OUTPUT,TAPE60=INPUT,TAPE61=OUTPUT)
      COMMON/ELMATX/ESM(6,6),X(3),Y(3),D(3,3),IELR
      COMMON/GRAD/B(3,6),AR2
      COMMON/MTL/EM,PR,TH
5     COMMON/TLE/TITLE(20)
      COMMON/AV/A(8500),JGF,JGSM,NP,NBW
      DIMENSION NEL(250,3),XC(200),YC(200)
      DIMENSION NS(6),U(6),STRA(3),STRE(6),ICK(250)
```

```
             DATA IN/60/,IO/61/,IFE/1/
10      C***************
        C***************
        C  DEFINITION OF THE INPUT VARIABLES FOR THIS PROGRAM
        C
        C    THIS LIST DOES NOT INCLUDE VALUES READ BY
15      C    THE SUBROUTINE MODIFY
        C
        C***************
        C***************
        C
20      C    TITLE AND PARAMETERS
        C
        C          TITLE - A DESCRIPTIVE STATEMENT OF THE PROBLEM
        C          NN - NUMBER OF NODES
        C          NE - NUMBER OF ELEMENTS
25      C          IPLVL - PRINT LEVEL
        C              0 - DO NOT WRITE THE ELEMENT MATRICES
        C              1 - WRITE THE ELEMENT MATRICES
        C
        C    MATERIAL PROPERTIES AND THICKNESS
30      C
        C          EM - MODULUS OF ELASTICITY
        C          PR - POISSON'S RATIO
        C          TH - THICKNESS OF THE REGION
        C
35      C    NODAL COORDNATES
        C
        C          XC(I) - X COORDINATES OF THE NODES IN NUMERICAL
        C                      SEQUENCE
        C          YC(I) - Y COORDINATES OF THE NODES IN NUMERICAL
40      C                      SEQUENCE
        C
        C    ELEMENT DATA
        C
        C          N - ELEMENT NUMBER
45      C          NEL(N,I) - NUMERICAL VALUE OF NODE I
        C          NEL(N,J) - NUMERICAL VALUE OF NODE J
        C          NEL(N,K) - NUMERICAL VALUE OF NODE K
        C
        C***************
50      C***************
        C
        C  INPUT SECTION OF THE PROGRAM
        C
        C***************
55      C***************
        C
        C  INPUT OF THE TITLE CARD AND PARAMETERS
        C
             READ(IN,3) TITLE
60         3 FORMAT(20A4)
             READ(IN,*) NN,NE,IPLVL
             NP=2*NN
             IF(IPLVL.GE.1) IPLVL=0
             IF(NE.GE.10) IPLVL=0
```

```
65          C
            C   COMPARISON CHECK OF NN AND NE WITH THE VALUES USED
            C       IN THE DIMENSION STATEMENTS
            C
                    ISTOP=0
70          C       CHECK NN
                    IF (NN.LE.200) GOTO6
                    WRITE (10,10)
                 10 FORMAT (10X,27HNUMBER OF NODES EXCEEDS 200/
                   +10X,20HEXECUTION TERMINATED)
75                  ISTOP-1
            C       CHECK NE
                  6 IF (NE.LE.250) GOTO1
                    WRITE (10,2)
                  2 FORMAT (10X,30HNUMBER OF ELEMENTS EXCEEDS 250/
80                 +10X,20HEXECUTION TERMINATED)
                    ISTOP=1
                  1 IF (ISTOP.EQ.1) STOP
            C
            C   INPUT OF THE MATERIAL PROPERTIES, THICKNESS
85          C     AND NODAL COORDINATES
            C
                    READ (IN,*) EM,PR,TH
                    READ (IN,*)  (XC (I) , I=1,NN)
                    READ (IN,*)  (YC (I) , I=1,NN)
90          C
            C   OUTPUT OF TITLE AND DATA HEADINGS
            C
                    WRITE (10,4) TITLE,NN,NE
                  4 FORMAT (1H1////10X,20A4//13X,5HNN  =16/13X,5HNE  =16)
95                  WRITE (10,16) EM,PR,TH
                 16 FORMAT (//10X,16HPARAMETER VALUES/13X,4HEM =,E15.5/13X,4HPR =
                   +E15.5/13X,4HTH =,E15.5)
                    WRITE (10,11)
                 11 FORMAT (//10X,17HNODAL COORDINATES/10X,4HNODE,5X,1HX,14X,1HY)
100                 WRITE (10,12)  (I,XC (I) ,YC (I) ,I=1,NN)
                 12     FORMAT (10X,I4,2E15.5)
            C
            C   INPUT AND ECHO PRINT OF ELEMENT DATA
            C     CHECK TO SEE IF THE ELEMENTS ARE IN SEQUENCE
105         C
                 28     WRITE (10,8) TITLE
                  8 FORMAT (1H1,///10X,20A4//10X,3HNEL,4X,12HNODE NUMBERS)
                    NID=0
                    DO9KK=1,NE
110                 READ (IN,*) N, (NEL (N,I) ,I=1,3)
                    IF ((N-1) .NE.NID) WRITE (10,17) N
                 17     FORMAT (10X,7HELEMENT,I4,16H NOT IN SEQUENCE)
                    NID=N
                  9     WRITE (10,5) N, (NEL (N,I) ,I=1,3)
115               5     FORMAT (10X,I3,2X,3I4)
            C*********
            C*********
            C
            C   ANALYSIS OF THE NODE NUMBERS
120         C
            C*********
```

```
        C*********
        C
        C   INITIALIZATION OF A CHECK VECTOR
125     C
              DO500I=1,NN
        500   ICK(I)=0
        C
        C   CHECK TO SEE IF ANY NODE NUMBER EXCEEDS NP
130     C
              DO501I=1,NE
              DO502J=1,3
              K=NEL(I,J)
              ICK(K)=1
135     502   IF(K.GT.NN) WRITE(10,503) J,I,NN
        503   FORMAT(/10X,4HNODE,I4,11H OF ELEMENT,I4,
             +13H EXCEEDS NN =,I4)
        501   CONTINUE
        C
140     C   CHECK TO SEE IF ALL NODE NUMBERS THROUGH
        C       NN ARE INCLUDED
        C
              DO505I=1,NN
        505   IF(ICK(I).EQ.0) WRITE(10,506) I
145     506   FORMAT(/10X,4HNODE,I4,15H DOES NOT EXIST)
        C************
        C************
        C
        C   CREATION AND INITIALIZATION OF THE A VECTOR
150     C
        C************
        C************
        C
        C   CALCULATION OF THE BANDWIDTH
155     C
              IEL=0
              INBW=0
              NBW=0
              DO20KK=1,NE
160           DO25I=1,3
        25    NS(I)=NEL(KK,I)
              DO21I=1,2
              IJ=I+1
              DO21J=IJ,3
165           NB=IABS(NS(I)-NS(J))
              IF(NB.EQ.0) WRITE(10,26) KK
        26    FORMAT(/10X,7HELEMENT,I3,
             +39HHAS TWO NODES WITH THE SAME NODE NUMBER)
              IF(NB.LE.NBW) GOTO21
170           INBW=KK
              NBW=NB
           21 CONTINUE
           20 CONTINUE
              NBW=(NBW+1)*2
175           WRITE(10,27) NBW,INBW
           27 FORMAT(10X,12HBANDWIDTH IS,I4,11H IN ELEMENT,I4)
        C
        C   INITIALIZATION OF THE COLUMN VECTOR A( )
        C
```

```
180              JGF=NP
                 JGSM=JGF+NP
                 JEND=JGSM+NP*NBW
                 IF(JEND.GT.8500) GOTO22
                 JL=JEND-JGF
185              DO241=1,JEND
              24 A(I)=0.0
                 GOTO30
              22 WRITE(10,23)
              23 FORMAT(10X,30HDIMENSION OF A VECTOR EXCEEDED/
190             +10X,20HEXECUTION TERMINATED)
                 STOP
        C*************
        C*************
        C
195     C  GENERATION OF THE SYSTEM OF EQUATIONS
        C
        C*************
        C*************
        C
200     C  GENERATION OF THE MATERIALS PROPERTY MATRIX D
        C
        30       R=EM/(1.-PR**2)
                 D(1,1)=R
                 D(2,2)=D(1,1)
205              D(3,3)=R*(1.0-PR)/2.
                 D(1,2)=PR*R
                 D(2,1)=D(1,2)
                 D(1,3)=0.0
                 D(3,1)=0.0
210              D(2,3)=0.0
                 D(3,2)=0.0
        C
        C  START OF THE LOOP FOR THE ELEMENT MATRICES
        C
215              IELR=0
                 KK=1
        C
        C  GENERATION OF THE NODAL DEGREES OF FREEDOM
        C     RETRIEVAL OF THE NODAL COORDNATES
220     C
        32       DO311=1,3
                 J=NEL(KK,I)
                 NS(2*I-1)=J*2-1
                 NS(2*I)=J*2
225              X(I)=XC(J)
              31 Y(I)=YC(J)
        C
        C  CALCULATION OF ELEMENT MATRICES
        C
230              CALL ELSTMX(KK,IPLVL)
        C
        C  DIRECT STIFFNESS PROCEDURE
        C
                 DO331=1,6
235              II=NS(I)
                 DO34J=1,6
```

```
                    JJ=NS(J)+1-II
                    IF(JJ.LE.O) GOTO34
                    J1=JGSM+(JJ-1)*NP+II-(JJ-1)*(JJ-2)/2
240                 A(J1)=A(J1)+ESM(I,J)
                 34 CONTINUE
                 33 CONTINUE
                    KK=KK+1
                    IF(KK.LE.NE) GOTO32
245     C*************
        C*************
        C
        C   MODIFICATION AND SOLUTION OF THE SYSTEM OF EQUATIONS
        C    DATA IS CALLED BY THE SUBROUTINE MODIFY
250     C
        C*************
        C*************
                    WRITE(IO,110) TITLE
            110     FORMAT(1H1///10X,20A4)
255                 CALL MODIFY(IFE)
                    CALL DCMPBD
                    CALL SLVBD
        C
        C   OUTPUT OF THE CALCULATED DISPLACEMENTS
260     C
                    WRITE(IO,112)
            112     FORMAT(///10X,25HNODAL DISPLACEMENT VALUES/
                   +/10X,4HNODE,6X,12HX DEFLECTION,6X,12HY DEFLECTION)
                    DO113I=1,NN
265         113     WRITE(IO,111) I,A(I*2-1),A(I*2)
            111     FORMAT(11X,I3,3X,E15.6,3X,E15.6)
        C*************
        C*************
        C
270     C   CALCULATION OF THE ELEMENT STRESS AND STRAIN COMPONENTS
        C      AND THE PRINCIPAL STRESS VALUES
        C
        C*************
        C*************
275                 IPLVL=0
                    IELR=1
                    WRITE(IO,110) TITLE
                    DO96KK=1,NE
                    IF(KK/9*9.EQ.KK) WRITE(IO,110) TITLE
280     C
        C   GENERATION OF THE NODAL DEGREES OF FREEDOM
        C     RETRIEVAL OF THE NODAL COORDINATES
        C
                    DO 51 I=1,3
285                 J=NEL(KK,I)
                    NS(2*I-1)=2*J-1
                    NS(2*I)=2*J
                    X(I)=XC(J)
             51     Y(I)=YC(J)
290     C
        C   RETRIEVAL OF THE ELEMENT NODAL DISPLACEMENTS
        C
                 65 DO73I=1,6,2
```

```
              NS1=NS (I)
295           NS2=NS (I+1)
              U (I) =A (NS1)
      73      U (I+1) =A (NS2)
      C
      C   CALCULATION OF THE STRAIN VECTOR,  (STRAIN) = (B) (U)
300   C
              CALL ELSTMX (KK, IPLVL)
              DO 1155 I=1,3
              STRA (I) =0.0
              DO1155K=1,6
305   1155    STRA (I) =STRA (I) +B (I,K) *U (K) /AR2
      C
      C   CALCULATION OF THE STRESS VECTOR,   (STRESS) = (D) (STRAIN)
      C
              DO58I=1,3
310           STRE (I) =0.0
              DO 58 K=1,3
      58      STRE (I) =STRE (I) +D (I,K) * (STRA (K) )
      C
      C   CALCULATION OF THE PRINCIPAL STRESSES
315   C
              AA= (STRE (1) +STRE (2) ) /2.
              AB=SQRT ( ( (STRE (1) -STRE (2) ) /2.) **2+STRE (3) **2)
              S1=AA+AB
              S2=AA-AB
320           TM=AB
              IF (ABS (STRE (1) -STRE (2) ) .LT.0.001) GOTO93
              AC=ATAN2 (2.*STRE (3) ,STRE (1) -STRE (2) )
              THM= ( (180.0/3.14159265) *AC) /2.0
              GO TO 94
325   93      THM=90.0
      C
      C   PRINTING OF THE RESULTS
      C
      94        WRITE (10,57) KK
330   57        FORMAT (/ 10X, 7HELEMENT, I4)
                WRITE (10,95) STRA (1) ,STRE (1) ,S1,STRA (2) ,STRE (2) ,S2,
              +STRA (3) ,STRE (3) ,TM,THM
      95        FORMAT (15X,5HEXX =,E12.5,5X,5HSXX =,E12.5,5X,5HS1 =,
              +E12.5/15X,5HEYY =,E12.5,5X,5HSYY =,E12.5,5X,5HS2 =,
335           +E12.5/15X,5HGXY =,E12.5,5X,5HTXY =,E12.5,4X,
              +6HTMAX =,E12.5/59X,5HANGLE,F8.2,4H DEG)
      96        CONTINUE
                STOP
                END


                SUBROUTINE ELSTMX (KK, IPLVL)
                COMMON/MTL/EM,PR,TH
                COMMON/GRAD/B (3,6) ,AR2
                COMMON/ELMATX/ESM (6,6) ,X (3) ,Y (3) ,D (3,3) ,IELR
5               DIMENSION C (6,3)
                IO=61
      C
      C   GENERATION OF THE B MATRIX
      C
```

```
10              DO 20 I=1,3
                DO 20 J=1,6
             20 B(I,J)=0.0
                B(1,1)=Y(2)-Y(3)
                B(1,3)=Y(3)-Y(1)
15              B(1,5)=Y(1)-Y(2)
                B(2,2)=X(3)-X(2)
                B(2,4)=X(1)-X(3)
                B(2,6)=X(2)-X(1)
                B(3,1)=B(2,2)
20              B(3,2)=B(1,1)
                B(3,3)=B(2,4)
                B(3,4)=B(1,3)
                B(3,5)=B(2,6)
                B(3,6)=B(1,5)
25              AR2=X(2)*Y(3)+X(3)*Y(1)+X(1)*Y(2)-X(2)*Y(1)-
                +X(3)*Y(2)-X(1)*Y(3)
                IF(IELR.EQ.1) RETURN
        C
        C   MATRIX MULTIPLICATION TO OBTAIN C = (BT)(D)
30      C
                DO 22 I=1,6
                DO 22 J=1,3
                C(I,J)=0.0
                DO22K=1,3
35           22 C(I,J)=C(I,J)+B(K,I)*D(K,J)
        C
        C   MATRIX MULTIPLICATION TO OBTAIN ESM
        C       ESM = (BT)(D)(B) = (C)(B)
        C
40              DO 27 I=1,6
                DO 27 J=1,6
                SUM=0.0
                DO 28 K=1,3
             28 SUM=SUM+C(I,K)*B(K,J)
45              ESM(I,J)=SUM*TH/(2.*AR2)
             27 CONTINUE
        C
        C   OUTPUT OF THE ELEMENT STIFFNESS MATRIX
        C
50              IF(IPLVL.EQ.0) RETURN
                WRITE(10,30) KK
             30 FORMAT(//5X,28HSTIFFNESS MATRIX FOR ELEMENT,I3)
                DO31I=1,6
             31 WRITE(10,32)  (ESM(I,J),J=1,6)
55           32 FORMAT(5X,6E15.5)
                RETURN
                END



                SUBROUTINE MODIFY(IFE)
                COMMON/AV/A(8500),JGF,JGSM,NP,NBW
                DATA IN/60/,IO/61/
        C*********
 5      C*********
        C
```

```
C    INPUT OF THE NODAL FORCE VALUES
C         FOR FIELD PROBLEMS
C              IB - NODE NUMBER
C              BV - SOURCE OR SINK VALUE
C         FOR SOLID MECHANICS PROBLEMS
C              IB - DEGREE OF FREEDOM OF THE FORCE
C              BV - VALUE OF THE FORCE
C
C     INPUT OF IB AND BV IS TERMINATED BY
C         INPUTTING A ZERO VALUE FOR IB
C
C*********
C*********
      NIW =0
202   READ(IN,*) IB
      IF(IB.LE.0) GOTO216
      IF(NIW.EQ.0.AND.IFE.EQ.0) WRITE(10,200)
      IF(NIW.EQ.0.AND.IFE.EQ.1) WRITE(10,201)
200   FORMAT(//10X,22HSOURCE AND SINK VALUES)
201   FORMAT(//10X,31HCONCENTRATED FORCES AND MOMENTS)
      NIW=1
      READ(IN,*) BV
      A(JGF+IB)=A(JGF+IB)+BV
      WRITE(10,203) IB,BV
203   FORMAT(10X,I3,E15.5)
      GOTO202
C*********
C********
C
C    INPUT OF THE PRESCRIBED NODAL VALUES
C         FOR FIELD PROBLEMS
C              IB - NODE NUMBER
C              BV - KNOWN VALUE OF PHI
C         FOR SOLID MECHANICS PROBLEMS
C              IB - DEGREE OF FREEDOM OF THE KNOWN DISPLACEMENT
C              BV - THE VALUE OF THE DISPLACEMENT
C
C     INPUT OF IB AND BV IS TERMINATED BY INPUTTING
C         A ZERO VAUE FOR IB
C
C********
C********
216   NIW=0
209   READ(IN,*) IB
      IF(IB.LE.0) RETURN
      IF(NIW.EQ.0.AND.IFE.EQ.0) WRITE(10,212)
      IF(NIW.EQ.0.AND.IFE.EQ.1) WRITE(10,208)
212   FORMAT(//10X,25HKNOWN NODAL VALUES OF PHI)
208   FORMAT(//10X,25HKNOWN DISPLACEMENT VALUES)
      NIW=1
      READ(IN,*) BV
C
C    MODIFICATION OF THE GLOBAL STIFFNESS MATRIX AND
C         THE GLOBAL FORCE VECTOR USING THE METHOD
C         OF DELETION OF ROWS AND COLUMNS .
C
```

```
             K=IB-1
             DO211J=2,NBW
65           M=IB+J-1
             IF(M.GT.NP) GOTO210
             IJ=JGSM+(J-1)*NP+IB-(J-1)*(J-2)/2
             A(JGF+M)=A(JGF+M)-A(IJ)*BV
             A(IJ)=0.0
70       210 IF(K.LE.0) GOTO 211
             KJ=JGSM+(J-1)*NP+K-(J-1)*(J-2)/2
             A(JGF+K)=A(JGF+K)-A(KJ)*BV
             A(KJ)=0.0
             K=K-1
75       211 CONTINUE
             A(JGF+IB)=A(JGSM+IB)*BV
         221 CONTINUE

             WRITE(IO,203) IB,BV
             GOTO209
80           END




             SUBROUTINE DCMPBD
             COMMON/AV/A(8500),JGF,JGSM,NP,NBW
             IO=61
     C*********
5    C*********
     C
     C  DECOMPOSITION OF A BANDED MATRIX INTO AN UPPER
     C      TRIANGULAR FORM USING GAUSSIAN ELIMINATION
     C
10   C*********
     C*********
             NP1=NP-1
             DO226I=1,NP1
             MJ=I+NBW-1
15           IF(MJ.GT.NP) MJ=NP
             NJ=I+1
             MK=NBW
             IF((NP-I+1).LT.NBW) MK=NP-I+1
             ND=0
20           DO225J=NJ,MJ
             MK=MK-1
             ND=ND+1
             NL=ND+1
             DO225K=1,MK
25           NK=ND+K
             JK=JGSM+(K-1)*NP+J-(K-1)*(K-2)/2
             INL=JGSM+(NL-1)*NP+I-(NL-1)*(NL-2)/2
             INK=JGSM+(NK-1)*NP+I-(NK-1)*(NK-2)/2
             II=JGSM+I
30       225 A(JK)=A(JK)-A(INL)*A(INK)/A(II)
         226 CONTINUE
             RETURN
             END
```

```
              SUBROUTINE SLVBD
              COMMON/AV/A(8500),JGF,JGSM,NP,NBW
              NP1=NP-1
       C*********
 5     C*********
       C
       C   DECOMPOSITION OF THE GLOBAL FORCE VECTOR
       C
       C*********
10     C*********
              DO250I=1,NP1
              MJ=I+NBW-1
              IF(MJ.GT.NP) MJ=NP
              NJ=I+1
15            L=1
              DO250J=NJ,MJ
              L=L+1
              IL=JGSM+(L-1)*NP+I-(L-1)*(L-2)/2
       250    A(JGF+J)=A(JGF+J)-A(IL)*A(JGF+I)/A(JGSM+I)
20     C*********
       C*********
       C
       C   BACKWARD SUBSTITUTION FOR DETERMINATION OF
       C       THE NODAL VALUES
25     C
       C*********
       C*********
              A(NP)=A(JGF+NP)/A(JGSM+NP)
              DO252K=1,NP1
30            I=NP-K
              MJ=NBW
              IF((I+NBW-1).GT.NP) MJ=NP-I+1
              SUM=0.0
              DO251J=2,MJ
35            N=I+J-1
              IJ=JGSM+(J-1)*NP+I-(J-1)*(J-2)/2
       251    SUM=SUM+A(IJ)*A(N)
       252    A(I)=(A(JGF+I)-SUM)/A(JGSM+I)
              RETURN
40            END
```

The computer program **STRESS** has four subroutines: **ELSTMX, MODIFY, DCMPBD** and **SLVBD**. The last two, **DCMPBD** and **SLVBD**, were discussed in Section 25.1. The subroutine **MODIFY** is identical to the subroutine **MODIFY** discussed in Section 16.2.

Subroutine ELSTMX. The subroutine **ELSTMX** evaluates the element stiffness matrix, $[k^{(e)}]$, for the linear triangular element using (23.22). The individual matrices, $[B]$ and $[D]$, are evaluated using (23.20) and (23.6), respectively. The matrix product $[B]^T[D][B]$ is evaluated within the subroutine. The subroutine also provides the option of printing the element stiffness matrix so that students can check a hand calculation of $[k^{(e)}]$. The subroutine also evaluates $[B]$ in the loop that calculates the stress components in each element.

25.4 AN EXAMPLE PROBLEM FOR STRESS

The input data and the computer output for a small two-dimensional elasticity problem are given in this section. The problem consists of a plate with a load applied over a segment of one side and continuously supported along the opposite side (Figure 25.3). The elements and node numbers are given in Figure 25.4.

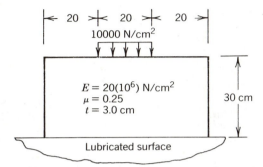

Figure 25.3. A thin plate with a distributed load over part of one side.

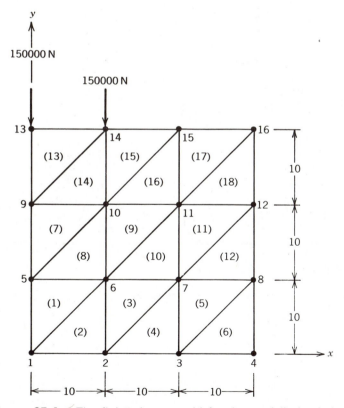

Figure 25.4. The finite element grid for the partially loaded plate.

The data for the program **STRESS** are given to the right of the solid vertical line in Table 25.2. The headings on the left of the line indicate the nature of the data.

Table 25.2 Computer Data for STRESS

Title	16	18	0				
Parameters							
Material properties	20000000.	0.25	3.0				
X-coordinates	0.	10.	20.	30.	0.	10.	20.	30.
	0.	10.	20.	30.	0.	10.	20.	30.
Y-coordinates	0.	0.	0.	0.	10.	10.	10.	10.
	20.	20.	20.	20.	30.	30.	30.	30.

Element data			
1	1	6	5
2	1	2	6
3	2	7	6
4	2	3	7
5	3	8	7
6	3	4	8
7	5	10	9
8	5	6	10
9	6	11	10
10	6	7	11
11	7	12	11
12	7	8	12
13	9	14	13
14	9	10	14
15	10	15	14
16	10	11	15
17	11	16	15
18	11	12	16

Known nodal forces	26	-15000.	28	-15000.	0

Known nodal displacements	1	0.	2	0.	4	0.	6	0.
	8	0.	9	0.	17	0.	25	0.	0

The computer output follows.

ELASTICITY EXAMPLE PROBLEM

```
        NN  =    16
        NE  =    18


    PARAMETER VALUES
        EM  =    .20000E+08
        PR  =    .25000E+00
        TH  =    .30000E+01
```

NODAL COORDINATES
```
NODE      X                    Y
  1    0.                   0.
  2     .10000E+02          0.
  3     .20000E+02          0.
  4     .30000E+02          0.
  5    0.                    .10000E+02
  6     .10000E+02           .10000E+02
  7     .20000E+02           .10000E+02
  8     .30000E+02           .10000E+02
  9    0.                    .20000E+02
 10     .10000E+02           .20000E+02
 11     .20000E+02           .20000E+02
 12     .30000E+02           .20000E+02
 13    0.                    .30000E+02
 14     .10000E+02           .30000E+02
 15     .20000E+02           .30000E+02
 16     .30000E+02           .30000E+02
```

ELASTICITY EXAMPLE PROBLEM

```
NEL     NODE NUMBERS
  1      1    6    5
  2      1    2    6
  3      2    7    6
  4      2    3    7
  5      3    8    7
  6      3    4    8
  7      5   10    9
  8      5    6   10
  9      6   11   10
 10      6    7   11
 11      7   12   11
 12      7    8   12
 13      9   14   13
 14      9   10   14
 15     10   15   14
 16     10   11   15
 17     11   16   15
 18     11   12   16
```
BANDWIDTH IS 12 IN ELEMENT   1

ELASTICITY EXAMPLE PROBLEM

CONCENTRATED FORCES AND MOMENTS
```
 26    -.15000E+06
 28    -.15000E+06
```

KNOWN DISPLACEMENT VALUES
```
  1    0.
  2    0.
```

```
              4    0.
              6    0.
              8    0.
              9    0.
             17    0.
             25    0.


         NODAL DISPLACEMENT VALUES

              NODE      X DEFLECTION        Y DEFLECTION
                1      0.                  0. .
                2       .111404E-02        0.
                3       .161945E-02        0.
                4       .176714E-02        0.
                5      0.                  -.288612E-02
                6       .105478E-02        -.222389E-02
                7       .158910E-02        -.117896E-02
                8       .178086E-02        -.570193E-03
                9      0.                  -.609005E-02
               10       .768601E-03        -.449951E-02
               11       .106851E-02        -.217900E-02
               12       .118223E-02        -.930698E-03
               13      0.                  -.961853E-02
               14      -.620767E-03        -.694166E-02
               15      -.377746E-03        -.272581E-02
               16      -.193326E-03        -.104652E-02


   ELASTICITY EXAMPLE PROBLEM

   ELEMENT   1
        EXX =  .10548E-03      SXX =  .71094E+03      S1   =  .75515E+03
        EYY = -.28861E-03      SYY = -.55945E+04      S2   = -.56387E+04
        GXY =  .66223E-04      TXY =  .52979E+03      TMAX =  .31969E+04
                                                      ANGLE    4.77 DEG

   ELEMENT   2
        EXX =  .11140E-03      SXX =  .11906E+04      S1   =  .11910E+04
        EYY = -.22239E-03      SYY = -.41501E+04      S2   = -.41506E+04
        GXY = -.59258E-05      TXY = -.47407E+02      TMAX =  .26708E+04
                                                      ANGLE   -.51 DEG

   ELEMENT   3
        EXX =  .53432E-04      SXX = -.46199E+02      S1   =  .90464E+02
        EYY = -.22239E-03      SYY = -.44593E+04      S2   = -.45960E+04
        GXY =  .98567E-04      TXY =  .78853E+03      TMAX =  .23432E+04
                                                      ANGLE    9.83 DEG

   ELEMENT   4
        EXX =  .50541E-04      SXX =  .44943E+03      S1   =  .44964E+03
        EYY = -.11790E-03      SYY = -.22456E+04      S2   = -.22458E+04
        GXY = -.30351E-05      TXY = -.24281E+02      TMAX =  .13477E+04
                                                      ANGLE   -.52 DEG
```

354

```
ELEMENT   5
    EXX =  .19176E-04      SXX = -.21969E+03      S1   = -.12605E+03
    EYY = -.11790E-03      SYY = -.24128E+04      S2   = -.25065E+04
    GXY =  .57841E-04      TXY =  .46273E+03      TMAX =  .11902E+04
                                                  ANGLE   11.44 DEG


ELEMENT   6
    EXX =  .14769E-04      SXX =  .10974E+02      S1   =  .11079E+02
    EYY = -.57019E-04      SYY = -.11376E+04      S2   = -.11377E+04
    GXY =  .13718E-05      TXY =  .10974E+02      TMAX =  .57441E+03
                                                  ANGLE    .55 DEG


ELEMENT   7
    EXX =  .76860E-04      SXX = -.69081E+02      S1   =  .17619E+03
    EYY = -.32039E-03      SYY = -.64251E+04      S2   = -.66704E+04
    GXY =  .15905E-03      TXY =  .12724E+04      TMAX =  .34233E+04
                                                  ANGLE   10.91 DEG


ELEMENT   8
    EXX =  .10548E-03      SXX =  .10365E+04      S1   =  .10535E+04
    EYY = -.22756E-03      SYY = -.42921E+04      S2   = -.43090E+04
    GXY =  .37605E-04      TXY =  .30084E+03      TMAX =  .26813E+04
                                                  ANGLE    3.22 DEG


ELASTICITY EXAMPLE PROBLEM

ELEMENT   9
    EXX =  .29990E-04      SXX = -.57387E+03      S1   = -.86572E+01
    EYY = -.22756E-03      SYY = -.46947E+04      S2   = -.52599E+04
    GXY =  .20343E-03      TXY =  .16275E+04      TMAX =  .26256E+04
                                                  ANGLE   19.15 DEG


ELEMENT  10
    EXX =  .53432E-04      SXX =  .60652E+03      S1   =  .67621E+03
    EYY = -.10000E-03      SYY = -.18485E+04      S2   = -.19182E+04
    GXY =  .52433E-04      TXY =  .41947E+03      TMAX =  .12972E+04
                                                  ANGLE    9.43 DEG


ELEMENT  11
    EXX =  .11372E-04      SXX = -.29075E+03      S1   = -.11742E+03
    EYY = -.10000E-03      SYY = -.20728E+04      S2   = -.22461E+04
    GXY =  .72771E-04      TXY =  .58217E+03      TMAX =  .10643E+04
                                                  ANGLE   16.58 DEG


ELEMENT  12
    EXX =  .19176E-04      SXX =  .21682E+03      S1   =  .21690E+03
    EYY = -.36050E-04      SYY = -.66680E+03      S2   = -.66688E+03
    GXY =  .10134E-05      TXY =  .81071E+01      TMAX =  .44189E+03
                                                  ANGLE    .53 DEG


ELEMENT  13
    EXX = -.62077E-04      SXX = -.32062E+04      S1   = -.23705E+04
    EYY = -.35285E-03      SYY = -.78585E+04      S2   = -.86941E+04
    GXY =  .26769E-03      TXY =  .21415E+04      TMAX =  .31618E+04
                                                  ANGLE   21.32 DEG
```

```
ELEMENT  14
     EXX =  .76860E-04      SXX =  .33720E+03      S1  =  .34224E+03
     EYY = -.24422E-03      SYY = -.48000E+04      S2  = -.48050E+04
     GXY =  .20117E-04      TXY =  .16094E+03      TMAX = .25736E+04
                                                   ANGLE   1.79 DEG

ELEMENT  15
     EXX =  .24302E-04      SXX = -.78404E+03      S1  =  .18671E+03
     EYY = -.24422E-03      SYY = -.50803E+04      S2  = -.60511E+04
     GXY =  .28265E-03      TXY =  .22612E+04      TMAX = .31189E+04
                                                   ANGLE  23.23 DEG

ELEMENT  16
     EXX =  .29990E-04      SXX =  .34816E+03      S1  =  .64444E+03
     EYY = -.54681E-04      SYY = -.10066E+04      S2  = -.13029E+04
     GXY =  .87425E-04      TXY =  .69940E+03      TMAX = .97365E+03
                                                   ANGLE  22.96 DEG

ELEMENT  17
     EXX =  .18442E-04      SXX =  .10180E+03      S1  =  .13079E+03
     EYY = -.54681E-04      SYY = -.10682E+04      S2  = -.10972E+04
     GXY =  .23304E-04      TXY =  .18643E+03      TMAX = .61397E+03
                                                   ANGLE   8.84 DEG


ELASTICITY EXAMPLE PROBLEM

ELEMENT  18
     EXX =  .11372E-04      SXX =  .18084E+03      S1  =  .20717E+03
     EYY = -.11582E-04      SYY = -.18643E+03      S2  = -.21276E+03
     GXY = -.12725E-04      TXY = -.10180E+03      TMAX = .20997E+03
                                                   ANGLE  -14.50 DEG
```

PART FOUR
LINEAR AND QUADRATIC ELEMENTS

Commercial finite element programs provide the option to use linear or quadratic elements. These program packages also use numerical integration techniques to evaluate the element matrices. General procedures for evaluating the element shape functions and numerically evaluating the integrals for the element matrices are discussed in this part of the book. The chapters that follow can be covered after one of the applications sections has been studied.

Chapter **26**

ELEMENT SHAPE FUNCTIONS

Elements with linear variations in the nodal values were used in all of the previous application chapters except those involving the beam element. The finite element method is not restricted to the use of linear elements. Most commercial finite element programs allow the user to select between elements with linear or quadratic interpolation functions.

Quadratic elements are useful because fewer elements are needed to obtain the same degree of accuracy in the nodal values; also, the two-dimensional quadratic elements can be shaped to model a curved boundary. The use of the quadratic elements, however, does not always lead to a reduction in the total computation time. Numerical integration techniques are used to evaluate the element matrices, and these techniques can involve a large number of calculations.

The shape functions for the one- and two-dimensional elements are developed in this chapter. The numerical integration techniques used to evaluate the element matrices are discussed in the next chapter. The computer implementation and the solution of some sample problems are discussed in Chapter 28.

26.1 LOCAL NODE NUMBERS

As the number of nodes associated with an element increases, it is no longer convenient to denote the nodes by letters. Some three-dimensional elements have 20 or more nodes. The standard procedure is to denote each element node using an integer. The numbering system for each element considered in this book is given in Figure 26.1.

The element node numbers are called local node numbers and should not be confused with the global node numbers. Consider the grid in Figure 26.2. The relationship between the local and global node numbers is given in Table 26.1. The equation for ϕ in each element is

$$\phi^{(1)} = N_1^{(1)}\Phi_4 + N_2^{(1)}\Phi_5 + N_3^{(1)}\Phi_2 + N_4^{(1)}\Phi_1$$
$$\phi^{(2)} = N_1^{(2)}\Phi_5 + N_2^{(2)}\Phi_6 + N_3^{(2)}\Phi_3 + N_4^{(2)}\Phi_2$$
$$\phi^{(3)} = N_1^{(3)}\Phi_7 + N_2^{(3)}\Phi_8 + N_3^{(3)}\Phi_5 + N_4^{(3)}\Phi_4 \qquad (26.1)$$

359

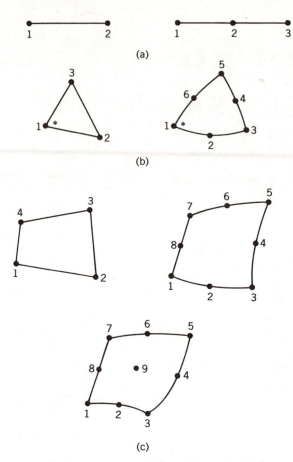

(a)

(b)

(c)

Figure 26.1. Local node numbers. (a) One-dimensional elements, (b) Triangular elements, (c) Quadrilateral elements.

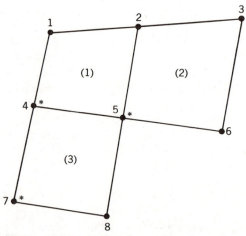

Figure 26.2. The global node numbers for a three-element grid.

Table 26.1 Global
Node Numbers for the
Elements in Figure 26.2

(e)	Global Node Numbers			
1	4	5	2	1
2	5	6	3	2
3	7	8	5	4

The nodal values, Φ, have a global node number as a subscript, but the shape functions retain their 1, 2, 3, or 4 value. Nothing is gained by referencing the shape functions to the global coordinate axis because all of the shape functions are developed relative to a natural coordinate system defined within an element.

26.2 EVALUATING THE SHAPE FUNCTIONS

The one-dimensional linear shape functions were evaluated in Chapter 2, and the shape functions for the triangular and rectangular elements were evaluated in Chapter 5. In each case, a system of equations was solved for some unknown coefficients. These coefficients were then substituted into the interpolation equation and the interpolation equation was rearranged to obtain the shape functions. This procedure becomes more difficult to apply as the number of unknown coefficients increases.

An alternate procedure for generating the shape functions is to assume that each shape function is a product of two functions

$$N_\beta = F_\beta G_\beta \qquad (26.2)$$

where F_β is a function that is zero at specific nodes and/or on specific sides, and G_β is selected so that N_β has the same powers of the coordinate variables as the interpolation function. The first function, F_β, is usually a product of two or more simple polynomials. The second function, G_β, contains unknown coefficients that are evaluated by requiring N_β to be one at its own node and zero at those nodes not included in F_β.

The procedure outlined above is clarified in the sections that follow. The method is based on the following three properties:

1. Each shape function has a value of one at its own node and is zero at each of the other nodes.
2. The shape functions for two-dimensional elements are zero along each side that the node does not touch.
3. Each shape function is a polynomial of the same degree as the interpolation equation.

All of the shape functions developed in this chapter are developed by using a natural coordinate system. The reason is that the most popular numerical integration techniques used to evaluate the element matrices are defined relative to natural coordinate systems.

26.3 THE ONE-DIMENSIONAL ELEMENT

The shape functions for the one-dimensional linear element relative to the ξ-coordinate system were developed in Chapter 6. The shape functions are

$$N_1 = \tfrac{1}{2}(1 - \xi) \quad \text{and} \quad N_2 = \tfrac{1}{2}(1 + \xi) \tag{26.3}$$

These shape functions have the general form of (26.2), when G is a constant.

The interpolation equation for the one-dimensional quadratic element (Figure 26.3a) is

$$\phi(\xi) = a_1 + a_2\xi + a_3\xi^2 \tag{26.4}$$

To evaluate N_1, we select $F_1(\xi) = \xi(\xi - 1)$ because this function is zero at nodes two and three. Note that it is a product of the two functions given in Figure 26.3b. Since $F_1 = \xi^2 - \xi$, that is, it contains the linear term ξ and the quadratic term ξ^2, G_β must be a constant. Therefore,

$$N_1 = C\xi(\xi - 1) \tag{26.5}$$

but $N_1 = 1$ at node one ($\xi = -1$); thus

$$1 = C(-1)(-1-1) = 2C$$

$$C = \frac{1}{2}$$

and

$$N_1(\xi) = \frac{\xi}{2}(\xi - 1) \tag{26.6}$$

It is left to the reader to verify that

$$F_2(\xi) = (\xi + 1)(\xi - 1) \tag{26.7}$$

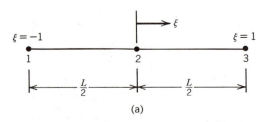

(a)

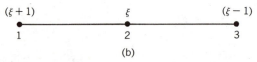

(b)

Figure 26.3. (a) The one-dimensional quadratic element. (b) Functions of ξ that are zero at the respective nodes.

and

$$F_3(\xi)=\xi(\xi+1) \tag{26.8}$$

lead to the shape function equations

$$N_2=-(\xi+1)(\xi-1) \quad \text{and} \quad N_3=\frac{\xi}{2}(\xi+1) \tag{26.9}$$

26.4   TRIANGULAR ELEMENTS

The shape functions for the linear triangular element written in terms of the area coordinates were developed in Chapter 6 and are

$$N_1=L_1, \quad N_2=L_2, \quad \text{and} \quad N_3=L_3 \tag{26.10}$$

The shape functions are the area coordinates. Recall, however, that each shape function is linear in x and y, that is, $N_1=a_1+b_1x+c_1y$. Thus each area coordinate is also linear in x and y.

The interpolation equation for the quadratic triangular element (Figure 26.1b) is

$$\phi(x, y)=a_1+a_2x+a_3y+a_4x^2+a_5xy+a_6y^2 \tag{26.11}$$

This equation is equivalent in form to

$$\phi(x, y)=C(L_\alpha-d_\alpha)(L_\delta-d_\delta) \tag{26.12}$$

because each term within the parentheses is linear in x and y and the product contains the x^2, xy, and y^2 terms that occur in (26.11). The equation $L_\alpha=d_\alpha$ represents a line of constant L_α, and the function $L_\alpha-d_\alpha$ is zero for all points on this line.

The shape function equations must have the same form as the interpolation equation; therefore,

$$N_\beta=C(L_\alpha-d_\alpha)(L_\delta-d_\delta) \tag{26.13}$$

where $(L_\alpha-d_\alpha)$ and $(L_\delta-d_\delta)$ represent two lines that pass through all of the nodes except node β. The constant C is evaluated by requiring that N_β be one at node β. The use of (26.13) is illustrated by evaluating a couple of shape functions. The evaluation of the other four is a very similar procedure and is left to the reader.

Consider the evaluation of N_1. We are looking for two lines that pass through all of the nodes except node one. These lines are shown in Figure 26.4 and are $L_1=0$ and $L_1=\frac{1}{2}$. The functions in (26.13), therefore, are (L_1-0) and $(L_1-\frac{1}{2})$ giving

$$N_1=C(L_1-0)(L_1-\tfrac{1}{2}) \tag{26.14}$$

The L_1 coordinate of node one is $L_1=1$; thus

$$1=C(1-0)\left(1-\frac{1}{2}\right)=\frac{C}{2}$$

and

$$N_1 = 2L_1\left(L_1 - \frac{1}{2}\right) = L_1(2L_1 - 1) \tag{26.15}$$

The two lines that pass through every node except node two are $L_1 = 0$ and $L_2 = 0$. Thus

$$N_2 = C(L_1 - 0)(L_2 - 0) = CL_1L_2 \tag{26.16}$$

The coordinates of node two are $L_1 = L_2 = \frac{1}{2}$ and

$$N_2 = 4L_1L_2 \tag{26.17}$$

The shape functions for the quadratic element are summarized in Figure 26.5. These functions are given in terms of two coordinates, L_1 and L_2, since L_3 is not an independent coordinate.

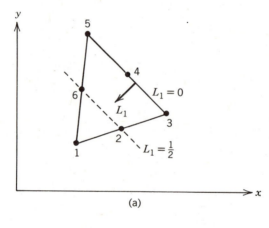

(a)

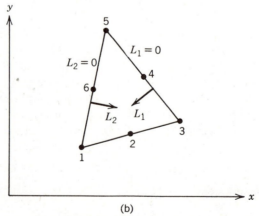

(b)

Figure 26.4. (a) Two lines passing through every node except node one. (b) Two lines passing through every node except node two.

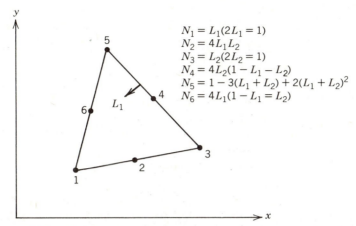

Figure 26.5. The quadratic shape functions for the triangular element.

26.5 QUADRILATERAL ELEMENTS

The two-dimensional quadrilateral elements are very useful and are one of the element choices in commercially available finite element computer programs. A special case of this element, the bilinear rectangle, was discussed in Chapters 5 and 6. Three quadrilateral elements are discussed in this section: the linear element and two versions of the quadratic element. All of the element shape functions are written in terms of the natural coordinates (ξ, η) because continuity exists between elements when this is done.

26.5.1 The Linear Quadrilateral Element

The shape functions for the four-node, linear quadrilateral element are generalizations of those developed for the rectangular element in Chapter 5 and given in natural coordinates as (6.19). The shape functions are

$$N_1 = \tfrac{1}{4}(1-\xi)(1-\eta), \qquad N_3 = \tfrac{1}{4}(1+\xi)(1+\eta)$$
$$N_2 = \tfrac{1}{4}(1+\xi)(1-\eta), \qquad N_4 = \tfrac{1}{4}(1-\xi)(1+\eta) \tag{26.18}$$

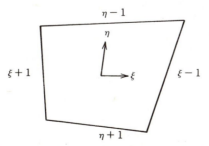

Figure 26.6. Functions that are zero along a side of the quadrilateral element.

Note that these functions have the form given in (26.2) when G_β and F_β consist of the product of simple functions that are zero on the sides of the elements. These functions are shown in Figure 26.6 and are used in conjunction with one of the quadratic elements.

26.5.2   The Lagrangian Element

One of the quadratic quadrilateral elements that is available for finite element computations is the nine-node element shown in Figure 26.7. The element is called the Lagrangian element because the shape functions are products of the one-dimensional Lagrangian interpolation functions. These Lagrangian functions are the same as (26.6) and (26.9).

The shape functions for the Lagrangian element match the method defined in (26.2) when F_β is the quadratic shape function written in terms of ξ, which has a value of one at the node while G_β has a similar property and is written in terms of η. For example,

$$N_1 = \frac{\xi}{2}(\xi-1)\frac{\eta}{2}(\eta-1) = \frac{\xi}{4}(\xi-1)(\eta-1) \tag{26.19}$$

The complete set of shape functions is

$$N_1 = \frac{\xi\eta}{4}(\xi-1)(\eta-1), \qquad N_2 = -\frac{\eta}{2}(\eta-1)(\xi^2-1)$$

$$N_3 = \frac{\xi\eta}{4}(\xi+1)(\eta-1), \qquad N_4 = -\frac{\xi}{2}(\xi+1)(\eta^2-1)$$

$$N_5 = \frac{\xi\eta}{4}(\xi+1)(\eta+1), \qquad N_6 = -\frac{\eta}{2}(\eta+1)(\xi^2-1)$$

$$N_7 = \frac{\xi\eta}{4}(\xi-1)(\eta+1), \qquad N_8 = -\frac{\xi}{2}(\xi-1)(\eta^2-1)$$

$$N_9 = (\xi^2-1)(\eta^2-1) \tag{26.20}$$

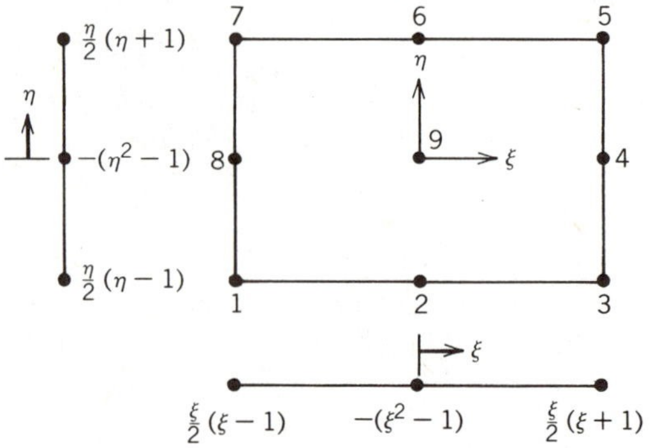

Figure 26.7.   The Lagrangian quadrilateral element and the one-dimensional quadratic shape functions.

Multiplication of the factors in each shape function and collecting terms shows that the general form for ϕ is

$$\phi = a_1 + a_2\xi + a_3\eta + a_4\xi\eta + a_5\xi^2 + a_6\eta^2 + a_7\xi^2\eta + a_8\xi\eta^2 + a_9\xi^2\eta^2 \quad (26.21)$$

26.5.3 The Eight-Node Quadratic Element

The interpolation equation for the eight-node quadratic quadrilateral element shown in Figure 26.8 is

$$\phi = a_1 + a_2\xi + a_3\eta + a_4\xi\eta + a_5\xi^2 + a_6\eta^2 + a_7\xi^2\eta + a_8\xi\eta^2 \quad (26.22)$$

which is identical to (26.21) except for the last term, which is deleted because there is one less node. The shape functions have the general form $N_\beta = F_\beta(\xi, \eta)G_\beta(\xi, \eta)$. The first function, $F_\beta(\xi, \eta)$, is selected such that N_β is zero on each side it does not touch. The second function, $G_\beta(\xi, \eta)$, is defined after $F_\beta(\xi, \eta)$ is known and contains those powers of ξ and η required to obtain the general form (26.22). The coefficients in $G_\beta(\xi, \eta)$ are determined by requiring N_β to be zero or one at those nodes not included in $F_\beta(\xi, \eta)$.

The procedure is illustrated by evaluating N_1. Since node one does not touch sides 3-4-5 or 5-6-7,

$$F_1(\xi, \eta) = (1 - \xi)(1 - \eta)$$
$$= 1 - \xi - \eta + \xi\eta \quad (26.23)$$

The function $G_1(\xi, \eta)$ must contain three terms because the conditions for N_1 at nodes one, two, and eight have not been satisfied. The equation for $G_1(\xi, \eta)$ is

$$G_1(\xi, \eta) = C_1 + C_2\xi + C_3\eta \quad (26.24)$$

which is linear in ξ and η so that the $F_1(\xi, \eta)G_1(\xi, \eta)$ product contains the correct powers of ξ and η. The three nodal conditions are

$$
\begin{array}{llll}
N_1 = 1 & \text{when} & \xi = -1, & \eta = -1 \\
N_1 = 0 & \text{when} & \xi = 0, & \eta = -1 \\
N_1 = 0 & \text{when} & \xi = -1, & \eta = 0
\end{array}
$$

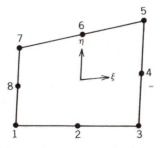

Figure 26.8. The eight-node quadratic quadrilateral element.

Substitution of these conditions into (26.24) produces

$$1 = 4C_1 - 4C_2 - 4C_3$$
$$0 = 2C_1 - 2C_3$$
$$0 = 2C_1 - 2C_2$$

which yields $C_1 = C_2 = C_3 = \frac{1}{4}$. The shape function is

$$N_1 = -\tfrac{1}{4}(1 - \xi)(1 - \eta)(1 + \xi + \eta) \tag{26.25}$$

The product functions $F_\beta(\xi, \eta)$ and $G_\beta(\xi, \eta)$ have the same general form for each corner node, and the other three evaluations are similar to the one just discussed. The four midside nodes are also similar in form, but the form is different from that observed for a corner node. It is instructive to consider one of the midside nodes.

The second shape function is given by

$$N_2 = F_2(\xi, \eta)G_2(\xi, \eta) \tag{26.26}$$

where

$$F_2(\xi, \eta) = (1 - \xi)(1 - \eta)(1 + \xi)$$
$$= 1 - \xi^2 - \eta + \eta\xi^2 \tag{26.27}$$

Equation (26.27) is zero at every node except node two; therefore, $G_2(\xi, \eta)$ consists of only one term because the only nodal condition to satisfy is $N_2 = 1$ at $(0, -1)$. If $G_2(\xi, \eta)$ contains either ξ or η, then terms of the type ξ^3, $\xi^3\eta$, or $\xi^2\eta^2$ occur in the $F_2(\xi, \eta)G_2(\xi, \eta)$ product. Since these terms are not in (26.22), we conclude that $G_2(\xi, \eta) = C$. Applying the nodal condition gives

$$1 = C(1 - 0^2 - (-1) + (-1)(0)) = 2C$$

or $C = \frac{1}{2}$, and the shape function is

$$N_2 = \tfrac{1}{2}(1 - \eta)(1 - \xi^2) \tag{26.28}$$

The complete set of shape functions for the eight node quadrilateral is

$$\begin{array}{ll}
N_1 = -\tfrac{1}{4}(1 - \xi)(1 - \eta)(1 + \xi + \eta), & N_2 = \tfrac{1}{2}(1 - \xi^2)(1 - \eta) \\
N_3 = \tfrac{1}{4}(1 + \xi)(1 - \eta)(\xi - \eta - 1), & N_4 = \tfrac{1}{2}(1 - \eta^2)(1 + \xi) \\
N_5 = \tfrac{1}{4}(1 + \xi)(1 + \eta)(\xi + \eta - 1), & N_6 = \tfrac{1}{2}(1 - \xi^2)(1 + \eta) \\
N_7 = -\tfrac{1}{4}(1 - \xi)(1 + \eta)(\xi - \eta + 1), & N_8 = \tfrac{1}{2}(1 - \eta^2)(1 - \xi)
\end{array} \tag{26.29}$$

PROBLEMS

26.1 Derive the one-dimensional linear shape functions given in (26.3) using the method defined by (26.2).

26.2 Develop the one-dimensional quadratic shape functions given in (26.9) using the definitions in (26.7) and (26.8).

26.3 Develop the one-dimensional quadratic shape functions for the s-coordinate system and node location shown in Figure P26.3. Use the method defined by (26.2).

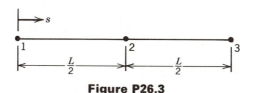

Figure P26.3

26.4 Develop the one-dimensional quadratic shape functions in terms of ξ for the node location shown in Figure P26.4. Use the method defined by (26.2).

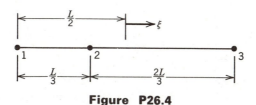

Figure P26.4

26.5 Develop the shape functions for the one-dimensional cubic element shown in Figure P26.5. Use the procedure defined by (26.2). The interpolation equation is

$$\phi(\xi)=a_1+a_2\xi+a_3\xi^2+a_4\xi^3$$

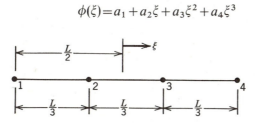

Figure P26.5

26.6 Evaluate N_3 and N_4 for the quadratic triangular element.

26.7 Evaluate N_5 and N_6 for the quadratic triangular element.

26.8 Verify that the set of shape functions given in Figure 26.5 sum to one.

26.9 The cubic triangular element is shown in Figure P26.9. The shape functions are given by the general equation

$$N_\beta=C(L_\alpha-a_\alpha)(L_\delta-a_\delta)(L_\mu-a_\mu)$$

Evaluate N_1, N_2, and N_3 for this element. The side nodes are equally spaced. Node 10 has the area coordinates $(\frac{1}{3}, \frac{1}{3}, \frac{1}{3})$. *Hint:* You need to find three straight lines that pass through all of the nodes except node β.

26.10 Evaluate N_4, N_5, and N_6 using the method and element discussed in Problem 26.9.

26.11 Evaluate N_7, N_8, and N_9 using the method and element discussed in Problem 26.9.

26.12 Evaluate N_3 and N_4 for the eight-node quadratic element in Figure 26.8.

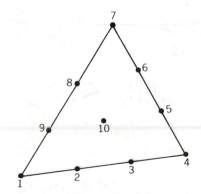

Figure P26.9.

26.13 Evaluate N_5 and N_6 for the eight-node quadratic element in Figure 26.8.
26.14 Evaluate N_7 and N_8 for the eight-node quadratic element in Figure 26.8.
26.15 Verify that the shape functions in (26.29) sum to one.
26.16 The temperature at three consecutive equally spaced points is known to be
$\Phi_1 = 30°C$, $\Phi_2 = 45°C$, and $\Phi_3 = 50°C$. Evaluate the temperature at a point
5 cm from the origin. The nodal locations are given in Figure P26.16.

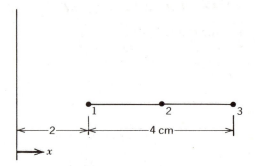

Figure P26.16

26.17 Do Problem 26.16 for a point 3 cm from the origin.
26.18 Develop an equation that gives the value of ϕ at the center of the eight-
node quadrilateral element ($\xi = \eta = 0$) in terms of the eight nodal values.
26.19 Show that the shape functions for one of the following two-dimensional
elements reduce to the one-dimensional shape functions along each edge of:
(a) The quadratic triangular element.
(b) The Lagrangian quadrilateral element.
(c) The eight-node quadrilateral element.

Chapter 27

ELEMENT MATRICES

When elements have curved boundaries, the integrals for the element matrices are most easily evaluated using a natural coordinate system. The two-dimensional elements become squares or triangles in a natural coordinate system and there is no need to find equations for the curved boundaries. Evaluating the integrals in a natural coordinate system, however, is not without its problems. The integration must be done numerically because explicit equations cannot be obtained for all of the steps. The objective of this chapter is to discuss the numerical techniques used to evaluate the integrals that give the element matrices.

27.1 CHANGING THE VARIABLES OF INTEGRATION

The integrals that define the element matrices are written with respect to dx or $dx\,dy$. A change in the integration variable(s) must be made if the integrals are to be evaluated in a natural coordinate system.

27.1.1 One-Dimensional Integrals

The equation for the change of variable in a one-dimensional integral was introduced and used in Chapter 6. The equation is

$$\int_{X_i}^{X_m} f(x)\,dx = \int_{p_1}^{p_2} g(p)\left(\frac{d(x(p))}{dp}\right)dp \tag{27.1}$$

where $x(p)$ is the equation that relates the two-coordinate systems and

$$\frac{d(x(p))}{dp} = [J] = J \tag{27.2}$$

is called the Jacobian matrix for the transformation equation. The matrix consists of one term in this case, which is denoted by J.

We wish to evaluate the integral using the natural coordinate ξ; therefore, (27.1) becomes

$$\int_{X_i}^{X_m} f(x)\,dx = \int_{-1}^{1} g(\xi)\left(\frac{d(x(\xi))}{d\xi}\right)d\xi \tag{27.3}$$

and we need an equation that gives x as a function of ξ. This transformation equation is established by using the element shape functions and writing

$$x = N_1(\xi)X_1 + \cdots + N_m(\xi)X_m \tag{27.4}$$

where $X_1, X_2, \ldots, X_m$ are the global coordinates for the local node numbers $1, 2, \ldots, m$.

Consider the linear element with two nodes. The transformation equation is

$$\begin{aligned} x(\xi) &= N_1(\xi)X_1 + N_2(\xi)X_2 \\ &= \tfrac{1}{2}(1-\xi)X_1 + \tfrac{1}{2}(1+\xi)X_2 \end{aligned} \tag{27.5}$$

using the shape functions given in (26.3). The Jacobian of the transformation is

$$\frac{dx(\xi)}{d\xi} = -\frac{X_1}{2} + \frac{X_2}{2} = \frac{X_2 - X_1}{2} \cdot \frac{L}{} = \frac{L}{2} \tag{26.6}$$

where L is the length of the element. The new integral is

$$\int_{X_i}^{X_m} f(x)\,dx = \frac{L}{2}\int_{-1}^{1} g(\xi)\,d\xi \tag{27.7}$$

Equation (27.7) also holds for the one-dimensional quadratic element.

27.1.2 Two-Dimensional Integrals

The change of variables equation for a double integral is

$$\int_A f(x, y)\,dx\,dy = \int_{-1}^{1}\int_{-1}^{1} g(\xi, \eta)\,|\det[J]|\,d\xi\,d\eta \tag{27.8}$$

when the new variables are the natural coordinates η and ξ and

$$\int_A f(x, y)\,dA = \int_0^1\int_0^{1-L_1} g(L_1, L_2)\,|\det[J]|\,dL_2\,dL_1 \tag{27.9}$$

when the new variables are area coordinates. In each case $[J]$ is the Jacobian matrix of the transformation equations. Two transformation equations exist for the two-dimensional problem because there are two possible local coordinate systems. The Jacobian matrices are

$$\begin{bmatrix} \dfrac{\partial x}{\partial \xi} & \dfrac{\partial y}{\partial \xi} \\[2ex] \dfrac{\partial x}{\partial \eta} & \dfrac{\partial y}{\partial \eta} \end{bmatrix} = [J] = \begin{bmatrix} \dfrac{\partial x}{\partial L_1} & \dfrac{\partial y}{\partial L_1} \\[2ex] \dfrac{\partial x}{\partial L_2} & \dfrac{\partial y}{\partial L_2} \end{bmatrix} \tag{27.10}$$

for the (ξ, η) and (L_1, L_2) systems, respectively.

The two-dimensional transformation equations are established using a procedure identical to that used for the one-dimensional problem; the equations are written using the element shape functions and the global coordinates for the nodes. For example, consider the linear triangular element. The transformation equations are

$$x(L_1, L_2) = N_1(L_1, L_2)X_1 + N_2(L_1, L_2)X_2 + N_3(L_1, L_2)X_3$$
$$y(L_1, L_2) = N_1(L_1, L_2)Y_1 + N_2(L_1, L_2)Y_2 + N_3(L_1, L_2)Y_3 \qquad (27.11)$$

where X_1, X_2, X_3, Y_1, Y_2, and Y_3 are the global coordinates of the three-element nodes. Replacing the shape functions by their area coordinate equivalents gives

$$x(L_1, L_2) = L_1 X_1 + L_2 X_2 + (1 - L_1 - L_2)X_3$$
$$y(L_1, L_2) = L_1 Y_1 + L_2 Y_2 + (1 - L_1 - L_2)Y_3 \qquad (27.12)$$

since $N_3 = L_3 = 1 - L_1 - L_2$. The components of the Jacobian matrix are

$$\frac{\partial x}{\partial L_1} = X_1 - X_3, \qquad \frac{\partial y}{\partial L_1} = Y_1 - Y_3$$

$$\frac{\partial x}{\partial L_2} = X_2 - X_3, \qquad \frac{\partial y}{\partial L_2} = Y_2 - Y_3 \qquad (27.13)$$

and the Jacobian matrix is

$$[J] = \begin{bmatrix} (X_1 - X_3) & (Y_1 - Y_3) \\ (X_2 - X_3) & (Y_2 - Y_3) \end{bmatrix} \qquad (27.14)$$

It is left to the reader to verify that $|\det[J]| = 2A$, where A is the area of the triangle.

Evaluation of $[J]$ becomes more difficult as the number of nodes increases because the shape functions are more complicated. The evaluation of $[J]$ for a linear quadrilateral element is done in the following example problem.

ILLUSTRATIVE EXAMPLE

Evaluate $[J]$ at $\xi = \eta = \frac{1}{2}$ for the linear quadrilateral element shown in Figure 27.1. The transformation equations are given by

$$x = N_1(\xi, \eta)X_1 + N_2(\xi, \eta)X_2 + N_3(\xi, \eta)X_3 + N_4(\xi, \eta)X_4$$
$$y = N_1(\xi, \eta)Y_1 + N_2(\xi, \eta)Y_2 + N_3(\xi, \eta)Y_3 + N_4(\xi, \eta)Y_4 \qquad (27.15)$$

where N_1, N_2, N_3, and N_4 are given by (26.18) and X_1, Y_1, and so on, are the global coordinates of the nodes. The Jacobian matrix can be written as the matrix product

$$[J] = \begin{bmatrix} \dfrac{\partial N_1}{\partial \xi} & \dfrac{\partial N_2}{\partial \xi} & \dfrac{\partial N_3}{\partial \xi} & \dfrac{\partial N_4}{\partial \xi} \\ \dfrac{\partial N_1}{\partial \eta} & \dfrac{\partial N_2}{\partial \eta} & \dfrac{\partial N_3}{\partial \eta} & \dfrac{\partial N_4}{\partial \eta} \end{bmatrix} \begin{bmatrix} X_1 & Y_1 \\ X_2 & Y_2 \\ X_3 & Y_3 \\ X_4 & Y_4 \end{bmatrix}. \qquad (27.16)$$

$$[J]^{-1} = \frac{adj[J]}{det[J]}$$

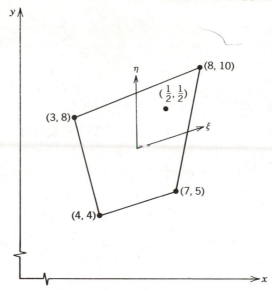

Figure 27.1.   The quadrilateral element for the example problem.

The respective derivative quantities are

$$\frac{\partial N_1}{\partial \xi} = -\frac{1}{4}(1-\eta), \qquad \frac{\partial N_1}{\partial \eta} = -\frac{1}{4}(1-\xi)$$

$$\frac{\partial N_2}{\partial \xi} = \frac{1}{4}(1-\eta), \qquad \frac{\partial N_2}{\partial \eta} = -\frac{1}{4}(1+\xi)$$

$$\frac{\partial N_3}{\partial \xi} = \frac{1}{4}(1+\eta), \qquad \frac{\partial N_3}{\partial \eta} = \frac{1}{4}(1+\xi)$$

$$\frac{\partial N_4}{\partial \xi} = -\frac{1}{4}(1+\eta), \qquad \frac{\partial N_4}{\partial \eta} = \frac{1}{4}(1-\xi) \tag{27.17}$$

Therefore,

$$[J] = \frac{1}{4}\begin{bmatrix} -(1-\eta) & (1-\eta) & (1+\eta) & -(1+\eta) \\ -(1-\xi) & -(1+\xi) & (1+\xi) & (1-\xi) \end{bmatrix}\begin{bmatrix} X_1 & Y_1 \\ X_2 & Y_2 \\ X_3 & Y_3 \\ X_4 & Y_4 \end{bmatrix} \tag{27.18}$$

Equation 27.18 is the general form of $[J]$ for the linear quadrilateral element. Substituting $\frac{1}{2}$ for ξ and η and the global coordinates gives

$$[J] = \frac{1}{8}\begin{bmatrix} -1 & 1 & 3 & -3 \\ -1 & -3 & 3 & 1 \end{bmatrix}\begin{bmatrix} 4 & 4 \\ 7 & 5 \\ 8 & 10 \\ 3 & 8 \end{bmatrix}$$

$$= \frac{1}{8}\begin{bmatrix} 18 & 7 \\ 2 & 19 \end{bmatrix}$$

and $|\det[J]| = [18(19) - (2)(7)]/64 = 41/8$.

27.2 NUMERICAL INTEGRATION TECHNIQUES

The numerical integration technique associated with the natural coordinate systems ξ and (ξ, η) relative to evaluating the element matrices is the Gauss–Legendre quadrature (Conte and deBoor, 1980). The Gauss–Legendre quadrature locates the sampling points to achieve the greatest accuracy. This means that for n sampling points, a polynomial of degree $(2n-1)$ can be integrated exactly. The sampling points and weighting coefficients are presented in Table 27.1. The sampling points are defined on the interval -1 to 1; the location of the points for $n=2$ and $n=3$ are shown in Figure 27.2.

Table 27.1   Location and Weights for Gauss–Legendre Quadrature

$n=1$	$\xi_i=0.0$	$W_i=2.0$
$n=2$	$\xi_i=\pm 0.577350$	$W_i=1.0$
$n=3$	$\xi_i=0.0$	$W_i=\frac{8}{9}$
	$\xi_i=\pm 0.774597$	$W_i=\frac{5}{9}$
$n=4$	$\xi_i=\pm 0.861136$	$W_i=0.347855$
	$\xi_i=\pm 0.339981$	$W_i=0.652145$

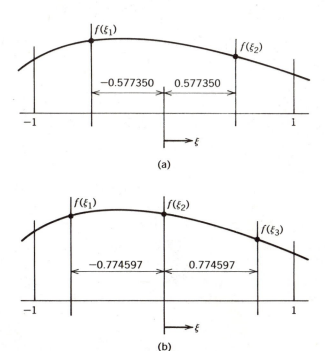

Figure 27.2.   Gauss-Legendre integration points for (A) n = 2, and (b) n = 3.

27.2.1   One-Dimensional Integrals

The one-dimensional integral is replaced by a summation, that is,

$$\frac{L}{2} \int_{-1}^{1} g(\xi)\, d\xi = \frac{L}{2} \sum_{i=1}^{n} g(\xi_i) W_i \tag{27.19}$$

where the functions $g(\xi_i)$ are evaluated at each of the sampling points and multiplied by the corresponding weighting coefficient, W_i. The number of sampling points, n, is determined by equating $(2n-1)$ to the highest power of ξ. As a simple example, consider $\int_{-1}^{1} (3\xi^2 + \xi^3)\, d\xi$. The highest power of ξ is 3; thus $(2n-1)=3$ and $n=2$. When n is a fraction, use the next highest integer value.

The numerical evaluation of the integral in the previous paragraph goes as follows

$$\int_{-1}^{1} (3\xi^2 + \xi^3)\, d\xi = \sum_{i=1}^{2} g(\xi_i) W_i$$

$$= [3(-0.577350)^2 + (-0.577350)^3]1.0 + [3(0.577350)^2$$
$$+ (0.577350)^3]1.0$$
$$= [6(0.333333) + 0]1.0 = 2.0$$

Since $n=2$, the sampling points are $\xi_1 = -0.577350$ and $\xi_2 = 0.577350$, whereas the weighting coefficient is 1.0 at each sampling point.

27.2.2   Quadrilateral Regions

The area integrals in the (ξ, η) coordinate system have the general form

$$\int_{-1}^{1} \int_{-1}^{1} f(\xi, \eta)\, d\eta\, d\xi \tag{27.20}$$

and are evaluated numerically by first evaluating the inner integral, keeping ξ constant, and then evaluating the outer integral. Evaluating the inner integral gives

$$\int_{-1}^{1} g(\xi, \eta)\, d\eta = \sum_{j=1}^{n} g(\xi, \eta_j) W_j = h(\xi) \tag{27.21}$$

where η_j and W_j are the Gauss–Legendre sampling points and weighting coefficients given in Table 27.1. The outer integral becomes

$$\int_{-1}^{1} h(\xi)\, d\xi = \sum_{i=1}^{m} h(\xi_i) W_i \tag{27.22}$$

Substituting (27.21) for $h(\xi)$ yields

$$\int_{-1}^{1} \int_{-1}^{1} g(\xi, \eta)\, d\eta\, d\xi = \sum_{i=1}^{m} \sum_{j=1}^{n} W_i W_j g(\xi_i, \eta_j) \tag{27.23}$$

The values of m and n are obtained by equating $(2m-1)$ to the highest power of ξ and $(2n-1)$ to the highest power of η. Equation (27.23) is usually implemented as a single sum over $K = n \times m$ sampling points with $W_i W_j$-type products being the weighting coefficient of a specific point.

27.2.3 Triangular Regions

The area integral over a triangular region is

$$\int_A f(x, y)\, dA = \int_0^1 \int_0^{1-L_2} g(L_1, L_2)\, dL_1\, dL_2 \qquad (27.24)$$

where $g(L_1, L_2)$ includes the $\left|\det[J]\right|$ term. The sampling points that have become associated with the finite element method were formulated by Hammer, et al. (1956). The location of the sampling points and the corresponding weighting functions are summarized in Table 27.2. The use of these sampling points is equivalent to replacing (27.24) by the single sum

$$\int_0^1 \int_0^{1-L_2} g(L_1, L_2)\, dL_1\, dL_2 = \sum_{i=1}^n g(L_{1i}, L_{2i}) W_i \qquad (27.25)$$

Table 27.2 Sampling Points and Weighting Coefficients for a Triangular Region

	n	Points	Coordinates L_1	Coordinates L_2	W_i	Error
	2	a	$\frac{1}{3}$	$\frac{1}{3}$	$\frac{1}{2}$	O(h)
	3	a	$\frac{1}{2}$	0	$\frac{1}{6}$	
		b	$\frac{1}{2}$	$\frac{1}{2}$	$\frac{1}{6}$	O(h^2)
		c	0	$\frac{1}{2}$	$\frac{1}{6}$	
	6	a	$\frac{1}{3}$	$\frac{1}{3}$	0.11250	
		b	α	β		
		c	β	β	$\left.\right\}$ 0.0661971	O(h^6)
		d	β	α		
		e	γ	γ		
		f	Δ	γ	$\left.\right\}$ 0.0629696	
		g	γ	Δ		

$\alpha = 0.0597159$
$\beta = 0.470142$
$\gamma = 0.101287$
$\Delta = 0.797427$

The value of n is determined by summing the powers of L_1 and L_2 in each term and using the largest sum. For example, to integrate the product $L_1^2 L_2$, n must be three or greater.

27.3   AN INTEGRATION EXAMPLE

There are three types of integrals associated with the element matrices, those involving $[N]^T$ and $[N]^t[N]$, those involving $[B]$ such as $[B]^T[D][B]$ and surface integrals. We will work through an example involving $[N]^T[N]$ for the one-dimensional quadratic element in this section. The integrals involving $[B]$ are discussed in the next section, and a discussion of surface integrals concludes this chapter.

We wish to numerically evaluate

$$\frac{L}{2} \int_{-1}^{1} [N]^T[N] \, d\xi \tag{27.26}$$

for the quadratic element. The shape functions are

$$N_1(\xi) = \frac{\xi}{2}(\xi - 1)$$

$$N_2(\xi) = -(\xi^2 - 1) \tag{27.27}$$

$$N_3(\xi) = \frac{\xi}{2}(\xi + 1)$$

The matrix product $[N]^T[N]$ contains products of the type $N_1 N_2$, N_2^2, and so on. Since each shape function has ξ^2 terms, the products have ξ^4 terms and $(2n - 1) = 4$ yields $n = \frac{5}{2}$. We must use three sampling points to evaluate the integral. The integral in (27.26) is replaced by the summation

$$\frac{L}{2} \int_{-1}^{1} [N]^T[N] \, d\xi = \frac{L}{2} \sum_{i=1}^{3} [N(\xi_i)]^T[N(\xi_i)] W_i \tag{27.28}$$

where ξ_i and W_i are given in Table 27.1.

The sampling points are $\xi_1 = -0.774597$, $\xi_2 = 0$, and $\xi_3 = 0.774597$. Starting with ξ_1 gives

$$N_1(\xi_1) = \frac{\xi_1}{2}(\xi_1 - 1) = \frac{-0.774597}{2}(-0.774597 - 1)$$

$$= 0.687299$$

$$N_2(\xi_1) = -(\xi_1^2 - 1) = 0.400000$$

$$N_3(\xi_1) = \frac{\xi_1}{2}(\xi_1 + 1) = -0.087298$$

and

$$g(\xi_1) = [N(\xi_1)]^T[N(\xi_1)]$$

$$= \left\{ \begin{array}{c} 0.687299 \\ 0.400000 \\ -0.087298 \end{array} \right\} \quad [0.687299 \quad 0.400000 \quad -0.087298]$$

$$= \left[\begin{array}{ccc} 0.472380 & 0.274920 & -0.060000 \\ 0.274920 & 0.160000 & -0.034919 \\ -0.060000 & -0.034919 & 0.007621 \end{array} \right]$$

Next consider $\xi_2 = 0.0$, which results in

$$N_1(\xi_2) = \frac{\xi_2}{2}(\xi_2 - 1) = 0$$

$$N_2(\xi_2) = -(\xi_2^2 - 1) = 1$$

$$N_3(\xi_2) = \frac{\xi_2}{2}(\xi_2 + 1) = 0$$

and

$$g(\xi_2) = [N(\xi_2)]^T[N(\xi_2)] = \left[\begin{array}{ccc} 0 & 0 & 0 \\ 0 & 1 & 0 \\ 0 & 0 & 0 \end{array} \right]$$

The final sampling point is $\xi_3 = 0.774597$. Substitution of this values into the shape function equations yields

$$N_1(\xi_3) = -0.087298$$
$$N_2(\xi_3) = 0.400000$$
$$N_3(\xi_3) = 0.687299$$

and

$$g(\xi_3) = [N(\xi_3)]^T[N(\xi_3)]$$

$$= \left[\begin{array}{ccc} 0.007621 & -0.034919 & -0.060000 \\ -0.034919 & 0.160000 & 0.274920 \\ -0.060000 & 0.274920 & 0.472380 \end{array} \right]$$

Since $W_1 = W_3$,

$$\frac{L}{2} \int_{-1}^{1} [N]^T[N]\, d\xi = \frac{L}{2}\{W_1[g(\xi_1) + g(\xi_3)] + W_2 g(\xi_2)\}$$

Substituting the matrices for $g(\xi_1)$, $g(\xi_2)$, and $g(\xi_3)$ into the above equation and performing the addition yields

$$\frac{L}{2} \int_{-1}^{1} [N]^T[N]\, d\xi$$

$$= \frac{L}{2}\left(\frac{5}{9} \left[\begin{array}{ccc} 0.480 & 0.240 & -0.120 \\ 0.240 & 0.320 & 0.240 \\ -0.120 & 0.240 & 0.480 \end{array} \right] + \frac{8}{9} \left[\begin{array}{ccc} 0 & 0 & 0 \\ 0 & 1 & 0 \\ 0 & 0 & 0 \end{array} \right] \right)$$

Evaluating the products and addition gives

$$\frac{L}{2} \int_{-1}^{1} [N]^T[N]\, d\xi = L \begin{bmatrix} 0.1333 & 0.0667 & -0.0333 \\ 0.0667 & 0.5333 & 0.0667 \\ -0.0333 & 0.0667 & 0.1333 \end{bmatrix} \tag{27.29}$$

An analytical integration gives

$$\frac{L}{2} \int_{-1}^{1} [N]^T[N]\, d\xi = \frac{L}{30} \begin{bmatrix} 4 & 2 & -1 \\ 2 & 16 & 2 \\ -1 & 2 & 4 \end{bmatrix} \tag{27.30}$$

which is equivalent to (27.29).

27.4   EVALUATING [B]

The numerical evaluation of the integrals involving $[B]$, namely $\int [B]^T[D][B]\, dA$ or $\int D[B]^T[B]\, dx$, proceeds in the same manner as illustrated in the previous section. The difficulty comes in determining the coefficients in $[B]$. To illustrate this, we consider the one-dimensional quadratic element where

$$[B] = \begin{bmatrix} \dfrac{dN_1}{dx} & \dfrac{dN_2}{dx} & \dfrac{dN_3}{dx} \end{bmatrix} \tag{27.31}$$

The row vector $[B]$ contains derivatives with respect to x, but these derivatives must be written in terms of ξ before the numerical integration can be performed. The change of variables requires that

$$\int_{X_i}^{X_m} [B(x)]^T[B(x)]\, dx = \frac{L}{2} \int_{-1}^{1} [B(\xi)]^T[B(\xi)]\, d\xi \tag{27.32}$$

where

$$[B(\xi)] = \begin{bmatrix} \dfrac{dN_1}{dx}(\xi) & \dfrac{dN_2}{dx}(\xi) & \dfrac{dN_3}{dx}(\xi) \end{bmatrix} \tag{27.33}$$

The derivatives in $[B]$ are obtained as a function of ξ by using the chain rule, which states that

$$\frac{dN_\beta}{d\xi}(\xi) = \frac{dN_\beta}{dx}(\xi)\frac{dx}{d\xi}(\xi) \tag{27.34}$$

We know $N_\beta(\xi)$ so that we can evaluate $dN_\beta(\xi)/d\xi$. We also know $x(\xi)$, (27.5), and $dx(\xi)/d\xi = L/2$, (27.6). Equation (27.34) can be solved for $dN_\beta(\xi)/dx$. The result is

$$\frac{dN_\beta}{dx}(\xi) = \frac{dN_\beta(\xi)}{d\xi}\frac{1}{\dfrac{dx(\xi)}{d\xi}} = \frac{2}{L}\frac{dN_\beta(\xi)}{d\xi} \tag{27.35}$$

Using the quadratic shape functions, (26.6) and (26.9), gives

$$\frac{dN_1}{dx}(\xi)=\frac{2}{L}\frac{d}{d\xi}\left(\frac{\xi}{2}(\xi-1)\right)=\frac{2}{L}\left(\xi-\frac{1}{2}\right)$$

$$\frac{dN_2}{dx}(\xi)=\frac{2}{L}\frac{d}{d\xi}(-(\xi^2-1))=-\frac{4\xi}{L} \qquad (27.36)$$

$$\frac{dN_3}{dx}(\xi)=\frac{2}{L}\frac{d}{d\xi}\left(\frac{\xi}{2}(\xi+1)\right)=\frac{2}{L}\left(\xi+\frac{1}{2}\right)$$

and

$$[B(\xi)]=\frac{2}{L}\left[\left(\xi-\frac{1}{2}\right) \quad -2\xi \quad \left(\xi+\frac{1}{2}\right)\right] \qquad (27.37)$$

It is re-emphasized that $[B(\xi)]$ contains the derivatives of the shape functions with respect to x but written in terms of ξ. Note that $dx(\xi)/d\xi=L/2$, (27.34), is also the Jacobian matrix, $[J]$, and that $1/(dx(\xi)/d\xi)$ is $[J]^{-1}$.

In the two-dimensional case, $[B]$ contains quantities related to the partial derivatives of the shape functions with respect to x or y. These derivatives, how-ever, must be functions of ξ and η or L_1 and L_2. The desired terms are again ob-tained by applying the chain rule for differentiation. Starting with N_1 and using the variables ξ and η, we obtain

$$\frac{\partial N_1(\xi, \eta)}{\partial \xi}=\frac{\partial N_1(\xi, \eta}{\partial x}\frac{\partial x(\xi, \eta)}{\partial \xi}+\frac{\partial N_1(\xi, \eta)}{\partial y}\frac{\partial y(\xi, \eta)}{\partial \xi}$$

$$\frac{\partial N_1(\xi, \eta)}{\partial \eta}=\frac{\partial N_1(\xi, \eta)}{\partial x}\frac{\partial x(\xi, \eta)}{\partial n}+\frac{\partial N_1(\xi, \eta)}{\partial y}\frac{\partial y(\xi, \eta)}{\partial n} \qquad (27.38)$$

This pair of equations can be written as

$$\begin{Bmatrix} \dfrac{\partial N_1}{\partial \xi} \\[2mm] \dfrac{\partial N_1}{\partial \eta} \end{Bmatrix}=\begin{bmatrix} \dfrac{\partial x}{\partial \xi} & \dfrac{\partial y}{\partial \xi} \\[2mm] \dfrac{\partial x}{\partial \eta} & \dfrac{\partial y}{\partial \eta} \end{bmatrix}\begin{Bmatrix} \dfrac{\partial N_1}{\partial x}(\xi, \eta) \\[2mm] \dfrac{\partial N_1}{\partial y}(\xi, \eta) \end{Bmatrix} \qquad (27.39)$$

The coefficient matrix is the Jacobian matrix, $[J]$, (27.10). Inverting $[J]$ gives

$$\begin{Bmatrix} \dfrac{\partial N_1}{\partial x}(\xi, \eta) \\[2mm] \dfrac{\partial N_1}{\partial y}(\xi, \eta) \end{Bmatrix}=[J(\xi, \eta)]^{-1}\begin{Bmatrix} \dfrac{\partial N_1}{\partial \xi}(\xi, \eta) \\[2mm] \dfrac{\partial N_1}{\partial \eta}(\xi, \eta) \end{Bmatrix} \qquad (27.40)$$

The Jacobian matrix and the column vector on the right-hand side of (27.40) are easily evaluated because both contain the element shape functions written in terms of ξ and η. The coefficients in $[J]$, however, are usually functions of ξ and η or L_1 and L_2, and an explicit inverse cannot be obtained. The derivatives must be numerically evaluated for each integration point. The inability to obtain an explicit inverse for $[J]$ is the reason why the element integrals must be evaluated numerically.

ILLUSTRATIVE EXAMPLE

Evaluate the first partial derivative of each shape function with respect to x and y for the linear quadrilateral element considered in the example in Section 27.1 (Figure 27.1). Evaluate the derivatives at $\xi = \eta = \frac{1}{2}$.

The Jacobian matrix and its inverse are

$$[J] = \frac{1}{8}\begin{bmatrix} 18 & 7 \\ 2 & 19 \end{bmatrix} \quad \text{and} \quad [J]^{-1} = \frac{1}{41}\begin{bmatrix} 19 & -7 \\ -2 & 18 \end{bmatrix}$$

The first derivatives of each shape function with respect to ξ and η are given in the previous example. Evaluating these derivatives at $\xi = \eta = \frac{1}{2}$ produces

$$\frac{\partial N_1}{\partial \xi} = -\frac{1}{8} \qquad \frac{\partial N_1}{\partial \eta} = -\frac{1}{8}$$

$$\frac{\partial N_2}{\partial \xi} = \frac{1}{8} \qquad \frac{\partial N_2}{\partial \eta} = -\frac{3}{8}$$

$$\frac{\partial N_3}{\partial \xi} = \frac{3}{8} \qquad \frac{\partial N_3}{\partial \eta} = \frac{3}{8}$$

$$\frac{\partial N_4}{\partial \xi} = -\frac{3}{8} \qquad \frac{\partial N_4}{\partial \eta} = \frac{1}{8}$$

The values of $\partial N_1/\partial x$ and $\partial N_1/\partial y$ are given by the matrix product in (27.40)

$$\begin{Bmatrix} \dfrac{\partial N_1}{\partial x} \\[2mm] \dfrac{\partial N_1}{\partial y} \end{Bmatrix} = \frac{1}{41}\begin{bmatrix} 19 & -7 \\ -2 & 18 \end{bmatrix}\begin{Bmatrix} -\dfrac{1}{8} \\[2mm] -\dfrac{1}{8} \end{Bmatrix} = \begin{Bmatrix} -\dfrac{12}{328} \\[2mm] -\dfrac{16}{328} \end{Bmatrix}$$

Similar products yield

$$\frac{\partial N_2}{\partial x} = \frac{40}{328} \qquad \frac{\partial N_2}{\partial y} = -\frac{56}{328}$$

$$\frac{\partial N_3}{\partial x} = \frac{36}{328} \qquad \frac{\partial N_3}{\partial y} = \frac{48}{328}$$

$$\frac{\partial N_4}{\partial x} = -\frac{50}{328} \qquad \frac{\partial N_4}{\partial y} = \frac{24}{328}$$

27.5 EVALUATING THE SURFACE INTEGRALS

Several types of surface integrals occur in the element matrices. These include

$$\int_{\Gamma} M[N]^T[N]\, d\Gamma, \qquad \text{and} \qquad \int_{\Gamma} [N]^T \begin{Bmatrix} p_x \\ p_y \end{Bmatrix} d\Gamma \qquad (27.41)$$

The evaluation of these integrals is relatively easy because each set of two-dimensional shape functions reduces to its one-dimensional counterparts along the edge of a triangular or quadrilateral element.

The fact that makes the evaluation of the surface integrals easy is that we have already evaluated each product that occurs in the integrals. Consider the integral $\int_\Gamma M[N]^T[N]\,d\Gamma$ along side 2–3 of a linear quadrilateral element. The integral is

$$\int_\Gamma M[N]^T[N]\,d\Gamma = \frac{L_{23}}{2}\int_{-1}^{1} M\begin{bmatrix} 0 & 0 & 0 & 0 \\ 0 & N_2^2 & N_2N_3 & 0 \\ 0 & N_2N_3 & N_3^2 & 0 \\ 0 & 0 & 0 & 0 \end{bmatrix}d\eta \qquad (27.42)$$

The rows and columns of zero occur because $N_1 = N_4 = 0$ along side 2–3. The shape functions N_2 and N_3, (26.18), reduce to

$$N_2 = \tfrac{1}{2}(1+\eta) \qquad \text{and} \qquad N_3 = \tfrac{1}{2}(1-\eta) \qquad (27.43)$$

because $\xi = 1$ along this edge. The products are easily integrated to yield

$$\int_\Gamma M[N]^T[N]\,d\Gamma = \frac{ML_{23}}{6}\begin{bmatrix} 0 & 0 & 0 & 0 \\ 0 & 2 & 1 & 0 \\ 0 & 1 & 2 & 0 \\ 0 & 0 & 0 & 0 \end{bmatrix} \qquad (27.44)$$

The nonzero coefficients in (27.44) are the coefficients associated with the integral

$$\int_{X_i}^{X_j}[N]^T[N]\,dx$$

for the one-dimensional element. A similar situation occurs for the quadratic elements. The nonzero coefficients in $\int_\Gamma M[N]^T[N]\,d\Gamma$ over any side of a quadratic element are those associated with $\int_{-1}^{1}[N]^T[N]\,d\xi$, which are given in (27.30). For example, $\int_\Gamma M[N]^T[N]\,d\Gamma$ over side 3–4–5 of an eight-node quadratic element is

$$\int_\Gamma M[N]^T[N]\,d\Gamma = \frac{ML_{345}}{30}\begin{bmatrix} 0 & 0 & 0 & 0 & 0 & 0 \\ 0 & 0 & 0 & 0 & 0 & 0 \\ 0 & 0 & 4 & 2 & -1 & 0 \\ 0 & 0 & 2 & 16 & 2 & 0 \\ 0 & 0 & -1 & 2 & 4 & 0 \\ 0 & 0 & 0 & 0 & 0 & 0 \end{bmatrix} \qquad (27.46)$$

where L_{345} is the length of side 3–4–5.

PROBLEMS

27.1 Write the transformation equation for the one-dimensional quadratic element and show that $J = L/2$.

27.2 Verify that $|\det[J]| = 2A$ for the linear triangular element. Recall that $[J]$ is given by (27.14).

27.3 Derive an equation for $|\det[J]|$ in terms of the element area when the linear quadrilateral element is a rectangle. Start with (27.18).

27.4 Numerically evaluate one of the following integrals:

(a) $\displaystyle\int_{-1}^{1} (6\xi + \xi^2)\, d\xi$ (b) $\displaystyle\int_{-1}^{1} (4\xi + \xi^4)\, d\xi$ (c) $\displaystyle\int_{-1}^{1}\int_{-1}^{1} 6\xi^4\eta^2\, d\eta\, d\xi$

(d) $\displaystyle\int_{-1}^{1}\int_{-1}^{1} 3\xi^3\eta^3\, d\eta\, d\xi$ (e) $\displaystyle\int_{0}^{1}\int_{0}^{1-L_1} L_1 L_2^2\, dL_2\, dL_1$

(f) $\displaystyle\int_{0}^{1}\int_{0}^{1-L_1} L_2 L_1\, dL_2\, dL_1$

27.5 Evaluate the surface integral $\int_\Gamma [N]^T\, d\Gamma$ along side 1–2–3 of a quadratic triangular element. Evaluate the integral analytically using area coordinates.

27.6–27.9 The corner coordinates for four rectangular elements are given in the following table. Evaluate $[B]$ for the integration point $\xi = \eta = 0.577350$ when eight nodes are used to define the quadratic interpolation surface. Use (7.27) for $[B]$.

Element Quantity	Problem Number			
	27.6	27.7	27.8	27.9
X_1	2.0	2.0	1.0	2.0
Y_1	1.0	2.0	1.0	5.0
X_3	5.0	6.0	3.0	4.0
Y_3	1.0	2.0	1.0	5.0
X_5	5.0	6.0	3.0	4.0
Y_5	3.0	5.0	5.0	7.0
X_7	2.0	2.0	1.0	2.0
Y_7	3.0	5.0	5.0	7.0

27.10–27.13 The corner coordinates for four triangular elements with straight sides are given in the following table. Evaluate $[B]$ for the integration point $L_1 = L_2 = \frac{1}{2}$ when six nodes are used to define the quadratic interpolation surface. Use (7.27) for $[B]$.

Element Quantity	Problem Number			
	27.10	27.11	27.12	27.13
X_1	0.0	1.0	0.0	3.0
Y_1	0.0	2.0	0.0	0.0
X_3	6.0	6.0	5.0	8.0
Y_3	1.0	0.0	0.0	5.0
X_5	2.0	7.0	0.0	0.0
Y_5	6.0	5.0	6.0	5.0

27.14 Numerically evaluate $\{f^{(e)}\} = \int_A Q[N]^T \, dA$ for the element in Problem 27.6. Assume a linear quadrilateral element.

27.15 Numerically evaluate the integral in Problem 27.14 for the element in Problem 27.10. Assume a quadratic variation for ϕ.

27.16 Numerically evaluate the integral in Problem 27.14 for the linear quadrilateral element in Figure P27.16.

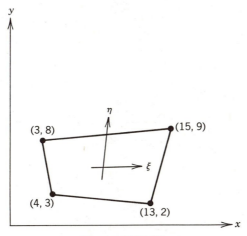

Figure P27.16

27.17 Numerically evaluate the integral in Problem 27.14 for the quadratic triangular element in Figure P27.17.

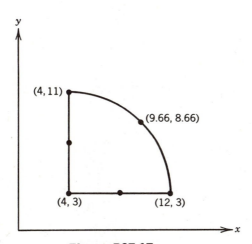

Figure P27.17

27.18 Numerically evaluate the integral in Problem 27.14 for the quadratic quadrilateral element in Figure P27.18. Use the linear shape functions to obtain $[J]$.

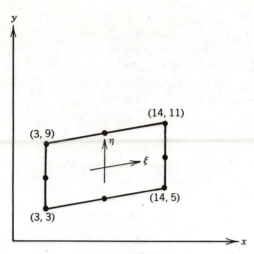

Figure P27.18

Chapter 28

ISOPARAMETRIC COMPUTER PROGRAMS

The flow of computations in an isoparametric computer program is similar to those for programs using the linear elements and shown in Figure 4.6. The primary change is the loops needed to numerically evaluate the element integrals and the gradient values at specific points within the element. A relatively simple isoparametric computer program is presented and discussed in this chapter as well as the flow of computations needed for evaluating $[k^{(e)}]$ and $\{f^{(e)}\}$.

28.1 COMPUTER EVALUATION OF $[k^{(e)}]$ and $\{f^{(e)}\}$

The flow of computations for the numerical integration of $[k^{(e)}]$ and $\{f^{(e)}\}$ is shown in Figure 28.1. The computation starts with the initialization of $[k^{(e)}]$ and $\{f^{(e)}\}$ with zeros. The next step is to determine the number of integration points needed, and to define their coordinates and the weighting coefficient at each point.

The outer DO loop in Figure 28.1 is for the number of integration points, M, which varies with the type of element and the level of integration. The arithmetic calculations within this loop include the evaluation of N_β, $\partial N_\beta(\xi, \eta)/\partial \xi$, and $\partial N_\beta(\xi, \eta)/\partial \eta$. These quantities are always evaluated in a separate subroutine that is written for a specific element. Once N_β and its derivatives are known, $[J]$ is constructed and $[J]^{-1}$ and $|\det[J]|$ are evaluated. This step is followed by the evaluation of $\partial N_\beta(\xi, \eta)/\partial x$ and $\partial N_\beta(\xi, \eta)/\partial y$, which are needed in $[B]$. Once these derivatives are known, the matrix product $[B]^T[D][B]$ is calculated for the integration point.

When solving field problems, the matrix product $G[N]^T[N]$ must also be evaluated as well as any contributions from derivative boundary conditions. The end result of this step is the numerical values for the matrix sums $[k_D^{(e)}] + [k_G^{(e)}] + [k_M^{(e)}]$ and $\{f_Q^{(e)}\} + \{f_S^{(e)}\}$ for the specific integration point. The coefficients in these sums are then multiplied by $|\det[J]|$ and WC(KK) and added to the previously calculated values of $[k^{(e)}]$ and $\{f^{(e)}\}$. These multiplications and additions are done in the DO loops on I and J shown in Figure 28.1. The variable NR is the number of rows in $\{f^{(e)}\}$ and $[k^{(e)}]$ and is a function of the number of element nodes.

It is important that N_β, $\partial N_\beta(\xi, \eta)/\partial \xi$ and $\partial N_\beta(\xi, \eta)/\partial \eta$ be evaluated in a separate subroutine because the rest of the calculations depend only on the number of

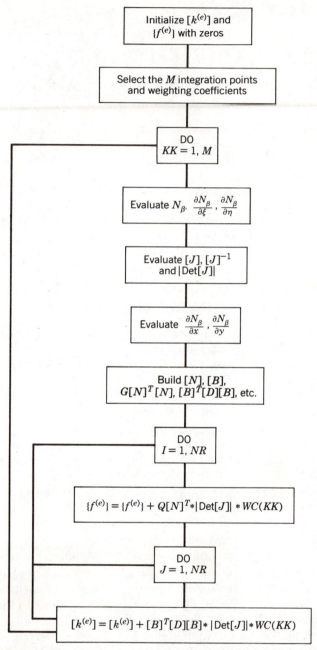

Figure 28.1. Flow chart for evaluating the element matrices.

rows and columns in $[k^{(e)}]$ and thus are applicable for all elements that might be used to solve a certain type of problem. Placing the element quantities in a separate subroutine provides for maximum flexibility. A new element can be introduced; all that needs to be written and checked is the element subroutine.

The above discussion applies equally well to triangular elements. The variables ξ and η are replaced by L_1 and L_2.

28.2 CURVED BOUNDARIES

The location of the nodes used to define the element shape and the determination of the nodal values in $\{f^{(e)}\}$ is not difficult when the element boundaries are straight. The existence of stresses or derivative boundary conditions on curved boundaries, however, complicates the determination of $\{f^{(e)}\}$ and $[k^{(e)}]$. The length of the boundary is needed for evaluating $\{f^{(e)}\}$ in both cases and $[k^{(e)}]$ when derivative boundary conditions occur. In addition, the midside node must divide the length into two equal segments. The determination of the length of the side and the location of the midside node is the subject of discussion in this section.

The mathematical determination of arc length involves the integral

$$\mathscr{L} = \int_a^b \sqrt{1 + \left(\frac{dy}{dx}\right)^2}\, dx \tag{28.1}$$

where $y = f(x)$ and the length is desired between points a and b. Equation 28.1 is a summation of lengths, $d\mathscr{L}$, each calculated using $d\mathscr{L}^2 = dx^2 + dy^2$. The integral in (28.1) can be approximated by calculating the length $d\mathscr{L}$ for very small increments in x and then summing all the increments.

The one-dimensional shape functions can be used to calculate the incremental length $d\mathscr{L}$. We start with the curve $y = f(x)$ defined on the interval $[a, b]$ (Figure 28.2). The curve is approximated using a quadratic polynomial; therefore, one additional node, equally spaced between a and b, is located. The values of y at these nodes, Y_1, Y_2, and Y_3, are determined and y is approximated using the one-dimensional quadratic shape functions:

$$y = [N_1 \quad N_2 \quad N_3] \begin{Bmatrix} Y_1 \\ Y_2 \\ Y_3 \end{Bmatrix} \tag{28.2}$$

The next step is to subdivide the x-axis into several increments (50 to 100) and calculate the length of the curve over each increment using

$$d\mathscr{L} = \sqrt{(X_j - X_i)^2 + (Y_j - Y_i)^2} \tag{28.3}$$

The addition of all the incremental lengths gives the total length. The coordinates of the point that divides the arc length into two equal segments is determined by storing a cumulative-length value as the individual increments in x are considered. Once the total length is known, it is easy to go back and find the two points on the x-axis between which the node is located. The x-coordinate of the midside node is

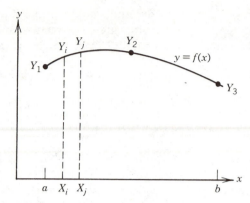

Figure 28.2. Approximating a curve using three values.

determined assuming a linear variation between these two points. The y-coordinate can be calculated using (28.2) once the x-coordinate is known.

28.3 SOLVING TIME-DEPENDENT FIELD PROBLEMS

The solution of time-dependent field problems using quadratic quadrilateral or triangular elements is often accompanied by increases or decreases in ϕ that violate the physical aspects of the problem. The difficulty arises because the quadratic elements do not satisfy the sign criteria discussed in Chapter 15. The diagonal coefficients in $[k^{(e)}]$ are positive, but not all of the off-diagonal values are negative. The element stiffness matrix for the equilateral triangle (Figure 28.3), is

$$
\begin{bmatrix}
57.7 & -38.4 & 9.6 & 0.0 & 9.6 & -38.4 \\
-38.4 & 230.0 & -38.4 & -76.9 & 0.0 & -76. \\
9.6 & -38.4 & 57.7 & -38.4 & 9.6 & 0.0 \\
0.0 & -76.9 & -38.4 & 230.0 & -38.4 & -76.9 \\
9.6 & 0.0 & 9.6 & -38.4 & 57.7 & -38.4 \\
-38.4 & -76.9 & 0.0 & -76.9 & -38.4 & 230.0
\end{bmatrix}
\quad (28.4)
$$

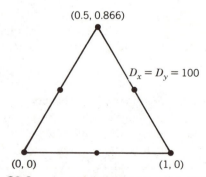

Figure 28.3. An equilateral triangular element.

Note that k_{13} as well as k_{15} and k_{35} is positive. The same type of situation occurs for a square quadratic element.

The square and the equilateral triangle are the best possible shapes for the two-dimensional quadratic elements. Since these two elements do not satisfy the positive diagonal, negative off-diagonal rule, the quadratic element should be used with caution. It is the author's opinion that quadratic elements should not be used to solve time-dependent field problems.

28.4 THE COMPUTER PROGRAM ISOFLD

A relatively simple isoparametric computer program is included in this book and is discussed in this section. The computer program ISOFLD uses the quadratic quadrilateral element to solve the general field equation, (7.1), with the derivative boundary conditions discussed in Chapter 9.

The program utilizes eight subroutines, three of which were used by the previously discussed programs STRESS and TDFIELD. These subroutines, MODIFY, DCMPBD, and SLVBD, were discussed in both Chapters 16 and 25. The new

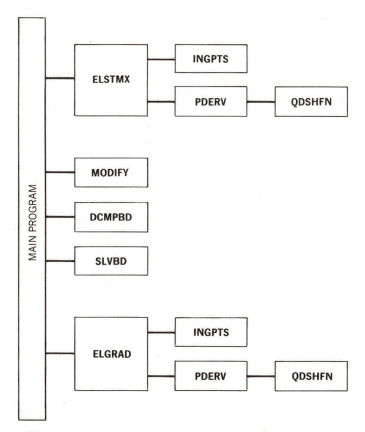

Figure 28.4. Subroutine hierarchy for isoparametric computer program.

subroutines are **ELSTMX, ELGRAD, INGPTS, PDERV,** and **QDSHFN.** Some of the subroutines are called by other subroutines. The subroutine hierarchy is illustrated in Figure 28.4. The objective of each new subroutine is discussed below.

Subroutine ELSTMX. The subroutine **ELSTMX** evaluates the element stiffness matrix and element force vector using Gauss–Legendre integration techniques. The numerical values of N_β and $\partial N_\beta/\partial x$ and $\partial N_\beta/\partial y$ are supplied by the subroutine **PDERV.** The integration coordinates and weighting coefficient are supplied by the subroutine **INGPTS.** The coefficients for the surface integrals are contained in the dimensioned arrays **H** and **FC.** The array **NSIDE** contains the local element node numbers for the four sides.

Subroutine ELGRAD. The subroutine **ELGRAD** evaluates the gradient related quantities $\pm D_x \, \partial\phi/\partial x$ and $\pm D_y \, \partial\phi/\partial y$ at the eight-element-node points and the center of the element. It evaluates $\partial\phi/\partial x$ and $\partial\phi/\partial y$ using the subroutines **PDERV** and **QDSHFN** and selects the appropriate sign using the integer variable **ITYP** and the values stored in the array **GRDC.**

Subroutine INGPTS. The subroutine **INGPTS** generates the ξ and η coordinates and the corresponding weighting coefficient for a nine-point Gauss–Legendre integration of the area integrals.

Subroutine PDERV. The subroutine **PDERV** evaluates the Jacobian transformation matrix, $[J]$, and uses $[J]^{-1}$ to obtain $\partial N_\beta/\partial x$ and $\partial N_\beta/\partial y$ as functions of ξ and η. The subroutine obtains $\partial N_\beta/\partial \xi$ and $\partial N_\beta/\partial \eta$ from the subroutine **QDSHFN.**

Subroutine QDSHFN. The subroutine **QDSHFN** returns the numerical values of N_β, $\partial N_\beta/\partial \xi$ and $\partial N_\beta/\partial \eta$ for the coordinate pair (ξ, η). The subroutine **QDSHFN** is called by the subroutine **PDERV.**

The listing of the statements in **ISOFLD** and the new subroutines follows.

```
          PROGRAM ISOPFLD(INPUT,OUTPUT,TAPE60=INPUT,TAPE61=OUTPUT)
          COMMON/ELMATX/ESM(8,8),EF(8),PHI(8),NS(8),VOL
          COMMON/MATL/DXE,DYE,GE,QE
          COMMON/HCV/IDBC(50,2),DBC(50,2),NDBC
5         COMMON/CRD/XC(200),YC(200)
          COMMON/TLE/TITLE(20)
          COMMON/AV/A(5000),JGF,JGSM,NP,NBW
          DIMENSION NEL(50,8),NMTL(50),ICK(50)
          DIMENSION DX(5),DY(5),G(5),Q(5)
10        DATA IN/60/,IO/61/,IFE/0/,VOL/0./
    C*********
    C*********
    C
    C  DEFINITION OF THE INPUT PARAMETERS
15  C
    C*********
    C*********
    C
    C  TITLE AND PARAMETERS
```

```
20          C
            C               TITLE - A DESCRIPTIVE STATEMENT OF THE
            C                         PROBLEM BEING SOLVED
            C               NP - NUMBER OF NODES
            C               NE - NUMBER OF ELEMENTS
25          C               NCOEF - NUMBER OF COEFFICIENT SETS
            C               NDBC - NUMBER OF ELEMENT SIDES WITH A DERIVATIVE
            C                      BOUNDARY CONDITION
            C               ITYP - TYPE OF FIELD PROBLEM BEING SOLVED
            C                        1 - TORSION PROBLEM
30          C                        2 - IDEAL FLUID, STREAMLINE FLOW
            C                        3 - IDEAL FLUID, POTENTIAL FLOW
            C                        4 - GROUNDWATER FLOW
            C                        5 - HEAT TRANSFER
            C
35          C   EQUATION COEFFICIENTS
            C        THE NUMBER OF SETS MUST EQUAL NCOEF
            C
            C          DX(I) - MATERIAL PROPERTY IN THE X DIRECTION
            C          DY(I) - MATERIAL PROPERTY IN THE Y DIRECTION
40          C          G(I) - COEFFICIENT MULTIPLYING PHI IN THE DIFF. EQUATION
            C          Q(I) - SOURCE COEFFICIENT IN THE DIFF. EQUATION
            C
            C   NODAL COORDNATES
            C
45          C          XC(I) - X COORDINATES OF THE NODES IN NUMERICAL
            C                     SEQUENCE
            C          YC(I) - Y COORDINATES OF THE NODES IN NUMERICAL
            C                     SEQUENCE
            C
50          C   ELEMENT DATA
            C
            C          N - ELEMENT NUMBER
            C          NMTL - COEFFICIENT SET FOR THE ELEMENT
            C          NEL(N,I) - THE ELEMENT NODE NUMBERS FOR ELEMENT N
55          C                      EIGHT NODE NUMBERS ARE SPECIFIED FOR EACH
            C                      ELEMENT
            C
            C   DERIVATIVE BOUNDARY CONDITION DATA
            C        THE NUMBER OF SETS OF DATA MUST EQUAL NDBC
60          C
            C               IDBC(I,1) - ELEMENT NUMBER WITH A DERIVATIVE
            C                              BOUNDARY CONDITION
            C               IDBC(I,2) - SIDE OF THE ELEMENT WITH THE
            C                              DERIVATIVE BOUNDARY CONDITION
65          C               DBC(I,1) -  (M COEFFICIENT)*(LENGTH OF THE SIDE)
            C               DBC(I,2) -  (S COEFFICIENT)*(LENGTH OF THE SIDE)
            C
            C
            C   DATA IS ALSO READ BY THE SUBROUTINE MODIFY
70          C
            C
            C*********
            C*********
            C
75          C   DATA INPUT SECTION OF THE PROGRAM
            C
            C*********
```

```
C*********
C
80    C    INPUT OF THE TITLE CARD AND PARAMETERS
      C
            READ(IN,3) TITLE
          3 FORMAT(20A4)
            READ(IN,*) NP,NE,NCOEF,NDBC,ITYP
85          IF(NP.LE.200) GOTO6
            WRITE(IO,10)
      10    FORMAT(IOX,27HNUMBER OF NODES EXCEEDS 200/
           +10X,20HEXECUTION TERMINATED)
            STOP
90        6 IF(NE.LE.50) GOTO1
            WRITE(IO,2)
          2 FORMAT(IOX,29HNUMBER OF ELEMENTS EXCEEDS 50/
           +10X,20HEXECUTION TERMINATED)
            STOP
95        1 IF(NDBC.LE.50) GOTO400
            WRITE(IO,70)
      70    FORMAT(IOX,34HDERIVATIVE BOUNDARY CONDITION DATA/
           +23HEXCEEDS DIMENSION OF 50/10X,16HINPUT TERMINATED)
            STOP
100   C
      C    INPUT OF THE EQUATON COEFFICIENTS AND THE NODAL COORDINATES
      C
      400   READ(IN,*)  (DX(I),DY(I),G(I),Q(I),I=1,NCOEF)
            READ(IN,*)  (XC(I),I=1,NP)
105         READ(IN,*)  (YC(I),I=1,NP)
      C
      C    OUTPUT OF THE TITLE AND PARAMETERS
      C
            WRITE(IO,4) TITLE,NP,NE,ITYP
110   4     FORMAT(1H1////10X,20A4//10X,5HNP  =,I6/10X,5HNE  =,I6/
           +10X,6HITYP =,I5)
      C
      C    OUTPUT OF THE EQUATION COEFFICIENTS
      C
115         WRITE(IO,48)
      48    FORMAT(//10X,21HEQUATION COEFFICIENTS,/10X,
           +8HMATERIAL/13X,3HSET,8X,2HDX,13X,2HDY,13X,1HG,14X,1HQ)
            WRITE(IO,16)  (I,DX(I),DY(I),G(I),Q(I),I=1,NMTL)
      16    FORMAT(14X,I2,4E15.5)
120   C
      C    OUTPUT OF THE NODAL COORDINATES
      C
            WRITE(IO,11)
      11    FORMAT(//10X,17HNODAL COORDINATES/10X,
125        +4HNODE,5X,1HX,14X,1HY)
            WRITE(IO,12)  (I,XC(I),YC(I),I=1,NP)
      12    FORMAT(10X,I4,2E15.5)
      C
      C    INPUT AND ECHO PRINT OF THE ELEMENT NODAL DATA
130   C
            WRITE(IO,8) TITLE
      8     FORMAT(1H1///10X,20A4//13X,1HN,2X,4HNMTL,
           +13X,12HNODE NUMBERS)
            NID=0
```

```
135            DO9KK=1,NE
               READ(IN,*) N,NMTL(N),(NEL(N,I),I=1,8)
               IF((N-1).NE.NID) WRITE(IO,71) N
        71     FORMAT(10X,7HELEMENT,I4,16H NOT IN SEQUENCE)
               NID=N
140     9      WRITE(IO,5) N,NMTL(N),(NEL(N,I),I=1,8)
        5      FORMAT(10X,2I5,3X,8I5)
        C
        C  INPUT AND ECHO PRINT OF THE DERIVATIVE
        C         BOUNDARY CONDITION DATA
145     C
               IF(NDBC.EQ.0) GOTO26
               WRITE(IO,18)
        18     FORMAT(//10X,34HDERIVATIVE BOUNDARY CONDITION DATA/
               +15X,7HELEMENT,4X,4HSIDE,7X,2HML,13X,2HSL)
150            DO17J=1,NDBC
               READ(IN,*) IDBC(I,1),IDBC(I,2),DBC(I,1),DBC(I,2)
        17     WRITE(IO,119) IDBC(I,1),IDBC(I,2),DBC(I,1),DBC(I,2)
        119    FORMAT(15X,I4,9X,I1,2E15.5)
        C*********
155     C*********
        C
        C  ANALYSIS OF THE NODE NUMBERS
        C
        C*********
160     C*********
        C
        C  INITIALIZATION OF A CHECK VECTOR
        C
               DO500I=1,NE
165     500    ICK(I)=0
        C
        C  CHECK TO SEE IF ANY NODE NUMBER EXCEEDS NP
        C
               DO501I=1,NE
170            DO502J=1,8
               K=NEL(I,J)
        502    IF(K.GT.NP) WRITE(IO,503) J,I,NP
        503    FORMAT(/10X,4HNODE,I4,11H OF ELEMENT,I4,
               +12HEXCEEDS NP =,I4)
175     501    ICK(K)=1
        C
        C  CHECK TO SEE IF ALL NODE NUMBERS THROUGH
        C     NP ARE INCLUDED
        C
180            DO505I=1,NE
        505    IF(ICK(I).EQ.0) WRITE(IO,506) I
        506    FORMAT(/10X,4HNODE,I4,15H DOES NOT EXIST)
        C*********
        C*********
185     C
        C  CREATION AND INITIALIZATION OF THE A VECTOR
        C
        C*********
        C*********
190     C
        C  CALCULATION OF THE BANDWIDTH
```

```
          C
            26 NBW=0
               DO20KK=1,NE
195            DO25I=1,8
            25 NS(I)=NEL(KK,I)
               LK=7
               DO21I=1,LK
               IJ=I+1
200            DO21J=IJ,8
               NB=IABS(NS(I)-NS(J))
               IF(NB.EQ.0) WRITE(IO,27) KK
            27 FORMAT(10X,7HELEMENT,I4,
               +40H HAS TWO NODES WITH THE SAME NODE NUMBER)
205            IF(NB.LE.NBW) GOTO21
               INBW=KK
               NBW=NB
            21 CONTINUE
            20 CONTINUE
210            NBW=NBW+1
               WRITE(IO,270) NBW,INBW
           270 FORMAT(//10X,12HBANDWIDTH IS,I4,11H IN ELEMENT,I4)
          C
          C  INITIALIZATION OF THE COLUMN VECTOR A()
215       C
               JGF=NP
               JGSM=JGF+NP
               JEND=JGSM+NP*NBW
               IF(JEND.GT.5000) GOTO22
220            JL=JEND-JGF
               DO24I=1,JEND
            24 A(I)=0.0
               GOTO30
            22 WRITE(IO,23)
225         23 FORMAT(10X,30HDIMENSION OF A VECTOR EXCEEDED/
               +10X,20HEXECUTION TERMINATED)
               STOP
          C*********
          C*********
230       C
          C  GENERATION OF THE SYSTEM OF EQUATIONS
          C
          C*********
          C*********
235       C
          C  START OF THE LOOP TO BUILD THE SYSTEM
          C      OF EQUATIONS
          C
            30    DO32KK=1,NE
240       C
          C  RETRIEVAL OF THE ELEMENT NODE NUMBERS
          C
               DO31I=1,8
            31    NS(I)=NEL(KK,I)
245       C
          C  ELEMENT COEFFICIENTS
          C
               J=NMTL(KK)
```

```
                    DXE=DX(J)
250                 DYE=DY(J)
                    GE=G(J)
                    QE=Q(J)
      C
      C  CALCULATION OF THE ELEMENT STIFFNESS MATRIX
255   C        AND THE ELEMENT FORCE VECTOR
      C
         CALL ELSTMX(KK)
      C
      C  DIRECT STIFFNESS PROCEDURE
260   C
         DO33I=1,8
            II=NS(I)
            A(JGF+II)=A(JGF+II)+EF(I)
            DO34J=1,8
265         JJ=NS(J)+1-II
            IF(JJ.LE.0) GOTO34
            J1=JGSM+(JJ-1)*NP+II-(JJ-1)*(JJ-2)/2
            A(J1)=A(J1)+ESM(I,J)
         34 CONTINUE
270      33 CONTINUE
         32 CONTINUE
      C********
      C********
      C
275   C  MODIFICATION AND SOLUTION OF THE SYSTEM OF EQUATIONS
      C        AND OUTPUT OF THE CALCULATED NODAL VALUES
      C
      C********
      C********
280      CALL MODIFY(IFE)
         CALL DCMPBD
         CALL SLVBD
      C
      C  OUTPUT OF THE CALCULATED VALUES
285   C
            WRITE(10,165)
      165    FORMAT(//10X,21HCALCULATED QUANTITIES/
            +12X,12HNODAL VALUES)
            WRITE(10,166)  (I,A(I),I=1,NP)
290   166    FORMAT(12X,I3,E14.5,3X,I3,E14.5,3X,I3,E14.5)
      C********
      C********
      C
      C  EVALUATION OF THE VOLUME UNDER THE PHI SURFACE
295   C     AND THE GRADIENT VALUES
      C
      C********
      C********
            ILINE=0
300         DO83KK=1,NE
            IF(ILINE.GT.0) GOTO110
      C
      C  OUTPUT OF THE CORRECT GRADIENT HEADING
      C
305         WRITE(10,43) TITLE
```

```
      43      FORMAT (1H1///10X,20A4)
              IF (ITYP.EQ.1) WRITE (10,44)
      44      FORMAT (//10X,7HELEMENT,4X,8HLOCATION,7X,7HTAU(ZX),
              +8X,7HTAU(ZY))
              IF (ITYP.NE.1.AND.ITYP.NE.5) WRITE (10,147)
      147     FORMAT (///10X,7HELEMENT,4X,8HLOCATION,8X,6HVEL(X),
              +10X,5X,6HVEL(Y))
              IF (ITYP.EQ.5) WRITE (10,146)
      146     FORMAT (///10X,7HELEMENT,4X,8HLOCATION,10X,4HQ(X),
              +11X,4HQ(Y))
      C
      C   RETRIEVAL OF THE NODE NUMBERS AND THE NODAL
      C      VALUES OF PHI FOR THE ELEMENT
      C
      110     DO40I=1,8
              NS(I)=NEL(KK,I)
              J=NS(I)
      40 PHI(I)=A(J)
      C
      C   ELEMENT COEFFICIENTS FOR CALCULATING
      C      THE GRADIENT VALUES
      C
              J=NMTL(KK)
              DXE=DX(J)
              DYE=DY(J)
      C
      C   EVALUATION OF  THE ELEMENT CONTRIBUTION TO
      C      THE VOLUME UNDER THE SURFACE.
      C      EVALUATION OF THE X AND Y GRADIENTS AT
      C      THE NODE POINTS
      C
        *   CALL ELGRAD(KK,ITYP)
      C
      C   LINE COUNTER
      C
              ILINE=ILINE+10
              IF (ILINE.GT.50) ILINE=0
      83      CONTINUE
      C*********
      C*********
      C
      C   OUTPUT OF THE INTEGRAL VALUE
      C
      C*********
      C*********
              VOL=VOL*2.
              IF (ITYP.EQ.1) WRITE (10,56) VOL
      56      FORMAT (//10X,29HTHE TORQUE FOR THE SECTION IS, E15.5)
              STOP
              END


        *   SUBROUTINE ELSTMX(KK)
      C*********
      C*********
      C
```

```
 5          C  THIS SUBROUTINE EVALUATES THE ELEMENT STIFFNESS
            C        MATRIX AND ELEMENT FORCE VECTOR USING
            C        NUMERICAL INTEGRATION TECHNIQUES
            C
            C*********
10          C*********
                  COMMON/ELMATX/ESM(8,8),EF(8),PHI(8),NS(8),VOL
                  COMMON/MATL/DXE,DYE,GE,QE
                  COMMON/HCV/IDBC(50,2),DBC(50,2),NDBC
                  COMMON/PDXY/VN(8),PNX(8),PNY(8),XX(8),YY(8),XD,YD,DET
15                COMMON/IPTS/VX(9),VY(9),WC(9)
                  COMMON/CRD/XC(200),YC(200)
                  DIMENSION H(3,3),FC(3),NT(3),NSIDE(3,4)
                  DATA NSIDE/1,2,3,3,4,5,5,6,7,7,8,1/
                  DATA H/4.,2.,-1.,2.,16.,2.,-1.,2.,4./, FC/1.,4.,1./
20          C*********
            C*********
            C
            C  INITIALIZATION OF THE ELEMENT MATRICES
            C
25          C*********
            C*********
                  DO 1 I=1,8
                  EF(I)=0.0
                  DO 1 J=1,8
30                1 ESM(I,J)=0.0
            C*********
            C*********
            C
            C  RETRFIEVAL OF THE ELEMENT NODAL COORDINATES
35          C
            C*********
            C*********
                  DO10I=1,8
                  J=NS(I)
40                XX(I)=XC(J)
            10    YY(I)=YC(J)
            C*********
            C*********
            C
45          C  EVALUALTION OF THE ELEMENT MATRICES WITHOUT
            C        A DERIVATIVE BOUNDARY CONDITION
            C
            C*********
            C*********
50              * CALL INGPTS
                  DO311=1,9
                * CALL PDERV(VX(II),VY(II))
                  DO2I=1,8
                  EF(I)=EF(I)+QE*VN(I)*DET*WC(II)
55                DO2J=1,8
                  A=(DXE*PNX(I)*PNX(J)+DYE*PNY(I)*PNY(J))*DET*WC(II)
                  B=GE*VN(I)*VN(J)*DET*WC(II)
                2 ESM(I,J)=ESM(I,J)+A+B
                3 CONTINUE
60                IF(NDBC.EQ.0) RETURN
            C*********
```

```
C*********
C
C   ADDITION OF THE DERIVATIVE BOUNDARY CONDITION
C
C*********
C*********
        DO7IJ=1,NDBC
        IF(IDBC(IJ,1).NE.KK) GOTO7
        K=IDBC(I,2)
        DO4I=1,3
4       NT(I)=NSIDE(I,K)
        DO5I=1,3
        II=NT(I)
        EF(II)=EF(II)+DBC(I,2)*FC(I)/6.
        DO5J=1,3
        JJ=NT(J)
5     ESM(II,JJ)=ESM(II,JJ)+DBC(I,1)*H(I,J)/30.
7       CONTINUE
        RETURN
        END




        SUBROUTINE ELGRAD(KK,ITYP)
C*********
C*********
C
C   THIS SUBROUTINE CALCULATES THE ELEMENT CONTRIBUTION
C       TO THE EVALUATION OF THE VOLUME UNDER THE PHI
C       SURFACE AND THE GRADIENT VALUES AT EACH NODE
C       POINT AND AT THE CENTER OF THE ELEMENT.
C
C*********
C*********
        COMMON/ELMATX/ESM(8,8),EF(8),PHI(8),NS(8),VOL
        COMMON/PDXY/VN(8),PNX(8),PNY(8),XX(8),YY(8),XD,YD,DET
        COMMON/MATL/DXE,DYE,GE,QE
        COMMON/IPTS/VX(9),VY(9),WC(9)
        COMMON/CRD/XC(200),YC(200)
        DIMENSION XQ(9),YQ(9),GDX(9),GDY(9),GRDC(5,2)
        DATA GRDC/-1.,-1.,1.,-1.,-1.,1.,1.,1.,-1.,-1./
        DATA XQ/-1.,0.,1.,1.,1.,0.,-1.,-1.,0/
        DATA YQ/-1.,-1.,-1.,0.,1.,1.,1.,0.,0./
        DATA IO/61/
C*********
C*********
C
C   RETRIEVAL OF THE ELEMENT NODAL COORDINATES
C
C*********
C*********
        DO1I=1,8
        J=NS(I)
        XX(I)=XC(J)
1       YY(I)=YC(J)
C*********
C*********
```

```
35          C
            C   EVALUATION OF THE VOLUME UNDER THE PHI SURFACE
            C
            C*********
            C*********
40              CALL INGPTS
                DO51 II=1,9
                CALL PDERV(VX(II),VY(II))
                DO6 I=1,8
              6 VOL=VOL+VN(I)*PHI(I)*WC(II)*DET
45            5 CONTINUE
            C*********
            C*********
            C
            C   EVALUATION OF THE GRADIENTS AT THE ELEMENT NODE
50          C       POINTS AND THE CENTER OF THE ELEMENT
            C
            C*********
            C*********
                DO10 I=1,9
55              GDX(I)=0.
                GDY(I)=0.
                CALL PDERV(XQ(I),YQ(I))
                DO11 J=1,8
                GDX(I)=GDX(I)+PNX(J)*PHI(J)
60           11 GDY(I)=GDY(I)+PNY(J)*PHI(J)
                GDX(I)=DXE*GDX(I)*GRDC(ITYP,1)
             12 GDY(I)=DYE*GDY(I)*GRDC(ITYP,2)
             10 CONTINUE
            C*********
65          C*********
            C
            C   OUTPUT OF THE GRADIENTS
            C
            C*********
70          C*********
                IF(ITYP.GE.3) GOTO50
            C
            C   GRADIENT OUTPUT FOR TROSION AND
            C       STREAMLINE FLOW
75          C
                WRITE(10,53) KK,NS(1),GDY(1),GDX(1)
             53 FORMAT(/13X,I3,5X,5HNODE ,I3,2E15.5)
                DO51 I=2,8
             51 WRITE(10,55) NS(I),GDY(I),GDX(I)
80           55 FORMAT(21X,5HNODE ,I3,2E15.5)
                WRITE(10,54) GDY(9),GDX(9)
             54 FORMAT(21X,6HCENTER,2X,2E15.5)
                RETURN
            C
85          C GRADIENT OUTPUT FOR POTENTIAL FLOW,
            C       GROUNDWATER FLOW AND HEAT TRANSFER
            C
             50 WRITE(10,53) KK,NS(1),GDX(1),GDY(1)
                DO52 I=2,8
90           52 WRITE(10,55) NS(I),GDX(I),GDY(I)
```

```
                        WRITE (10,54) GDX (9) ,GDY (9)
                        RETURN
                        END



                        SUBROUTINE INGPTS
                        COMMON/IPTS/VX (9) ,VY (9) ,WC (9)
                        DIMENSION A (3) ,B (3)
                        DATA  A/-0.774597,0.0,0.774597/
      5                 DATA B/5.,  8.,  5./
            C*********
            C*********
            C
            C  GENERATION OF THE NINE INTEGRATION POINTS
     10     C       FOR THE FQUADRATIC QUADRILATERAL
            C       ELEMENT
            C
            C*********
            C*********
     15                 N=0
                        DO1I=1,3
                        DO1J=1,3
                        N=N+1
                        VX (N) =A (I)
     20                 VY (N) =A (J)
            1           WC (N) =B (I) *B (J) /81.
                        RETURN
                        END



                        SUBROUTINE PDERV (X1,X2)
            C*********
            C*********
            C
      5     C   THIS SUBROUTING EVALUATES THE JACOBIAN
            C       TRANSFORMATION MATRIX AND USES
            C       THE INVERSE JACOBIAN MATRIX TO
            C       OBTAIN THE DERIVATIVES OF THE SHAPE
            C       FUNCTIONS WITH RESPECT TO X AND Y
     10     C
            C*********
            C*********
                        COMMON/PDXY/VNN (8) ,PNX (8) ,PNY (8) ,X (8) ,Y (8) ,XD,YD,DET
                        COMMON/DERV/VN (8) ,PNS (8) ,PNE (8)
     15                 REAL JOCB (2,2)
            C*********
            C*********
            C
            C  INITIALIZATION OF THE JACOBIAN MATRIX
     20     C
            C*********
            C*********
                        DO1I=1,2
                        DO1J=1,2
     25     1           JOCB (I,J) =0.0
```

```
                    XD=0.0
                    YD=0.0
            C*********
            C*********
30          C
            C   EVALUATION OF THE SHAPE FUNCTIONS AND
            C      THEIR DERIVATIVES
            C*********
            C*********
35             ⚡ CALL QDSHFN(X1,X2)
            C*********
            C*********
            C
            C   CALCULATION OF THE COORDINATES OF THE
40          C         INTEGRATION POINT (XD,YD)
            C   AND THE JOCABIAN MATRIX
            C
            C*********
            C*********
45                  DO2I=1,8
                    XD=XD+VN(I)*X(I)
                    YD=YD+VN(I)*Y(I)
                    JOCB(1,1)=JOCB(1,1)+PNS(I)*X(I)
                    JOCB(1,2)=JOCB(1,2)+PNS(I)*Y(I)
50                  JOCB(2,1)=JOCB(2,1)+PNE(I)*X(I)
                  2 JOCB(2,2)=JOCB(2,2)+PNE(I)*Y(I)
            C*********
            C*********
            C
55          C   CALCULATION OF THE INVERSE OF THE JOCABIAN MATRIX
            C
            C*********
            C*********
                    A=JOCB(1,1)*JOCB(2,2)-JOCB(2,1)*JOCB(1,2)
60                  B=JOCB(1,1)
                    JOCB(1,1)=JOCB(2,2)/A
                    JOCB(1,2)=-JOCB(1,2)/A
                    JOCB(2,1)=-JOCB(2,1)/A
                    JOCB(2,2)=B/A
65                  DET=ABS(A)
            C*********
            C*********
            C
            C   CALCULATION OF THE PARTIAL DERIVATIVES
70          C      WITH RESPECT TO X AND Y
            C
            C*********
            C*********
                    DO3I=1,8
75                  VNN(I)=VN(I)
                    PNX(I)=JOCB(1,1)*PNS(I)+JOCB(1,2)*PNE(I)
                  3 PNY(I)=JOCB(2,1)*PNS(I)+JOCB(2,2)*PNE(I)
                    RETURN
                    END
```

```fortran
      SUBROUTINE QDSHFN(SI,ETA)
C*********
C*********
C
C   THIS SUBROUTINE CALCULATES THE VALUE OF THE SHAPE
C       FUNCTIONS AND THEIR DERIVATIVES FOR THE
C       QUADRATIC QUADRILATERAL ELEMENT GIVEN
C       SPECIFIC VALUES OF KSI AND ETA
C         VN - VALUE OF THE SHAPE FUNCTION N
C         PNS - PARTIAL DERIVATIVE WITH RESPECT TO KSI
C         PNE - PARTIAL DERIVATIVE WITH RESPECT TO ETA
C
C*********
C*********
      COMMON/DERV/VN(8),PNS(8),PNE(8)
      DIMENSION SQ(8),EQ(8)
      DATA SQ/-1.,0.,1.,1.,1.,0.,-1.,-1./, EQ/-1.,-1.,-1.,
     10.,1.,1.,1.,0./
C*********
C*********
C
C    CORNER NODES
C
C*********
C*********
      DO5I=1,7,2
      VN(I)=0.25*(1.+SI*SQ(I))*(1.+ETA*EQ(I))
     +*(SI*SQ(I)+ETA*EQ(I)-1.)
      PNS(I)=0.25*(1.+ETA*EQ(I))*(2.*SI+ETA*EQ(I)*SQ(I))
    5 PNE(I)=0.25*(1.+SI*SQ(I))*(2.*ETA+SI*SQ(I)*EQ(I))
C*********
C*********
C
C    MIDSIDE NODES, KSI EQUAL ZERO
C
C*********
C*********
      DO6I=2,6,4
      VN(I)=0.5*(1.-SI**2)*(1.+ETA*EQ(I))
      PNS(I)=-SI*(1.+ETA*EQ(I))
    6 PNE(I)=0.5*EQ(I)*(1.-SI**2)
C*********
C*********
C
C    MIDSIDE NODES, ETA EQUAL ZERO
C
C*********
C*********
      DO7I=4,8,4
      VN(I)=0.5*(1.+SI*SQ(I))*(1.-ETA**2)
      PNS(I)=0.50*SQ(I)*(1.-ETA**2)
    7 PNE(I)=-ETA*(1.+SI*SQ(I))
      RETURN
      END
```

28.5 A COMPUTER EXAMPLE

Data input for the computer program ISOFLD discussed in the previous section is illustrated by solving the convection heat transfer problem given in Problem 11.16 and shown in Figure 28.5. The body has a vertical axis of symmetry and a known temperature distribution on the inside boundary. Convection heat transfer occurs on the outer boundary.

The heat transfer problem is solved using the four-element grid shown in Figure 28.6. The nodes 1, 4, 6, 9, 11, 14, 16, 19, and 21 divide the innercircular boundary into eight equal arc lengths (22.5° increments); nodes 3, 5, 8, 10, 13, 15, 18, 20, and 23 do the same for the outer circular boundary. The four-element grid is probably too coarse for this problem, but it is sufficient to illustrate all aspects of the data input.

The input data for ISOFLD are given to the right of the vertical line in Table 28.1. General comments about the data are given to the left of the line. The data required for a solution are very similar to those required by TDFIELD, which used the linear triangular and bilinear rectangular elements.

The calculated nodal temperatures are shown in Figure 28.7. The calculated nodal values are probably fairly good values because the heat flow perpendicular to surface 1-2-3 is 2.97 W/cm^2 and 1.91 W/cm^2 perpendicular to surface 21-22-23. The values of q_x at the nodes are shown in Figure 28.8. The total heat flow was obtained using Simpson's quadrature. The values of 1.91 and 2.97 are close to the theoretical values of zero and would decrease if the number of elements was increased.

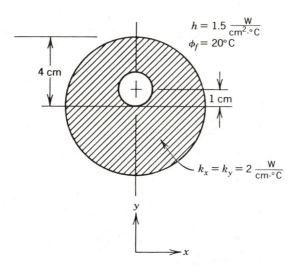

$$h = 1.5 \frac{W}{cm^2 \cdot °C}$$
$$\phi_f = 20°C$$

4 cm

1 cm

$$k_x = k_y = 2 \frac{W}{cm \cdot °C}$$

Boundary of the inner cylinder is at 140°C
Diameter of the inner cylinder, 2 cm
Diameter of the outer cylinder, 8 cm

Figure 28.5. The physical configuration of the heat transfer example.

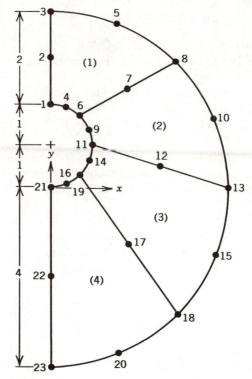

Figure 28.6. The isoparametric finite element grid.

The computer output follows.

```
ISOPARAMETRIC COMPUTER EXAMPLE

NP   =    23
NE   =    4
ITYP =    5


EQUATION COEFFICIENTS
MATERIAL
     SET       DX             DY              G              Q
      1      .20000E+01     .20000E+01     0.             0.


NODAL COORDINATES
NODE     X              Y
  1    0.             .20000E+01
  2    0.             .30000E+01
  3    0.             .40000E+01
  4     .38300E+00    .19240E+01
  5     .15310E+01    .36960E+01
  6     .70700E+00    .17070E+01
  7     .17680E+01    .22680E+01
```

```
 8      .28280E+01       .28280E+01
 9      .92400E+00       .13830E+01
10      .36960E+01       .15310E+01
11      .10000E+01       .10000E+01
12      .25000E+01       .50000E+00
13      .40000E+01      0.
14      .92400E+00       .61700E+00
15      .36960E+01      -.15310E+01
16      .70700E+00       .29300E+00
17      .17680E+01      -.12680E+01
18      .28280E+01      -.28280E+01
19      .38300E+00       .76000E-01
20      .15310E+01      -.36960E+01
21     0.               0.
22     0.               -.20000E+01
23     0.               -.40000E+01
```

ISOPARAMETRIC COMPUTER EXAMPLE

N	NMTL			NODE NUMBERS				
1	1	6	7	8	5	3	2	1	4
2	1	11	12	13	10	8	7	6	9
3	1	16	17	18	15	13	12	11	14
4	1	21	22	23	20	18	17	16	19

DERIVATIVE BOUNDARY CONDITION DATA

ELEMENT	SIDE	ML	SL
1	2	.47100E+01	.94200E+02
2	2	.47100E+01	.94200E+02
3	2	.47100E+01	.94200E+02
4	2	.47100E+01	.94200E+02

BANDWIDTH IS   8 IN ELEMENT   1

KNOWN NODAL VALUES OF PHI
```
 1      .14000E+03
 4      .14000E+03
 6      .14000E+03
 9      .14000E+03
11      .14000E+03
14      .14000E+03
16      .14000E+03
19      .14000E+03
21      .14000E+03
```

CALCULATED QUANTITIES
  NODAL VALUES

1	.14000E+03	2	.88474E+02	3	.55705E+02
4	.14000E+03	5	.54188E+02	6	.14000E+03
7	.83889E+02	8	.50207E+02	9	.14000E+03
10	.45870E+02	11	.14000E+03	12	.76059E+02

13	.42352E+02	14	.14000E+03	15	.39383E+02
16	.14000E+03	17	.70721E+02	18	.37913E+02
19	.14000E+03	20	.36557E+02	21	.14000E+03
22	.68966E+02	23	.36580E+02		

ISOPARAMETRIC COMPUTER EXAMPLE

ELEMENT	LOCATION		Q(X)	Q(Y)
1	NODE	6	.82271E+02	.84274E+02
	NODE	7	.58253E+02	.49984E+02
	NODE	8	.30612E+02	.22346E+02
	NODE	5	.19532E+02	.37748E+02
	NODE	3	.97118E+00	.46782E+02
	NODE	2	.31631E+01	.84295E+02
	NODE	1	.16241E+01	.12181E+03
	NODE	4	.46800E+02	.11293E+03
	CENTER		.35709E+02	.73720E+02
2	NODE	11	.10588E+03	.14117E+01
	NODE	12	.63106E+02	-.59765E+01
	NODE	13	.23635E+02	-.34566E+01
	NODE	10	.29575E+02	.67015E+01
	NODE	8	.30806E+02	.21978E+02
	NODE	7	.59983E+02	.46710E+02
	NODE	6	.83652E+02	.81663E+02
	NODE	9	.10357E+03	.42921E+02
	CENTER		.66909E+02	.18845E+02
3	NODE	16	.67686E+02	-.66077E+02
	NODE	17	.36409E+02	-.40677E+02
	NODE	18	.10561E+02	-.11514E+02
	NODE	15	.16687E+02	-.10055E+02
	NODE	13	.23180E+02	-.48218E+01
	NODE	12	.62614E+02	-.74546E+01
	NODE	11	.10494E+03	-.13992E+01
	NODE	14	.91342E+02	-.37855E+02
	CENTER		.51419E+02	-.27327E+02
4	NODE	21	.12048E+01	-.90358E+02
	NODE	22	.31342E+00	-.51710E+02
	NODE	23	.10408E+01	-.13061E+02
	NODE	20	.46919E+01	-.13595E+02
	NODE	18	.88717E+01	-.12662E+02
	NODE	17	.35788E+02	-.41098E+02
	NODE	16	.65775E+02	-.67376E+02
	NODE	19	.34831E+02	-.84046E+02
	CENTER		.18404E+02	-.49247E+02

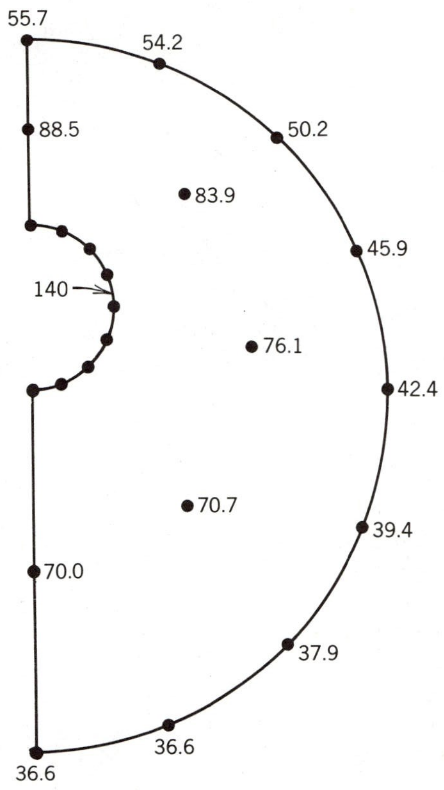

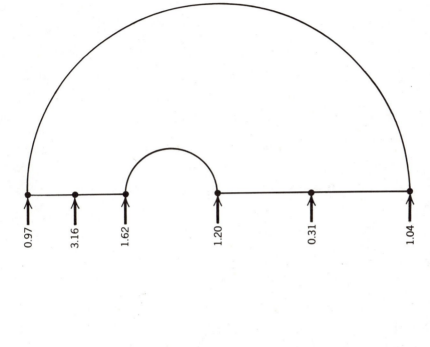

Figure 28.7. Nodal temperature values for the heat transfer example.

Figure 28.8. Heat flow along the axis of symmetry in the heat transfer example.

Table 28.1 Computer Data for ISOFLD

Title	ISOPARAMETRIC COMPUTER EXAMPLE				
Parameters	23 4 1 4 5				
Equation coefficients	2.0 2.0 0. 0.				
X-coordinates	0.	0.	0.	0.383	1.531
	0.707	1.768	2.828	0.924	3.696
	1.0	2.50	4.0	0.924	3.696
	0.707	1.768	2.828	0.383	1.531
	0.	0.	0.		
Y-coordinates	2.	3.	4.	1.924	3.696
	1.707	2.268	2.828	1.383	1.531
	1.	0.50	0.	0.617	-1.531
	0.293	-1.268	-2.828	0.076	-3.696
	0.	-2.0	-4.0		

Element data									
1	1	6	7	8	5	3	2	1	4
2	1	11	12	13	10	8	7	6	9
3	1	16	17	18	15	13	12	11	14
4	1	21	22	23	20	18	17	16	19

Derivative boundary conditions			
1	2	4.71	94.2
2	2	4.71	94.2
3	2	4.71	94.2
4	2	4.71	94.2

Source and sink	0

Known values of ϕ									
1	140.	4	140.	6	140.	9	140.	11	140.
14	140.	16	140.	19	140.	21	140.	0	

REFERENCES

Abramowitz, Milton, and Irene A. Stegun, 1964. *Handbook of Mathematical Functions with Formulas, Graphs, and Mathematical Tables*, National Bureau of Standards.

Bathe, Klaüs-Jurgen, and Edward L. Wilson, 1976. *Numerical Methods in Finite Element Analysis*, Prentice-Hall, Englewood Cliffs, N.J.

Conte, S. D. and Carl deBoor, 1980. *Elementary Numerical Analysis: An Algorithmic Approach*, 3rd ed., McGraw-Hill, New York.

Cook, Robert D., 1981. *Concepts and Applications of Finite Element Analysis*, 2nd ed., John Wiley, New York.

Eisenberg, Martin A., and Lawrence E. Malvern, 1973. "On Finite Element Integration in Natural Coordinates," *International Journal for Numerical Methods in Engineering*, Vol. 7, pp. 574–575.

Fried, Isaac, 1979. *Numerical Solution of Differential Equations*, Academic Press, New York.

Fung, Y. C., 1965. *Foundations of Solid Mechanics*, Prentice-Hall, Englewood Cliffs, N.J.

Hammer, P. C., O. P. Marlowe, and A. H. Stroud, 1956. "Numerical Integration Over Simplexes and Cones," *Mathematics Tables Aids Computation*, Vol. 10, pp. 130–137.

Higdon, Archie, Edward H. Olsen, William B. Stiles, John A. Weese, and William F. Riley, 1976. *Mechanics of Materials*, 3rd ed., John Wiley, New York.

Hildebrand, Francis B., 1965. *Methods of Applied Mathematics*, 2nd edition, Prentice-Hall, Englewood Cliffs, N.J.

Kaplan, Wilfred, 1962. *Operational Methods for Linear Systems*, Addison-Wesley, Reading, Mass.

Kreith, Frank, 1973. *Principles of Heat Transfer*, 3rd ed., Intext Educational Publishers, New York.

Langhaar, Henry L., 1962. *Energy Methods in Applied Mechanics*, John Wiley, New York.

Maadooliat, Reza, 1983. Element and Time Step Criteria for Solving Time-Dependent Field Problems Using the Finite Element Method, unpublished Ph.D. Dissertation, Michigan State University, East Lansing, Mich.

Myers, Glen E., 1971. *Analytical Methods in Conduction Heat Transfer*, McGraw-Hill, New York.

Olmstead, John M. H., 1961. *Advanced Calculus*, Prentice-Hall, Englewood Cliffs, N.J.

Popov, E. P., 1976. *Mechanics of Materials*, 2nd ed., Prentice-Hall, Englewood Cliffs, N.J.

Shuku, T., and K. Ishihara, 1973. "The Analysis of the Acoustic Field in Irregularly Shaped Rooms by the Finite Element Method," *Journal of Sound and Vibrations*, Vol. 29, pp. 67–76.

Timoshenko, S. P., and J. N. Goodier, 1970. *Theory of Elasticity*, McGraw-Hill, New York.

Visser, W., 1965. "A Finite Element Method for the Determination of Non-Stationary Temperature Distribution of Non-Stationary Temperature Distribution and Thermal Deformations," *Proceedings of Conference on Matrix Methods in Structural Mechanics*, Air Force Institute of Technology, Wright Patterson Air Force Base, Dayton, Ohio.

Appendix **1**

MATRIX NOTATION

The principles of the finite element method are presented using matrix notation. Matrices are often denoted in printed material using boldface type such as **B**, **D**, **N**, and so on. The difficulty with this notation is that the boldface type is very difficult to duplicate in a classroom situation. It is much easier to identify matrices and vectors on a chalkboard when they are enclosed in brackets. Since this book is a textbook and the written material will be presented in a lecture format, all of the matrices and vectors have been enclosed in brackets. The following notational rules are used throughout this book.

1. All column vectors are enclosed in pointed brackets, that is,

$$\{F\} = \begin{Bmatrix} F_1 \\ F_2 \\ \vdots \\ F_p \end{Bmatrix}.$$

2. All row vectors and matrices are enclosed in regular brackets, that is,

$$[N] = [N_1 \quad N_2 \cdots N_p]$$

$$[B] = \begin{bmatrix} b_{11} & b_{12} & b_{13} \\ b_{21} & b_{22} & b_{23} \end{bmatrix}$$

The two rules given above should make it easier to distinguish between row vectors and column vectors. The use of brackets to denote matrices allows instructors to use the same notation during lectures that is used within this book.

Appendix **2**

DIFFERENTIATION OF MATRIX EQUATIONS

The minimization procedure discussed in Chapter 18 required the differentiation of matrix products $\{U\}^T\{C\}$ and $\{U\}^T[A]\{U\}$ with respect to $\{U\}$, where $\{C\}$ is a column vector and $[A]$ is a square symmetric matrix. These differentiations are relatively easy to perform but, since the results are not given in many references for matrix algebra, we consider them in this appendix.

Starting with the quantity

$$\Lambda = \{U\}^T\{C\} \tag{1}$$

or

$$\Lambda = [U_1 \ \ U_2 \cdots U_r] \begin{Bmatrix} C_1 \\ C_2 \\ \vdots \\ C_r \end{Bmatrix} \tag{2}$$

we find that the desired derivative is $\partial\Lambda/\partial\{U\}$. This derivative is equivalent to the column vector of derivatives

$$\frac{\partial\Lambda}{\partial\{U\}} = \begin{Bmatrix} \dfrac{\partial\Lambda}{\partial U_1} \\[2mm] \dfrac{\partial\Lambda}{\partial U_2} \\[2mm] \vdots \\[2mm] \dfrac{\partial\Lambda}{\partial U_r} \end{Bmatrix} \tag{3}$$

The derivatives in (3) can be evaluated after the product in (2) has been evaluated. In this case

$$\Lambda = U_1 C_1 + U_2 C_2 + \cdots + U_r C_r$$

The derivatives become

$$\frac{\partial\Lambda}{\partial U_1} = C_1, \qquad \frac{\partial\Lambda}{\partial U_2} = C_2, \ldots, \frac{\partial\Lambda}{\partial U_r} = C_r \tag{4}$$

Substituting these derivatives into (3) produces

$$\frac{\partial \Lambda}{\partial \{U\}} = \left\{\begin{array}{c} C_1 \\ C_2 \\ \vdots \\ C_r \end{array}\right\} = \{C\} \tag{5}$$

and

$$\frac{\partial(\{U\}^T\{C\})}{\partial\{U\}} = \{C\} \tag{6}$$

The other derivative of interest is

$$\frac{\partial(\{U\}^T[A]\{U\})}{\partial\{U\}} \tag{7}$$

The expansion of $[A]\{U\}$ gives the column vector

$$[A]\{U\} = \left\{\begin{array}{c} A_{11}U_1 + A_{12}U_2 + \cdots + A_{1r}U_r \\ A_{21}U_1 + A_{22}U_2 + \cdots + A_{2r}U_r \\ \vdots \qquad \vdots \qquad\qquad \vdots \\ A_{r1}U_1 + A_{r2}U_2 + \cdots + A_{rr}U_r \end{array}\right\} \tag{8}$$

Completion of the product $\{U\}^T[A]\{U\}$ gives the single term

$$\begin{aligned}\Lambda = &\{U\}^T[A]\{U\} \\ = &A_{11}U_1^2 + (A_{12} + A_{21})U_1U_2 + (A_{13} + A_{31})U_1U_3 + \cdots \\ &+ (A_{1r} + A_{r1})U_rU_1 + \cdots + (A_{nm} + A_{mn})U_nU_m + \cdots + A_{rr}U_r^2 \end{aligned} \tag{9}$$

The evaluation of the derivatives yields

$$\frac{\partial \Lambda}{\partial U_1} = 2A_{11}U_1 + (A_{12} + A_{21})U_2 + \cdots + (A_{1r} + A_{r1})U_r$$

$$\frac{\partial \Lambda}{\partial U_2} = (A_{21} + A_{12})U_1 + 2A_{22}U_2 + \cdots + (A_{2r} + A_{r2})U_r$$

$$\vdots$$

$$\frac{\partial \Lambda}{\partial U_r} = (A_{r1} + A_{1r})U_1 + (A_{r2} + A_{2r})U_2 + \cdots + 2A_{rr}U_r$$

Each of the derivatives can be separated into two sums as follows

$$\frac{\partial \Lambda}{\partial U_1} = (A_{11}U_1 + A_{12}U_2 + \cdots + A_{1r}U_r) + (A_{11}U_1 + A_{21}U_2 + \cdots + A_{r1}U_r)$$

$$\frac{\partial \Lambda}{\partial U_2} = (A_{21}U_1 + A_{22}U_2 + \cdots + A_{2r}U_r) + (A_{12}U_1 + A_{22}U_2 + \cdots + A_{r2}U_r)$$

$$\vdots$$

$$\frac{\partial \Lambda}{\partial U_r} = (A_{r1}U_1 + A_{r2}U_2 + \cdots + A_{rr}U_r) + (A_{1r}U_1 + A_{2r}U_2 + \cdots + A_{r2}U_r)$$

The above equations are equivalent to

$$\frac{\partial(\{U\}^T[A]\{U\})}{\partial\{U\}} = [A]\{U\} + [A]^T\{U\} \tag{10}$$

When $[A]$ is symmetric,

$$\frac{\partial(\{U\}^T[A]\{U\})}{\partial\{U\}} = 2[A]\{U\} \tag{11}$$

Appendix **3**

MODIFYING THE SYSTEM OF EQUATIONS

The system of equations

$$[K]\{\Phi\} = \{F\}$$

or

$$[K]\{U\} = \{F\} + \{P\}$$

obtained by using the direct stiffness procedure must be modified whenever some of the values in $\{\Phi\}$ or $\{U\}$ are known. All field problems except some problems involving convection heat transfer must have some of the boundary values specified and all solid mechanics problems must have displacements specified to eliminate rigid body motion. Therefore, the modification of the system of equations to incorporate known nodal conditions is more the rule than the exception.

Our objective here is to discuss and then illustrate a systematic procedure for modifying $[K]$ and $\{F\}$ such that we satisfy two criteria. First, we must obtain the correct answers for all values in $\{\Phi\}$ or $\{U\}$. Second, we do not want to change the size of $[K]$, $\{F\}$, and $\{P\}$ because this leads to programming difficulties. We shall consider the steady-state situation first and then discuss the modification of equations associated with time-dependent field problems.

III.1 STEADY-STATE EQUATIONS

The modification of the system of equations $[K]\{\Phi\} = \{F\}$ is a two-step procedure once the subscript of the known nodal parameter is available. For example, suppose that Φ_5 has a known value. The modification proceeds as follows.

1. All of the coefficients in row five are set equal to zero except the diagonal term, which is left unaltered. In equation form, $K_{5j} = 0, j = 1, \ldots, n$ and $j \neq 5$. The associated term in the column vector $\{F\}$, F_5, is replaced by the product $K_{55}\Phi_5$.
2. All of the remaining equations are modified by subtracting the product $K_{j5}\Phi_5$ from F_j and then setting $K_{j5} = 0, j = 1, \ldots, n, j \neq 5$.

ILLUSTRATIVE EXAMPLE

Modify the following system of equations when $\Phi_1 = 150$ and $\Phi_5 = 40$.

$$
\begin{bmatrix}
55 & -46 & 0 & 0 & 0 \\
-46 & 140 & -46 & 0 & 0 \\
4 & -46 & 110 & -46 & 4 \\
0 & 0 & -46 & 142 & -46 \\
0 & 0 & 4 & -46 & 65
\end{bmatrix}
\begin{Bmatrix}
\Phi_1 \\ \Phi_2 \\ \Phi_3 \\ \Phi_4 \\ \Phi_5
\end{Bmatrix}
=
\begin{Bmatrix}
500 \\ 2000 \\ 1000 \\ 2000 \\ 900
\end{Bmatrix}
$$

To implement step one, we set all of the coefficients in rows one and five to zero except the diagonal terms, which are left unaltered. The corresponding terms in $\{F\}$, F_1 and F_5, are then replaced by $F_1 = K_{11}\Phi_1$ and $F_5 = K_{55}\Phi_5$, respectively. This step yields

$$
\begin{bmatrix}
55 & 0 & 0 & 0 & 0 \\
-46 & 140 & -46 & 0 & 0 \\
4 & -46 & 110 & -46 & 4 \\
0 & 0 & -46 & 142 & -46 \\
0 & 0 & 0 & 0 & 65
\end{bmatrix}
\begin{Bmatrix}
\Phi_1 \\ \Phi_2 \\ \Phi_3 \\ \Phi_4 \\ \Phi_5
\end{Bmatrix}
=
\begin{Bmatrix}
8250 \\ 2000 \\ 1000 \\ 2000 \\ 2600
\end{Bmatrix}
$$

The second step involves the elimination of the columns of coefficients that multiply Φ_1 and Φ_5. This is accomplished by transferring the coefficients involving Φ_1 and Φ_5 to the right-hand side. For example, F_2 becomes $2000 + 46\Phi_1$, or 8900. Completion of this step gives

$$
\begin{bmatrix}
55 & 0 & 0 & 0 & 0 \\
0 & 140 & -46 & 0 & 0 \\
0 & -46 & 110 & -46 & 0 \\
0 & 0 & -46 & 142 & 0 \\
0 & 0 & 0 & 0 & 65
\end{bmatrix}
\begin{Bmatrix}
\Phi_1 \\ \Phi_2 \\ \Phi_3 \\ \Phi_4 \\ \Phi_5
\end{Bmatrix}
=
\begin{Bmatrix}
8250 \\ 8900 \\ 240 \\ 3840 \\ 2600
\end{Bmatrix}
$$

III.2 TIME-DEPENDENT EQUATIONS

The incorporation of specified nodal values in time-dependent problems is more complicated because the solution procedure involves combinations of $[C]$ and $[K]$, namely $[A]$ and $[P]$. We shall place the same requirement on the time-dependent solution that was placed on the steady-state solution. We want to keep the dimensions of $[C]$ and $[K]$ and thus $[A]$ and $[P]$ the same after modification as they were before modification.

The algorithm for modifying $[C]$ and $[K]$ is more easily understood once we have looked at a specific problem. Let us reconsider the problem in Section 14.5 without the heat source at node one. Instead we assume that $\Phi_1 = 40°C$ for all time values. The vector of initial conditions $\{\Phi\}_a$ becomes $\{\Phi\}_a^T = [40 \quad 0 \quad 0]$.

Our desire is to maintain $[A]$ and $[P]$ as 3×3 matrices with the new property that $\Phi_1 = 40°C$ for all of the calculated solutions.

If we use a lumped capacitance matrix with the $[K]$ that was obtained for the example problem in Section 14.5, the system of differential equations is

$$12\frac{d\Phi_1}{dt} + 2\Phi_1 - 2\Phi_2 = 0$$

$$24\frac{d\Phi_2}{dt} - 2\Phi_1 + 4\Phi_2 - 2\Phi_3 = 0$$

$$12\frac{d\Phi_3}{dt} - 2\Phi_2 + 2\Phi_3 = 0 \qquad (1)$$

The first equation of (1) comes from $R_1 = 0$. Since Φ_1 has the fixed value of 40, the first equation should not be included. We must eliminate the residual equation for node one because Φ_1 is known. The correct system of differential equations is

$$24\frac{d\Phi_2}{dt} + 4\Phi_2 - 2\Phi_3 - 80 = 0$$

$$12\frac{d\Phi_3}{dt} - 2\Phi_2 + 2\Phi_3 = 0 \qquad (2)$$

The value -80 in the first equation of (2) comes from substituting $\Phi_1 = 40$ into $-2\Phi_1$ in the original equation.

The equivalent matrix form is

$$\begin{bmatrix} 24 & 0 \\ 0 & 12 \end{bmatrix}\frac{d\{\Phi^*\}}{dt} + \begin{bmatrix} 4 & -2 \\ -2 & 2 \end{bmatrix}\{\Phi^*\} - \begin{Bmatrix} 80 \\ 0 \end{Bmatrix} = \begin{Bmatrix} 0 \\ 0 \end{Bmatrix} \qquad (3)$$

where $\{\Phi^*\}^T = [\Phi_2 \quad \Phi_3]$.

A central difference solution of (3) using $\Delta t = 1$ is

$$\begin{bmatrix} 26 & -1 \\ -1 & 13 \end{bmatrix}\{\Phi\}_b = \begin{bmatrix} 22 & -1 \\ -1 & 11 \end{bmatrix}\{\Phi\}_a + \begin{Bmatrix} 80 \\ 0 \end{Bmatrix} \qquad (4)$$

Our objective is to have a system of equations that includes (4) but which also gives the correct value of Φ_1 for each time step. One way of achieving this objective is to expand (4) into a larger system as follows.

$$\begin{bmatrix} 12 & 0 & 0 \\ 0 & 26 & -1 \\ 0 & -1 & 13 \end{bmatrix}\{\Phi\}_b = \begin{bmatrix} 12 & 0 & 0 \\ 0 & 22 & -1 \\ 0 & -1 & 11 \end{bmatrix}\{\Phi\}_a + \begin{Bmatrix} 0 \\ 80 \\ 0 \end{Bmatrix} \qquad (5)$$

The diagonal values A_{11} and P_{11} are set equal to C_{11}. All of the other coefficients in the first row of $[A]$ and $[P]$ are zero. When (5) is solved, Φ_1 at $t = b$ will be the same as Φ_1 at $t = a$.

The modification of a system of differential equations when some of the nodal values are known can be accomplished by the following algorithm. Assume that

Φ_i is the known nodal value.

1. Add the products $K_{ji}\Phi_i$, $j=1,\ldots,n$ to the corresponding coefficient in $\{F\}$, that is, F_j.
2. Replace the coefficients in row i and column i of $[K]$ by zeros.
3. Set $F_i=0$.
4. If $[C]$ comes from the consistent formulation, sum the coefficients in row i and replace C_{ii} with this sum. Set all of the off-diagonal coefficients in the row to zero.

When these three steps are completed, $[A]$ and $[P]$ have properties similar to the matrices in (6). The diagonal coefficients, A_{ii} and P_{ii}, are the same as C_{ii}; thus, Φ_i at time b is the same as Φ_i at time a.

Answers to Selected Problems

1.1 $y = -\dfrac{4wH^4}{\pi^5 EI} \sin \dfrac{\pi x}{H}$

1.3 $y = -\dfrac{0.01326wH^4}{EI} \sin \dfrac{\pi x}{H}$

1.6 $y = -\dfrac{2M_0 H^2}{\pi^3 EI} \sin \dfrac{\pi x}{H}$

1.9 $y = -\dfrac{0.0645 M_0 H^2}{EI} \sin \dfrac{\pi x}{H}$

1.11 $y = -\dfrac{PH^3}{16\pi EI} \sin \dfrac{\pi x}{H}$

1.14 $y = -\dfrac{4M_0 H^2}{\pi^3 EI}$

$$\times \left(\sin \dfrac{\pi x}{H} + \dfrac{1}{27} \sin \dfrac{3\pi x}{H} \right)$$

1.16 $x/H = 0.3008$

2.1 (a) 50.9, (c) 55.8

2.2 (a) -11.3, (c) -12.0

2.5 $N_i(s) = 1 - \dfrac{s}{L}, \ N_j(s) = \dfrac{s}{L}$

2.8 $N_i = \left(1 - \dfrac{s}{L_s} \right)\left(1 - \dfrac{t}{L_t} \right),$

$N_j = \dfrac{s}{L_s}\left(1 - \dfrac{t}{L_t} \right), \ N_k = \dfrac{st}{L_s L_t},$

$N_m = \dfrac{t}{L_t}\left(1 - \dfrac{s}{L_s} \right)$

3.2 $\Phi_2 = 19.1, \ \Phi_3 = -4.48$

3.4 (a) $\Phi_2 = 2.25, \ \Phi_3 = 3.50, \ \Phi_4 = 3.75$

3.4 (c) $\Phi_2 = 0.50, \ \Phi_3 = -0.25,$
$\Phi_4 = -0.25$

3.8 $Y_2 = -7.33, \ Y_3 = -10.67,$
$Y_4 = -7.33$

5.5 $N_i + N_j + \cdots + N_m = 1$

5.7 (a) $\phi = 171.7$, (b) (0.21, 0.04),
(0.202, 0.088),

(c) $\dfrac{\partial \phi}{\partial x} = -230.8, \ \dfrac{\partial \phi}{\partial y} = -38.41$

5.9 (a) $\phi = 170.9$, (b) (0.183, 0.13),
(0.166, 0.166),

(c) $\dfrac{\partial \phi}{\partial x} = -283.3, \ \dfrac{\partial \phi}{\partial y} = -133.3$

5.12 (a) $\phi = 92.47$, (b) (0.368, 0.18),
(0.346, 0.25),

(c) $\dfrac{\partial \phi}{\partial x} = -420.4, \ \dfrac{\partial \phi}{\partial y} = -134.7$

5.14 (a) $\phi = 140.3$, (b) (0.259, 0.060),
(0.250, 0.095),

(c) $\dfrac{\partial \phi}{\partial x} = -438.1, \ \dfrac{\partial \phi}{\partial y} = -92.9$

6.4 $N_i = \dfrac{2}{3} - \dfrac{r}{L}, \ N_j = \dfrac{1}{3} + \dfrac{r}{L}$

6.6 (a) $\dfrac{L}{15}$, (c) $\dfrac{2L}{15}$

6.9 (a) $\dfrac{A}{30}$, (c) $\dfrac{A}{2}$

7.1 (a) $\dfrac{QA}{4}$ **7.2** (a) $\dfrac{2GA}{36}$

7.4 $\{f^{(e)}\} = \dfrac{A}{12}\begin{bmatrix} 2 & 1 & 1 \\ 1 & 2 & 1 \\ 1 & 1 & 2 \end{bmatrix}\begin{Bmatrix} Q_i \\ Q_j \\ Q_k \end{Bmatrix}$

7.9 (a)

$[k^{(e)}]$

$= \begin{bmatrix} 0.6238 & -0.2891 & -0.3243 \\ -0.2891 & 0.5469 & -0.2474 \\ -0.3243 & -0.2474 & 0.5821 \end{bmatrix},$

$\{f^{(e)}\}^T = [0.0130 \quad 0.0130 \quad 0.0130]$

7.10 (b) $[k^{(e)}] = \begin{bmatrix} 0.3420 & -0.1207 & -0.1625 & -0.0433 \\ -0.1207 & 0.3420 & -0.0433 & -0.1675 \\ -0.1625 & -0.0433 & 0.3420 & -0.1207 \\ -0.0433 & -0.1675 & -0.1207 & 0.3420 \end{bmatrix},$

$\{f^{(e)}\}^T = [0.0420 \quad 0.0420 \quad 0.0420 \quad 0.0420]$

8.1 $\Phi_1 = 218.0, \Phi_2 = 159.8, \Phi_4 = 123.5$

8.4 $\tau_{zx} = -133.3, \tau_{zy} = 283.3,$
$T^{(e)} = 1.190$

8.6 $\tau_{zx} = -38.5, \tau_{zy} = 230.8, T^{(e)} = 2.782$

8.10 $\tau_{zx} = -192.8, \tau_{zy} = 375.0,$
$T^{(e)} = 0.507$

8.12 $\tau_{zx} = -125.0, \tau_{zy} = 450.0,$
$T^{(e)} = 0.440$

9.1 (a) $M = -6, S = -3$

9.4 $\{f_Q^{(e)}\}^T = [8.96 \quad 23.4 \quad 7.72]$

9.7 $\{f_Q^{(e)}\}^T = [16.6 \quad 6.2 \quad 17.2]$

9.10 $\{f_Q^{(e)}\}^T = [14.3 \quad 2.86 \quad 3.81 \quad 19.0]$

9.12 $\{f_Q^{(e)}\}^T = [15.2 \quad 11.4 \quad 5.71 \quad 7.62]$

11.2 $[k_M^{(e)}] = \begin{bmatrix} hA & 0 \\ 0 & 0 \end{bmatrix}, \{f_s^{(e)}\} = \begin{Bmatrix} hA\phi_f \\ 0 \end{Bmatrix}$

11.3 $\{\Phi\}^T = [80 \quad 64.2 \quad 45.3 \quad 35.2 \quad 29.9]$

11.5 $\{\Phi\}^T = [50 \quad 44.0 \quad 17.0 \quad 11.1]$

11.8 $\Phi_1 = 5.38$

11.9 $[k^{(e)}]$

$= \begin{bmatrix} 1.886 & -0.752 & -0.722 \\ -0.752 & 1.386 & -0.222 \\ -0.722 & -0.222 & 0.944 \end{bmatrix},$

$\{f^{(e)}\}^T = [13.83 \quad 13.83 \quad 1.67]$

11.11 $[k^{(e)}]$

$= \begin{bmatrix} 1.422 & -0.500 & -0.289 \\ -0.500 & 1.250 & -0.750 \\ -0.289 & -0.750 & 1.672 \end{bmatrix}$

$\{f^{(e)}\}^T = [9.35 \quad 0 \quad 9.35]$

12.1 $Z_1 = 0, Z_2 = 2$

12.2 $Z_1 = 0, Z_2 = \frac{1}{5}, Z_3 = 1, Z_4 = 2$

12.8 $Z_1 = 0.108, Z_2 = 0.500, Z_3 = 1.32$

13.7 $\{f_Q^{(e)}\}^T = [461 \quad 545 \quad 503]$

13.9 $\{f_Q^{(e)}\}^T = [9549 \quad 10555 \quad 10052]$

13.12 $[k_M^{(e)}] = \begin{bmatrix} 62.8 & 37.7 & 0 \\ 37.7 & 88.0 & 0 \\ 0 & 0 & 0 \end{bmatrix}$

13.14 $[k_M^{(e)}] = \begin{bmatrix} 0 & 0 & 0 \\ 0 & 1591 & 754 \\ 0 & 754 & 1424 \end{bmatrix}$

14.1 $\{\Phi\}\dfrac{T}{2} = [0.380 \quad 0.0149 \quad 0.0011]$

14.4 $\begin{bmatrix} 16 & 1 & 0 \\ 1 & 16 & 1 \\ 0 & 1 & 16 \end{bmatrix}\begin{Bmatrix} \Phi_2 \\ \Phi_3 \\ \Phi_4 \end{Bmatrix}_b$

$$= \begin{bmatrix} 10 & 4 & 0 \\ 4 & 10 & 4 \\ 0 & 4 & 10 \end{bmatrix} \begin{Bmatrix} \Phi_2 \\ \Phi_3 \\ \Phi_4 \end{Bmatrix}_a + \begin{Bmatrix} 30 \\ 0 \\ 30 \end{Bmatrix}$$

14.6 $\{\Phi\}_2^T$

$$= [10 \quad 43.9 \quad 49.5 \quad 43.9 \quad 10]$$

14.8
$$\begin{bmatrix} 8 & 2 & 0 & 0 \\ 2 & 8 & 2 & 0 \\ 0 & 2 & 8 & 2 \\ 0 & 0 & 2 & 4 \end{bmatrix} \begin{Bmatrix} \Phi_2 \\ \Phi_3 \\ \Phi_4 \\ \Phi_5 \end{Bmatrix}_b$$

$$= \begin{bmatrix} 6 & 3 & 0 & 0 \\ 3 & 6 & 3 & 0 \\ 0 & 3 & 6 & 3 \\ 0 & 0 & 3 & 3 \end{bmatrix} \begin{Bmatrix} \Phi_2 \\ \Phi_3 \\ \Phi_4 \\ \Phi_5 \end{Bmatrix}_a + \begin{Bmatrix} 5 \\ 0 \\ 0 \\ 0 \end{Bmatrix}$$

14.10 $\{\Phi\}_2^T = [5 \quad 29.9 \quad 30 \quad 30 \quad 30]$

14.12 $\{\Phi\}_2^T = [10 \quad 1.41 \quad 0.243]$

15.1 (a) $\Delta t \leqslant 0.1867/(1-\theta)$

15.4 $\Delta t \leqslant \lambda A/18D(1-\theta)$

17.1 $U_2 = 0.002679$

17.3 $U_2 = U_3 = 0.00500$

19.1 $U_5 = -0.001869$, $U_6 = -0.06906$

19.3 $U_3 = 0.100$, $U_4 = -0.3121$, $U_5 = 0.200$

19.5 $U_1 = -0.00004270$, $U_2 = -0.005390$

19.7 $S_2^{(1)} = 283000(T)$, $S_2^{(3)} = 199800(C)$, $S_4^{(4)} = 200000(T)$

19.10 $S_3^{(1)} = 33300(T)$, $S_3^{(3)} = 60340(C)$

20.1 $U_2 = -0.027780$, $U_3 = -8.8890$, $U_4 = -0.01111$, $U_6 = 0.02222$

20.2 $U_3 = -1.9753$, $U_4 = -0.003704$

20.4 $U_3 = -13.021$, $U_4 = -0.03906$

20.8 $U_3 = -2.8483$, $U_4 = -0.002441$, $U_6 = 0.009766$

20.12 $U_3 = -0.001758$, $U_4 = -0.00001465$, $U_6 = 0.00001172$

10.13 (a) $-6EI/L^2$

21.2 $U_4 = 5.325$, $U_5 = -5.341$, $U_6 = -0.02828$

23.1 $\{\sigma\}^T = [22800 \quad 1200 \quad -3000]$

23.3 $\{\sigma\}^T = [-2821 \quad -11284 \quad -5053]$

23.5 $\{\sigma\}^T = [2660 \quad -9340 \quad 19600]$

23.8 $\{\sigma\}^T = [-2400 \quad -6000 \quad 9960]$

23.11 $\{f_p^{(e)}\}^T$

$$= \frac{t p_0 L_{jk}}{6} [0 \quad 0 \quad 1 \quad 0 \quad 0 \quad 2]$$

24.1 $[\bar{B}] = \begin{bmatrix} -0.500 & 0 & 0.500 & 0 & 0 & 0 \\ 0.143 & 0 & 0.143 & 0 & 0.143 & 0 \\ 0 & 0 & 0 & -0.333 & 0 & 0.333 \\ 0 & -0.500 & -0.333 & 0.500 & 0.333 & 0 \end{bmatrix}$

24.3 $[\bar{B}] = \begin{bmatrix} -0.500 & 0 & 0.500 & 0 & 0 & 0 \\ 0.111 & 0 & 0.111 & 0 & 0.111 & 0 \\ 0 & -0.250 & 0 & -0.250 & 0 & 0.500 \\ 0 & -0.500 & -0.250 & 0.500 & 0.500 & 0 \end{bmatrix}$

24.7 $\{f_T^{(e)}\}^T = [-2741 \quad 1002 \quad 3262 \quad -3000 \quad 261 \quad 1998]$

24.8 $\{f_b^{(e)}\}^T = [0 \quad 115.2 \quad 0 \quad 136.1 \quad 0 \quad 125.7]$

24.11 $P_4 = -29845$, $P_6 = -84823$,
$P_8 = -54978$

26.16 48.75   **27.3** $[J] = \dfrac{A}{4}$

24.12 $\{f_p^{(e)}\}^T = \dfrac{2\pi p_z L_{jk}}{12}$

27.4 (e) 0.01666485

$[0 \quad 0 \quad 0 \quad (3R_j + R_k) \quad 0 \quad (R_j + R_k)]$

24.15 $\{\sigma\}^T$
$= [-24472 \quad 1256 \quad -19144 \quad 18664]$

26.4 $N_1 = -\frac{3}{4}(\frac{1}{3} + \xi)(1 - \xi)$,
$N_2 = \frac{9}{8}(1 + \xi)(1 - \xi)$,
$N_3 = \frac{3}{8}(\frac{1}{3} + \xi)(1 + \xi)$

26.9 $N_1 = \frac{9}{2}L_1(L_1 - \frac{1}{3})(L_1 - \frac{2}{3})$,
$N_2 = \frac{27}{2}L_1 L_2(L_1 - \frac{1}{3})$,
$N_3 = \frac{27}{2}L_1 L_2(L_2 - \frac{1}{3})$

27.7 $[B]^T =$
$$
\begin{bmatrix}
0.091507 & 0.122009 \\
-0.122009 & -0.222222 \\
0.030502 & 0.151781 \\
0.166667 & -0.607122 \\
0.341506 & 0.455341 \\
-0.455342 & 0.222222 \\
0.113836 & 0.040669 \\
0.166667 & -0.162678
\end{bmatrix}
$$

27.10 $[B] = \dfrac{1}{34}\begin{bmatrix} -5 & 2 & 6 & -2 & 1 & -2 \\ -4 & -12 & -2 & 12 & -6 & 12 \end{bmatrix}$

27.15 $\{f_Q^{(e)}\}^T = [0 \quad 5.67 \quad 0 \quad 5.67 \quad 0 \quad 5.67]$

27.17 $\{f_Q^{(e)}\}^T = [0 \quad 15.09 \quad 0 \quad 19.52 \quad 0 \quad 15.09]$

27.18 $\{f_Q^{(e)}\}^T = [-5.50 \quad 22.0 \quad -5.50 \quad 31.53 \quad -5.50 \quad 22.0 \quad -5.50 \quad 12.47]$

INDEX